T0255024

LONDON MATHEMATICAL SOCIETY STUDENT TEXTS

Managing editor: Professor D. Benson,
Department of Mathematics, University of Aberdeen, UK

15 Presentations of groups: Second edition, D. L. JOHNSON
17 Aspects of quantum field theory in curved spacetime, S. A. FULLING
18 Braids and coverings: selected topics, VAGN LUNDSGAARD HANSEN
20 Communication theory, C. M. GOLDIE & R. G. E. PINCH
21 Representations of finite groups of Lie type, FRANCOIS DIGNE & JEAN MICHEL
22 Designs, graphs, codes, and their links, P. J. CAMERON & J. H. VAN LINT
23 Complex algebraic curves, FRANCES KIRWAN
24 Lectures on elliptic curves, J. W. S. CASSELS
26 An introduction to the theory of L-functions and Eisenstein series, H. HIDA
27 Hilbert Space; compact operators and the trace theorem, J. R. RETHERFORD
28 Potential theory in the complex plane, T. RANSFORD
29 Undergraduate commutative algebra, M. REID
31 The Laplacian on a Riemannian manifold, S. ROSENBERG
32 Lectures on Lie groups and Lie algebras, R. CARTER, G. SEGAL & I. MACDONALD
33 A primer of algebraic D-modules, S. C. COUTINHO
34 Complex algebraic surfaces, A. BEAUVILLE
35 Young tableaux, W. FULTON
37 A mathematical introduction to wavelets, P. WOJTASZCZYK
38 Harmonic maps, loop groups, and integrable systems, M. GUEST
39 Set theory for the working mathematician, K. CIESIELSKI
40 Ergodic theory and dynamical systems, M. POLLICOTT & M. YURI
41 The algorithmic resolution of diophantine equations, N. P. SMART
42 Equilibrium states in ergodic theory, G. KELLER
43 Fourier analysis on finite groups and applications, AUDREY TERRAS
44 Classical invariant theory, PETER J. OLVER
45 Permutation groups, P. J. CAMERON
46 Riemann surfaces: A primer, A. BEARDON
47 Introductory lectures on rings and modules, J. BEACHY
48 Set theory, A HAJNAL & P. HAMBURGER
49 K-theory for C*-algebras, M. RORDAM, F. LARSEN & N. LAUSTSEN
50 A brief guide to algebraic number theory, H. P. F. SWINNERTON-DYER
51 Steps in commutative algebra: Second edition, R. Y. SHARP
52 Finite Markov chains and algorithmic applications, O. HÄGGSTRÖM
53 The prime number theorem, G. J. O. JAMESON
54 Topics in graph automorphisms and reconstruction, J. LAURI & R. SCAPELLATO
55 Elementary number theory, group theory, and Ramanujan graphs, G. DAVIDOFF, P. SARNAK & A. VALETTE
56 Logic, induction and sets, T. FORSTER
57 Introduction to Banach algebras and harmonic analysis, H. G. DALES *et al.*
58 Computational algebraic geometry, HAL SCHENCK
59 Frobenius algebras and 2-D topological quantum field theories, J. KOCK
60 Linear operators and linear systems, J. R. PARTINGTON
61 An introduction to noncommutative Noetherian rings, K. R. GOODEARL & R. B. WARFIELD
62 Topics from one dimensional dynamics, K. M. BRUCKS & H. BRUIN
63 Singularities of plane curves, C. T. C. WALL
64 A short course on Banach space theory, N. L. CAROTHERS
65 Elements of the representation theory of associative algebras Volume I, I. ASSEM, A. SKOWROŃSKI & D. SIMSON
66 An introduction to sieve methods and their applications, A. C. COJOCARU & M. R. MURTY
67 Elliptic functions, V. ARMITAGE & W. F. EBERLEIN
68 Hyperbolic geometry from a local viewpoint, L. KEEN & N. LAKIC
69 Lectures on Kähler Geometry, A. MORIANU
70 Dependence logic, J. VÄÄNÄNEN

London Mathematical Society Student Texts 71

Elements of the Representation Theory of Associative Algebras

Volume 2 Tubes and Concealed Algebras of Euclidean type

DANIEL SIMSON
Nicolaus Copernicus University

ANDRZEJ SKOWROŃSKI
Nicolaus Copernicus University

CAMBRIDGE
UNIVERSITY PRESS

CAMBRIDGE
UNIVERSITY PRESS

University Printing House, Cambridge CB2 8BS, United Kingdom

One Liberty Plaza, 20th Floor, New York, NY 10006, USA

477 Williamstown Road, Port Melbourne, VIC 3207, Australia

314-321, 3rd Floor, Plot 3, Splendor Forum, Jasola District Centre, New Delhi - 110025, India

103 Penang Road, #05-06/07, Visioncrest Commercial, Singapore 238467

Cambridge University Press is part of the University of Cambridge.

It furthers the University's mission by disseminating knowledge in the pursuit of education, learning and research at the highest international levels of excellence.

www.cambridge.org
Information on this title: www.cambridge.org/9780521836104

First published 2007

A catalogue record for this publication is available from the British Library

ISBN   978-0-521-83610-4   Hardback
ISBN   978-0-521-54420-7   Paperback

To our Wives

Sabina and Mirosława

Contents

Introduction   *page* ix

X.   **Tubes**   1
  X.1.   Stable tubes   2
  X.2.   Standard stable tubes   13
  X.3.   Generalised standard components   32
  X.4.   Generalised standard stable tubes   35
  X.5.   Exercises   46

XI.   **Module categories over concealed algebras of Euclidean
         type**   51
  XI.1. The Coxeter matrix and the defect of a hereditary algebra
         of Euclidean type   52
  XI.2. The category of regular modules over a hereditary algebra
         of Euclidean type   60
  XI.3. The category of regular modules over a concealed algebra of
         Euclidean type   69
  XI.4. The category of modules over the Kronecker algebra   77
  XI.5. A characterisation of concealed algebras of Euclidean type .   84
  XI.6. Exercises   87

XII.   **Regular modules and tubes over concealed algebras
         of Euclidean type**   91
  XII.1. Canonical algebras of Euclidean type   92
  XII.2. Regular modules and tubes over canonical algebras
         of Euclidean type   106
  XII.3. A separating family of tubes over a concealed algebra
         of Euclidean type   124
  XII.4. A controlled property of the Euler form of a concealed
         algebra of Euclidean type   133
  XII.5. Exercises   139

**XIII.   Indecomposable modules and tubes over hereditary
            algebras of Euclidean type   143**
XIII.1. Canonically oriented Euclidean quivers, their Coxeter
         matrices and the defect 145
XIII.2. Tubes and simple regular modules over hereditary algebras
         of Euclidean type 153
XIII.3. Four subspace problem 197
XIII.4. Exercises 223

XIV.   Minimal representation-infinite algebras   227
XIV.1. Critical integral quadratic forms 228
XIV.2. Minimal representation-infinite algebras 231
XIV.3. A criterion for the infinite representation type of algebras 239
XIV.4. A classification of concealed algebras of Euclidean type . . 244
XIV.5. Exercises 278

        Bibliography   285

        Index   305

        List of symbols   307

Introduction

The first volume serves as a general introduction to some of the techniques most commonly used in representation theory. The quiver technique, the Auslander–Reiten theory and the tilting theory were presented with some application to finite dimensional algebras over a fixed algebraically closed field.

In particular, a complete classification of those hereditary algebras that are representation-finite (that is, admit only finitely many isomorphism classes of indecomposable modules) is given. The result, known as Gabriel's theorem, asserts that a basic connected hereditary algebra A is representation-finite if and only if the quiver Q_A of A is a Dynkin quiver, that is, the underlying non-oriented graph $\overline{Q}_A$ of Q_A is one of the Dynkin diagrams

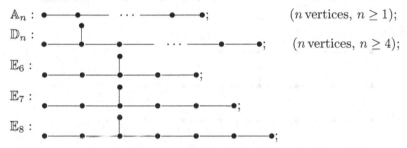

We also study in Volume 1 the class of hereditary algebras that are representation-infinite. It is shown in Chapter VIII that if B is a representation-infinite hereditary algebra, or B is a tilted algebra of the form

$$B = \operatorname{End} T_{KQ},$$

where KQ is a representation-infinite hereditary algebra and T_{KQ} is a postprojective tilting KQ-module, then B is representation-infinite and the Auslander–Reiten quiver $\Gamma(\operatorname{mod} B)$ of B has the shape

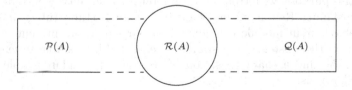

where $\operatorname{mod} B$ is the category of finite dimensional right B-modules, $\mathcal{P}(B)$ is the unique postprojective component of $\Gamma(\operatorname{mod} B)$ containing all the indecomposable projective B-modules, $\mathcal{Q}(B)$ is the unique preinjective component of $\Gamma(\operatorname{mod} B)$ containing all the indecomposable injective B-modules, and $\mathcal{R}(B)$ is the (non-empty) regular part consisting of the remaining components of $\Gamma(\operatorname{mod} B)$.

A prominent rôle in the representation theory is played by the class of hereditary algebras that are representation-infinite and minimal with respect to this property. They are just the hereditary algebras of Euclidean type, that is, the path algebras KQ, where Q is a connected acyclic quiver whose underlying non-oriented graph $\overline{Q}$ is one of the following Euclidean diagrams

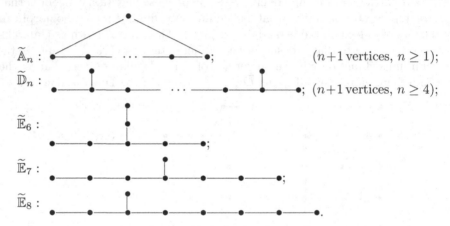

$$\widetilde{\mathbb{A}}_n : \qquad (n+1 \text{ vertices}, \ n \geq 1);$$
$$\widetilde{\mathbb{D}}_n : \qquad (n+1 \text{ vertices}, \ n \geq 4);$$
$$\widetilde{\mathbb{E}}_6 :$$
$$\widetilde{\mathbb{E}}_7 :$$
$$\widetilde{\mathbb{E}}_8 :$$

It is shown in Chapter VII that the underlying graph $\overline{Q}$ of a finite connected quiver $Q = (Q_0, Q_1)$ is a Dynkin diagram, or a Euclidean diagram, if and only if the associated quadratic form $q_Q : \mathbb{Z}^{|Q_0|} \longrightarrow \mathbb{Z}$ is positive definite, or positive semidefinite and not positive definite, respectively.

The main aim of Volumes 2 and 3 is to study the representation-infinite tilted algebras $B = \operatorname{End} T_{KQ}$ of a Euclidean type Q and, in particular, to give a fairly complete description of their indecomposable modules, their module categories $\operatorname{mod} B$, and the Auslander–Reiten quivers $\Gamma(\operatorname{mod} B)$.

For this purpose, we introduce in Chapter X a special type of components in the Auslander–Reiten quivers of algebras, namely stable tubes, and study their behaviour in module categories. In particular, we present a handy criterion on the existence of a standard self-hereditary stable tube, due to Ringel [215], and a characterisation of generalised standard stable tubes, due to Skowroński [246], [247], [254].

In Chapters XI and XII, we present a detailed description and properties of the regular part $\mathcal{R}(B)$ of the Auslander–Reiten quiver $\Gamma(\bmod B)$ of any concealed algebra B of Euclidean type, that is, a tilted algebra

$$B = \operatorname{End} T_{KQ}$$

of a Euclidean type Q defined by a postprojective tilting KQ-module T_{KQ}. In particular, it is shown that:

- the regular part $\mathcal{R}(B)$ of the Auslander–Reiten quiver $\Gamma(\bmod B)$ is a disjoint union of the $\mathbb{P}_1(K)$-family

$$\boldsymbol{\mathcal{T}}^B = \{\mathcal{T}_\lambda^B\}_{\lambda \in \mathbb{P}_1(K)}$$

  of pairwise orthogonal standard stable tubes $\mathcal{T}_\lambda^B$, where $\mathbb{P}_1(K)$ is the projective line over K,
- the family $\boldsymbol{\mathcal{T}}^B$ separates the postprojective component $\mathcal{P}(B)$ from the preinjective component $\mathcal{Q}(B)$,
- the module category $\bmod B$ is controlled by the Euler quadratic form $q_B : K_0(B) \longrightarrow \mathbb{Z}$ of the algebra B.

A crucial rôle in the investigation is played by the canonical algebras of Euclidean type, introduced by Ringel [215]. As an application of the developed theory, we present in Chapter XIII a complete list of indecomposable regular KQ-modules over any path algebra KQ of a canonically oriented Euclidean quiver Q, and we show how a simple tilting process allows us to construct the indecomposable regular modules over any path algebra KQ of a Euclidean type Q.

In Chapter XIV, we give the Happel–Vossieck [112] characterisation of the minimal representation-infinite algebras B having a postprojective component in the Auslander–Reiten quiver $\Gamma(\bmod B)$. As a consequence, we get a finite representation type criterion for algebras. We also present a complete classification, by means of quivers with relations, of all concealed algebras of Euclidean type, due independently by Bongartz [29] and Happel–Vossieck [112].

In Volume 3, we introduce some concepts and tools that allow us to give there a complete description of arbitrary representation-infinite tilted algebras B of Euclidean type and the module category $\bmod B$, due to Ringel [215]. We also investigate the wild hereditary algebras $A = KQ$, where Q is an acyclic quiver such that the underlying graph is neither a Dynkin nor a Euclidean diagram. We describe the shape of the components of the regular part $\mathcal{R}(A)$ of $\Gamma(\bmod A)$ and we establish a wild behaviour of the category $\bmod A$, for any such an algebra A. Finally, we introduce in Volume 3 the concepts of tame representation type and of wild representation type for algebras, and we discuss the tame and the wild nature of module categories mod B. Also, we present (without proofs) selected results of the representation theory of finite dimensional algebras that are related to the material discussed in the book.

It was not possible to be encyclopedic in this work. Therefore many important topics from the theory have been left out. Among the most notable omissions are covering techniques, the use of derived categories and partially ordered sets. Some other aspects of the theory presented here are discussed in the books [10], [15], [16], [91], [121], [235], and especially [215].

We assume that the reader is familiar with Volume 1, but otherwise the exposition is reasonably self-contained, making it suitable either for courses and seminars or for self-study. The text includes many illustrative examples and a large number of exercises at the end of each of the Chapters X-XIV.

The book is addressed to graduate students, advanced undergraduates, and mathematicians and scientists working in representation theory, ring and module theory, commutative algebra, abelian group theory, and combinatorics. It should also, we hope, be of interest to mathematicians working in other fields.

Throughout this book we use freely the terminology and notation introduced in Volume 1. We denote by K a fixed algebraically closed field. The symbols $\mathbb{N}$, $\mathbb{Z}$, $\mathbb{Q}$, $\mathbb{R}$, and $\mathbb{C}$ mean the sets of natural numbers, integers, rational, real, and complex numbers. The cardinality of a set X is denoted by $|X|$. Given a finite dimensional K-algebra A, the A-module means a finite dimensional right A-module. We denote by $\operatorname{Mod} A$ the category of all right A-modules, by $\operatorname{mod} A$ the category of finite dimensional right A-modules, and by $\Gamma(\operatorname{mod} A)$ the Auslander–Reiten translation quiver of A. The ordinary quiver of an algebra A is denoted by Q_A. Given a matrix $C = [c_{ij}]$, we denote by C^t the transpose of C.

A finite quiver $Q = (Q_0, Q_1)$ is called a **Euclidean quiver** if the underlying graph $\overline{Q}$ of Q is any of the Euclidean diagrams $\widetilde{\mathbb{A}}_m$, with $m \geq 1$, $\widetilde{\mathbb{D}}_m$, with $m \geq 4$, $\widetilde{\mathbb{E}}_6$, $\widetilde{\mathbb{E}}_7$, and $\widetilde{\mathbb{E}}_8$. Analogously, Q is called a **Dynkin quiver** if the underlying graph $\overline{Q}$ of Q is any of the Dynkin diagrams $\mathbb{A}_m$, with $m \geq 1$, $\mathbb{D}_m$, with $m \geq 4$, $\mathbb{E}_6$, $\mathbb{E}_7$, and $\mathbb{E}_8$.

We take pleasure in thanking all our colleagues and students who helped us with their comments and suggestions. We wish particularly to express our appreciation to Ibrahim Assem, Sheila Brenner, Otto Kerner, and Kunio Yamagata for their helpful discussions and suggestions. Particular thanks are due to Dr. Jerzy Białkowski and Dr. Rafał Bocian for their help in preparing a print-ready copy of the manuscript.

Chapter X

Tubes

In Chapter VIII of Volume 1, we have started to study the Auslander–Reiten quiver $\Gamma(\operatorname{mod} A)$ of any hereditary K-algebra A of Euclidean type, that is, the path algebra $A = KQ$ of an acyclic quiver Q whose underlying graph $\overline{Q}$ is one of the Euclidean diagrams $\widetilde{\mathbb{A}}_m$, with $m \geq 1$, $\widetilde{\mathbb{D}}_m$, with $m \geq 4$, $\widetilde{\mathbb{E}}_6$, $\widetilde{\mathbb{E}}_7$, and $\widetilde{\mathbb{E}}_8$. We recall that any such an algebra A is representation-infinite.

We have shown in (VIII.2.3) that the quiver $\Gamma(\operatorname{mod} A)$ contains a unique postprojective component $\mathcal{P}(A)$ containing all the indecomposable projective A-modules, a unique preinjective component $\mathcal{Q}(A)$ containing all the indecomposable injective A-modules, and the family $\mathcal{R}(A)$ of the remaining components being called regular (see (VIII.2.12)). This means that $\Gamma(\operatorname{mod} A)$ has the disjoint union form

$$\Gamma(\operatorname{mod} A) = \mathcal{P}(A) \cup \mathcal{R}(A) \cup \mathcal{Q}(A).$$

The indecomposable modules in $\mathcal{R}(A)$ are called regular. We have shown in (VIII.4.5) that there is a similar structure of $\Gamma(\operatorname{mod} B)$, for any concealed algebra B of Euclidean type, that is, the endomorphism algebra

$$B = \operatorname{End} T_A$$

of a postprojective tilting module T_A over a hereditary algebra $A = KQ$ of Euclidean type. The algebra B is representation-infinite.

The objective of Chapters XI-XIII is to describe the structure of regular components of the Auslander–Reiten quiver $\Gamma(\operatorname{mod} B)$ of any concealed algebra B of Euclidean type.

We introduce in this chapter a special type of a translation quiver, which we call a stable tube. The main aim of Section 1 is to describe special properties of irreducible morphisms between indecomposable modules in stable tubes of the Auslander–Reiten quiver $\Gamma(\operatorname{mod} B)$ of an algebra B and their compositions with arbitrary homomorphisms in the module category $\operatorname{mod} B$. In particular, some relevant properties of the radical rad_B and the infinite radical $\operatorname{rad}_B^\infty$ of the category $\operatorname{mod} B$ of finite dimensional right B-modules are described.

In Section 2, we introduce the important concept of a standard component and we prove Ringel's handy criterion on the existence of a standard self-hereditary stable tube in the Auslander–Reiten quiver $\Gamma(\operatorname{mod} B)$ of any algebra B. By applying the criterion, we show in Chapter XI that the regular components of any (representation-infinite) concealed algebra B of Euclidean type are self-hereditary standard stable tubes.

In Section 3, we introduce the concept of a generalised standard compo-
nent of $\Gamma(\operatorname{mod} B)$, invoking the infinite radical $\operatorname{rad}_B^\infty$ of the category $\operatorname{mod} B$,
and exhibit basic examples of generalised standard components. The main
result of Section 4 is a characterisation of (generalised) standard stable
tubes obtained by Skowroński in [246], [247], and [254]. It asserts that, for
a stable tube $\mathcal{T}$ in the Auslander–Reiten quiver $\Gamma(\operatorname{mod} B)$ of any algebra
B, the following three statements are equivalent:

- $\mathcal{T}$ is a standard stable tube,
- the mouth of $\mathcal{T}$ consists of pairwise orthogonal bricks, and
- $\mathcal{T}$ is a generalised standard stable tube.

It is also shown that $\operatorname{pd} X = 1$ and $\operatorname{id} X = 1$, for any indecomposable B-
module lying in a faithful generalised standard stable tube $\mathcal{T}$ of $\Gamma(\operatorname{mod} B)$.

Throughout, we assume that K is an algebraically closed field, and by
an algebra we mean a finite dimensional K-algebra. Given a finite quiver
$Q = (Q_0, Q_1)$, we denote by KQ the path K-algebra of Q. We recall that
the dimension $\dim_K KQ$ of KQ is finite if and only if the quiver Q is **acyclic**,
that is, there is no oriented cycle in Q, see Chapters II and III.

X.1. Stable tubes

We have defined in (VIII.1.1) the translation quiver $\mathbb{Z}\Sigma$, for Σ being a
connected and acyclic quiver. Thus, letting Σ be the infinite quiver

$$\mathbb{A}_\infty : \quad \underset{1}{\circ} \longrightarrow \underset{2}{\circ} \longrightarrow \underset{3}{\circ} \longrightarrow \underset{4}{\circ} \longrightarrow \cdots \longrightarrow \underset{m}{\circ} \longrightarrow \underset{m+1}{\circ} \longrightarrow \cdots$$

we obtain the **infinite translation quiver**

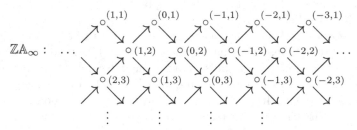

where $\tau(n, i) = (n + 1, i)$, for $n \in \mathbb{Z}$ and $i \geq 1$. Thus, by definition,
τ is an automorphism of $\mathbb{Z}\mathbb{A}_\infty$, and hence so is any power τ^r of τ (with
$r \in \mathbb{Z}$). For a fixed $r \geq 1$, let (τ^r) denote the infinite cyclic group of
automorphisms of $\mathbb{Z}\mathbb{A}_\infty$ generated by τ^r, and let $\mathbb{Z}\mathbb{A}_\infty/(\tau^r)$ denote the
orbit space of $\mathbb{Z}\mathbb{A}_\infty$ under the action of (τ^r). That is, $\mathbb{Z}\mathbb{A}_\infty/(\tau^r)$ is the
translation quiver obtained from $\mathbb{Z}\mathbb{A}_\infty$ by identifying each point (n, i) of
$\mathbb{Z}\mathbb{A}_\infty$ with the point $\tau^r(n, i) = (n + r, i)$, and each arrow $\alpha : x \to y$ in $\mathbb{Z}\mathbb{A}_\infty$
with the arrow $\tau^r\alpha : \tau^r x \to \tau^r y$. We are thus led to the following definition.

1.1. Definition. Let $(\mathcal{T}, \tau)$ be a translation quiver.

(a) $(\mathcal{T}, \tau)$ is defined to be a **stable tube of rank** $r = r_{\mathcal{T}} \geq 1$ if there is an isomorphism of translation quivers $\mathcal{T} \cong \mathbb{Z}\mathbb{A}_{\infty}/(\tau^r)$.

(b) A stable tube of rank $r = 1$ is defined to be a **homogeneous tube**.

(c) Let $(\mathcal{T}, \tau)$ be a stable tube of rank $r \geq 1$. A sequence $(x_1, \ldots, x_r)$ of points of $\mathcal{T}$ is said to be a τ-**cycle** if $\tau x_1 = x_r, \tau x_2 = x_1, \ldots, \tau x_r = x_{r-1}$.

For example, a stable tube of rank 3 is obtained from the quiver

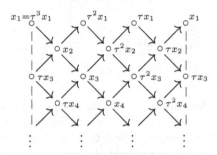

by identifying along the vertical dotted lines, thus giving the following

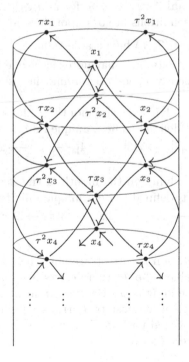

Similarly, a homogeneous tube has the following shape

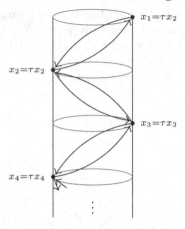

We observe that the translation τ still acts as an automorphism over a stable tube of rank r (that is the reason why such tubes are called stable), and that τ^r acts as the identity. The latter fact is expressed by saying that any point of $\mathbb{Z}\mathbb{A}_\infty/(\tau^r)$ is τ-**periodic** of period r.

We recall from Section IX.2 that a path $x_0 \to \cdots \to x_t$ in a translation quiver is called **sectional** if $\tau x_i \not\cong x_{i-2}$, for all $i \in \{2, \dots, t\}$.

The following two definitions are of importance in the theory.

1.2. Definition. Let $(\mathcal{T}, \tau)$ be a stable tube.

(a) The set of all points in $\mathcal{T}$ having exactly one immediate predecessor (or, equivalently, exactly one immediate successor) is called the **mouth** of $\mathcal{T}$.

(b) Given a point x lying on the mouth of the stable tube $\mathcal{T}$, a **ray** starting at x is defined to be a unique infinite sectional path
$$ x = x[1] \longrightarrow x[2] \longrightarrow x[3] \longrightarrow x[4] \longrightarrow \dots \longrightarrow x[m] \longrightarrow \dots $$
in the tube $\mathcal{T}$.

(c) Given a point x lying on the mouth of the stable tube $\mathcal{T}$, a **coray** ending with x is defined to be a unique infinite sectional path
$$ \dots \longrightarrow [m]x \longrightarrow \dots \longrightarrow [4]x \longrightarrow [3]x \longrightarrow [2]x \longrightarrow [1]x = x $$
in the tube $\mathcal{T}$.

To see that the definition is correct, we note that, for each point x lying on the mouth of a stable tube $\mathcal{T}$, there exists a unique arrow starting at x and a unique arrow ending at x. Because an arbitrary point in $\mathcal{T}$ is the source (and the target) of at most two arrows, this implies the existence of a unique infinite sectional path in $\mathcal{T}$ starting at x, and a unique infinite sectional path in $\mathcal{T}$ ending with x.

1.3. Definition. Let A be an algebra and C be a component of the Auslander–Reiten quiver $\Gamma(\mathrm{mod}\,A)$ of A.

(a) A **ray point** of C is defined to be a point X in C such that there exists an infinite sectional path in C

$$X = X[1] \longrightarrow X[2] \longrightarrow X[3] \longrightarrow X[4] \longrightarrow \ldots \longrightarrow X[m] \longrightarrow \ldots$$

starting at X and containing all sectional paths starting at X. The corresponding A-module X is called a **ray module**. The unique infinite sectional path starting at X is called the **ray** starting at X.

(b) A **coray point** of C is defined to be a point X in C such that there exists an infinite sectional path in C

$$\ldots \longrightarrow [m]X \longrightarrow \ldots \longrightarrow [4]X \longrightarrow [3]X \longrightarrow [2]X \longrightarrow [1]X = X$$

ending with X and containing all sectional paths ending with X. The corresponding A-module X is called the **coray module**. The unique infinite sectional path ending with X is called the **coray** ending with X.

A **ray point of a stable tube** $\mathcal{T}$ and a **coray point of a stable tube** $\mathcal{T}$ are defined analogously.

It is easy to see that if $(\mathcal{T}, \tau)$ is a stable tube and x is a point x in $\mathcal{T}$, then the following three statements are equivalent:

- x is a ray point of the tube $\mathcal{T}$,
- x is a coray point of $\mathcal{T}$, and
- x lies on the mouth of the tube $\mathcal{T}$.

Now we collect basic facts on the structure of any stable tube of $\Gamma(\mathrm{mod}\,A)$.

1.4. Lemma. *Let A be an algebra, and $\mathcal{T}$ a stable tube of rank $r = r_{\mathcal{T}} \geq 1$ of the Auslander–Reiten quiver $\Gamma(\mathrm{mod}\,A)$ of A. Assume that $(X_1, \ldots, X_r)$ is a τ_A-cycle of (indecomposable) mouth modules of mouth A-modules of the tube $\mathcal{T}$, that is, the modules $X_1, \ldots, X_r$ lie on the mouth of $\mathcal{T}$ and satisfy $\tau_A X_1 \cong X_r, \tau_A X_2 = X_1, \ldots, \tau_A X_r = X_{r-1}$.*

(a) *For each $i \in \{1, \ldots, r\}$, there exists a unique ray*

$$(\mathfrak{r}_i) \quad X_i = X_i[1] \longrightarrow X_i[2] \longrightarrow X_i[3] \longrightarrow \ldots \longrightarrow X_i[m] \longrightarrow X_i[m+1] \longrightarrow \ldots$$

in $\mathcal{T}$ starting at X_i, and a unique coray

$$(\mathfrak{c}_i) \quad \ldots \longrightarrow [m+1]X_i \longrightarrow [m]X_i \longrightarrow \ldots \longrightarrow [3]X_i \longrightarrow [2]X_i \longrightarrow [1]X_i = X_i$$

in $\mathcal{T}$ ending with X_i.

(b) *Every indecomposable A-module M in $\mathcal{T}$ is of the form $M \cong X_i[m]$, for some $i \in \{1, \ldots, r\}$ and $m \geq 1$.*

(c) *Every indecomposable A-module M in $\mathcal{T}$ is of the form $M \cong [m]X_s$, for some $s \in \{1, \ldots, r\}$ and $m \geq 1$.*

(d) *$[m]X_s \cong X_{s-m+1}[m]$ and $X_s[m] \cong [m]X_{s+m-1}$, for each $s \in \{1, \ldots, r\}$ and $m \geq 1$, where $s - m + 1$ is reduced modulo $r - 1$ if $s - m + 1 \leq 0$ or $s - m + 1 \geq r$.*

(e) *Under the isomorphisms of A-modules*

$$[1]X_i \cong X_i[1], \quad [2]X_i \cong X_{i-1}[2], \quad \ldots, \quad [m]X_i \cong X_{i+m-1}[m], \ldots$$

the coray $(\mathfrak{c}_i)$ has the form

$(\mathfrak{c}_i)$ $\ldots \longrightarrow X_{i-m}[m+1] \longrightarrow X_{i-m+1}[m] \longrightarrow \ldots \longrightarrow X_{i-1}[2] \longrightarrow X_i[1] = X_i.$

(f) *For any $i \in \{1, \ldots, r\}$ and $m \geq 1$, there exists an almost split sequence*

$$0 \longrightarrow X_i[m] \xrightarrow{\begin{bmatrix} f_{i,m+1} \\ g_{i,m} \end{bmatrix}} X_i[m+1] \oplus X_{i+1}[m-1] \xrightarrow{[g_{i,m+1} \; f_{i+1,m}]} X_{i+1}[m] \longrightarrow 0$$

in mod A, where we set $X_i[0] = 0$ and $X_{i+kr}[j] = X_i[j]$, for all $i \in \{1, \ldots, r\}$, $j \geq 1$, and $k \in \mathbb{Z}$.

Proof. Assume that $(X_1, \ldots, X_r)$ is a τ_A-cycle of mouth modules of the tube $\mathcal{T}$ of rank $r \geq 1$. Then $X_1, \ldots, X_r$ are indecomposable, lie on the mouth of the tube $\mathcal{T}$, and there are isomorphisms $\tau_A X_1 \cong X_r, \tau_A X_2 = X_1, \ldots, \tau_A X_r = X_{r-1}$. Because the tube $\mathcal{T}$ is stable then there is a surjective morphism $f : \mathbb{Z}\mathbb{A}_\infty \longrightarrow \mathcal{T}$ of translation quivers such that $f(-1, 1) = X_1, f(-2, 1) = X_2, \ldots, f(-r, 1) = X_r$ and the induced morphism $\bar{f} : \mathbb{Z}\mathbb{A}_\infty/(\tau^r) \xrightarrow{\simeq} \mathcal{T}$ is an isomorphism. It is clear that the conditions (a), (b), and (c) are satisfied in the translation quiver $(\mathbb{Z}\mathbb{A}_\infty, \tau)$. Hence we easily conclude that (a), (b), and (c) hold in $\mathcal{T}$, if we set $X_i[m] = f(-i, m)$ and $[m]X_i = f(-i+m-1, m)$.

The statement (d) follows from (a), (b), and (c) by an easy induction on $m \geq 1$.

Now we prove (e). It follows from (d) that, given $i \in \{1, \ldots, r\}$ and $m \geq 1$, there are isomorphisms $X_{i-m}[m+1] \cong [m+1]X_i$ and $X_{i-m+1}[m] \cong [m]X_i$. Hence, the following arrow in the coray $(\mathfrak{c}_i)$

$$X_{i-m}[m+1] \cong [m+1]X_i \longrightarrow [m]X_i \cong X_{i-m+1}[m]$$

corresponds to an irreducible morphism $X_{i-m}[m+1] \longrightarrow X_{i-m+1}[m]$ in mod A. To prove (f), we note that in view of the shape of the stable tube $\mathcal{T}$, each of its vertices is a source of at most two arrows, and the arrows correspond to some irreducible morphisms in mod A; thus yield a required almost split sequence. The proof of the lemma is then complete.　　□

In the remaining part of this section we investigate properties of irreducible morphisms between indecomposable modules in a stable tube $\mathcal{T}$ of the Auslander–Reiten quiver $\Gamma(\operatorname{mod} A)$ of an algebra A. The investigation needs some new concepts and preliminary results.

Assume that A is an arbitrary algebra. We recall from Section A.3 of Volume 1 that the **radical** $\operatorname{rad}_A = \operatorname{rad}(\operatorname{mod} A)$ of the category $\operatorname{mod} A$ is the two-sided ideal of $\operatorname{mod} A$ defined by the formula

$$\operatorname{rad}_A(X,Y) = \{h \in \operatorname{Hom}_A(X,Y); \ 1_X - gh \text{ is invertible, for any } g : Y \to X\},$$

for each pair of modules X and Y in $\operatorname{mod} A$. If the A-modules X and Y are indecomposable then $\operatorname{rad}_A(X,Y)$ consists of all non-isomorphisms $h : X \longrightarrow Y$ in $\operatorname{mod} A$, see (A.3.4) and (A.3.5) of Volume 1. In particular, $\operatorname{rad}_A(X,X)$ is just the radical of the local algebra $\operatorname{End} X$, for any indecomposable A-module X.

In other words, the radical rad_A of the category $\operatorname{mod} A$ is the two-sided ideal of $\operatorname{mod} A$ generated by all non-isomorphisms between indecomposable A-modules.

Given $m \geq 1$, the mth power $\operatorname{rad}_A^m \subseteq \operatorname{rad}_A$ of rad_A is the two-sided ideal of $\operatorname{mod} A$ such that, for A-modules X and Y, $\operatorname{rad}_A^m(X,Y)$ is the subspace of $\operatorname{rad}_A(X,Y)$ consisting of all finite sums of composite homomorphisms of the form

$$X = X_0 \xrightarrow{h_1} X_1 \xrightarrow{h_2} X_2 \longrightarrow \ \cdots \ \longrightarrow X_{m-1} \xrightarrow{h_m} X_m = Y,$$

where $h_j \in \operatorname{rad}_A(X_{j-1}, X_j)$, for any $j \in \{1, 2, \ldots, m-1, m\}$. For $m = 0$, we set $\operatorname{rad}_A^m(X,Y) = \operatorname{Hom}_A(X,Y)$. The intersection

$$\operatorname{rad}_A^\infty = \bigcap_{m=1}^{\infty} \operatorname{rad}_A^m$$

of all powers rad_A^m of rad_A is called the **infinite radical** of $\operatorname{mod} A$.

We recall from (IV.1.6) that a homomorphism $h : X \longrightarrow Y$ between two indecomposable modules X and Y in $\operatorname{mod} A$ is an irreducible morphism if and only if $h \in \operatorname{rad}_A(X,Y) \setminus \operatorname{rad}_A^2(X,Y)$. It follows that the radical rad_A of the category $\operatorname{mod} A$ is the two-sided ideal of $\operatorname{mod} A$ generated by the irreducible morphisms in $\operatorname{mod} A$, as a left ideal and as a right ideal.

Throughout, we need the following two preliminary lemmata.

1.5. Lemma. *Let A be an algebra, and M, N be a pair of modules in* $\operatorname{mod} A$.

(a) *There exists an integer $m \geq 0$ such that* $\operatorname{rad}_A^\infty(M,N) = \operatorname{rad}_A^m(M,N)$.
(b) $\operatorname{rad}_A^\infty(M,N) = \operatorname{rad}_A(M,N)$, *if the modules M and N are indecomposable and lie in two different components $\mathcal{C}_1$ and $\mathcal{C}_2$ of the Auslander–Reiten quiver $\Gamma(\operatorname{mod} A)$ of A.*

Proof. (a) Because the vector space $\operatorname{Hom}_A(M,N)$ is finite dimensional, the descending chain

$\operatorname{Hom}_A(M,N) \supseteq \operatorname{rad}_A(M,N) \supseteq \operatorname{rad}_A^2(M,N) \supseteq \ldots \supseteq \operatorname{rad}_A^m(M,N) \supseteq \ldots$

terminates, that is, $\operatorname{rad}_A^m(M,N) = \operatorname{rad}_A^{m+1}(M,N) = \operatorname{rad}_A^{m+2}(M,N) = \ldots$, for some $m \geq 0$. It follows that $\operatorname{rad}_A^\infty(M,N) = \operatorname{rad}_A^m(M,N)$.

The statement (b) is a consequence of (IV.5.1). $\qquad\square$

1.6. Lemma. *Let A be an algebra, M an A-module, and*

$$0 \longrightarrow Z \xrightarrow{\left[\begin{smallmatrix} g \\ g' \end{smallmatrix}\right]} X \oplus X' \xrightarrow{[f,\,f']} Y \longrightarrow 0$$

an almost split sequence in $\operatorname{mod} A$, where X, X', Y, and Z are indecomposable modules.

(a) *Let $\ell \geq 1$ be an integer and $h : M \longrightarrow X$ be a homomorphism in $\operatorname{mod} A$ such that $h \notin \operatorname{rad}_A^\ell(M,X)$ and $fh \in \operatorname{rad}_A^{\ell+1}(M,Y)$. Then there exists a homomorphism $h' : M \longrightarrow Z$ in $\operatorname{mod} A$ such that $h' \notin \operatorname{rad}_A^{\ell-1}(M,Z)$ and $g'h' \in \operatorname{rad}_A^\ell(M,X')$.*

(b) *Let $\ell \geq 1$ be an integer and $t : X \longrightarrow M$ be a homomorphism in $\operatorname{mod} A$ such that $t \notin \operatorname{rad}_A^\ell(X,M)$ and $tg \in \operatorname{rad}_A^{\ell+1}(Z,M)$. Then there exists a homomorphism $t' : Y \longrightarrow M$ in $\operatorname{mod} A$ such that $t' \notin \operatorname{rad}_A^{\ell-1}(Y,M)$ and $t'f' \in \operatorname{rad}_A^\ell(X',M)$.*

Proof. We only prove the statement (a), because the proof of (b) is similar.

Assume that $h \notin \operatorname{rad}_A^\ell(M,X)$ and $fh \in \operatorname{rad}_A^{\ell+1}(M,Y)$. Then there exists an A-module N and two homomorphisms $w \in \operatorname{rad}_A^\ell(M,N)$ and $v \in \operatorname{rad}_A(N,Y)$ such that $fh = vw$. Then v is not a retraction and, hence, there exists a homomorphism $\left[\begin{smallmatrix} u \\ u' \end{smallmatrix}\right] : N \longrightarrow X \oplus X'$ such that $v = [f, f'] \cdot \left[\begin{smallmatrix} u \\ u' \end{smallmatrix}\right] = fu + f'u'$. Consider the homomorphisms

$$\left[\begin{smallmatrix} uw-h \\ u'w \end{smallmatrix}\right] : M \longrightarrow X \oplus X'$$

and note that

$$[f, f'] \cdot \left[\begin{smallmatrix} uw-h \\ u'w \end{smallmatrix}\right] = fuw - fh + f'u'w = (fu + f'u')w - fh = vw - fh = 0.$$

It follows that there exists a homomorphism $h' : N \longrightarrow Z$ such that

$$\left[\begin{smallmatrix} uw-h \\ u'w \end{smallmatrix}\right] = \left[\begin{smallmatrix} g \\ g' \end{smallmatrix}\right] \cdot h',$$

and therefore $g'h' = u'w \in \operatorname{rad}_A^\ell(M,X')$, because $\left[\begin{smallmatrix} g \\ g' \end{smallmatrix}\right]$ is the kernel of $[f, f']$. Note also that $h' \notin \operatorname{rad}_A^{\ell-1}(M,Z)$, because otherwise we get the contradiction $h = uw - gh' \in \operatorname{rad}_A^\ell(M,X)$. This finishes the proof. $\qquad\square$

1.7. Proposition. *Let A be an algebra, $\mathcal{T}$ a stable tube of rank $r = r_{\mathcal{T}} \geq 1$ of the Auslander–Reiten quiver $\Gamma(\operatorname{mod} A)$ of A, and $(X_1, \ldots, X_r)$ a τ_A-cycle of mouth modules of the tube $\mathcal{T}$. In the notation of (1.4), let $(\mathfrak{r}_i)$*

and $(\mathfrak{c}_i)$ be the ray starting at X_i and the coray ending with X_i in $\mathcal{T}$, for each $i \in \{1, \dots, r\}$.

  (a) For each $i \in \{1, \dots, r\}$ and $m \geq 2$, any irreducible morphism $f_{i,m} :$
       $X_i[m{-}1] \longrightarrow X_i[m]$ corresponding to the arrow $X_i[m{-}1] \longrightarrow X_i[m]$
       in the ray $(\mathfrak{r}_i)$ of $\mathcal{T}$ starting at X_i is a monomorphism.
  (b) For each $i \in \{1, \dots, r\}$ and $m \geq 2$, any irreducible morphism $g_{i,m} :$
       $[m]X_i \longrightarrow [m{-}1]X_i$ corresponding to the arrow $[m]X_i \longrightarrow [m{-}1]X_i$
       in the coray $(\mathfrak{c}_i)$ of $\mathcal{T}$ ending with X_i is an epimorphism.
  (c) For each $i \in \{1, \dots, r\}$ and $m \geq 2$, there exist irreducible morphisms

$$v_{i,m} : X_i[m{-}1] \longrightarrow X_i[m] \quad \text{and} \quad q_{i,m} : X_i[m] \longrightarrow X_{i+1}[m{-}1]$$

in $\operatorname{mod} A$ such that
       (c1) $q_{i,2}v_{i,2} \in \operatorname{rad}_A^3(X_i[1], X_{i+1}[1])$, and
       (c2) $v_{i+1,m}q_{i,m} + q_{i,m+1}v_{i,m+1} \in \operatorname{rad}_A^3(X_i[m], X_{i+1}[m])$.

Proof. Assume that $\mathcal{T}$ is a stable tube of rank $r = r_{\mathcal{T}} \geq 1$ of $\Gamma(\operatorname{mod} A)$ and $(X_1, \dots, X_r)$ is a τ_A-cycle of mouth modules of the tube $\mathcal{T}$, that is, the modules $X_1, \dots, X_r$ are indecomposable, lie on the mouth of the tube $\mathcal{T}$, and there are isomorphisms $\tau_A X_1 \cong X_r, \tau_A X_2 = X_1, \dots, \tau_A X_r = X_{r-1}$. Then (1.4) applies, and we freely use the notation introduced there. Clearly, for each $i \in \{1, \dots, r\}$ and $m \geq 2$, the arrow $X_i[m{-}1] \longrightarrow X_i[m]$ in the ray $(\mathfrak{r}_i)$ of $\mathcal{T}$ starting from X_i corresponds to an irreducible morphism $f_{i,m} : X_i[m{-}1] \longrightarrow X_i[m]$ in $\operatorname{mod} A$. By (1.4)(d), there is an isomorphism $[m]X_s \cong X_{s-m+1}[m]$, for each $s \in \{1, \dots, r\}$ and $m \geq 1$, where $s - m + 1$ is reduced modulo $r - 1$ if $s - m + 1 \leq 0$ or $s - m + 1 \geq r$. It follows that, given $i \in \{1, \dots, r\}$ and $m \geq 1$, there are isomorphisms $X_i[m] \cong [m]X_{i+m-1}$ and $X_{i+1}[m{-}1] \cong [m{-}1]X_{i+m-1}$. Hence, the arrow

$$X_i[m] \cong [m]X_{i+m-1} \longrightarrow [m{-}1]X_{i+m-1} \cong X_{i+1}[m{-}1]$$

in the coray $(\mathfrak{c}_i)$ of $\mathcal{T}$ ending with X_i corresponds to an irreducible morphism $g_{i,m} : X_i[m] \longrightarrow X_{i+1}[m{-}1]$ in $\operatorname{mod} A$. In view of the shape of the stable tube $\mathcal{T}$, each of its vertices is a source of at most two arrows. It follows that, for any $i \in \{1, \dots, r\}$ and $m \geq 1$, there exists an almost split sequence

$$0 \longrightarrow X_i[m] \xrightarrow{\left[\begin{smallmatrix} f_{i,m+1} \\ g_{i,m} \end{smallmatrix}\right]} X_i[m{+}1] \oplus X_{i+1}[m{-}1] \xrightarrow{[g_{i,m+1}\ f_{i+1,m}]} X_{i+1}[m] \longrightarrow 0$$

in $\operatorname{mod} A$, where we set $X_i[0] = 0$ and $X_{i+kr}[j] = X_i[j]$, for all $i \in \{1, \dots, r\}$, $j \geq 1$, and $k \in \mathbb{Z}$. Counting the dimensions we get

$$\dim_K X_i[m] + \dim_K X_{i+1}[m] = \dim_K X_i[m{+}1] + \dim_K X_i[m{-}1].$$

Hence, we conclude the inequalities $\dim_K X_i[1] < \dim_K X_i[2] > \dim_K X_{i+1}[1]$, and an easy induction on $m \geq 1$ shows that

$$\dim_K X_i[m] < \dim_K X_i[m{+}1] > \dim_K X_{i+1}[m],$$

for all $i \in \{1, \dots, r\}$ and $m \geq 1$. Hence we easily conclude that any irreducible morphism $f_{i,m} : X_i[m{-}1] \longrightarrow X_i[m]$ is a monomorphism and any irreducible morphism $g_{i,m} : [m]X_i \longrightarrow [m{-}1]E_i$ is an epimorphism, because we know from (IV.1.4) that every irreducible morphism in $\mod A$ is a proper monomorphism or a proper epimorphism. This finishes the proof of (a) and (b).

Now we prove (c). First we observe that if $f, g : X \longrightarrow Y$ are irreducible morphisms in $\mod A$ and X, Y are indecomposable modules lying in the tube $\mathcal{T}$ then

$$f + \operatorname{rad}_A^2(X, Y) = \lambda \cdot g + \operatorname{rad}_A^2(X, Y),$$

for some $\lambda \in K \setminus \{0\}$, because

$$\dim_K \operatorname{rad}_A(X, Y)/\operatorname{rad}_A^2(X, Y) = 1,$$

by (IV.1.6), (IV.4.6), and the definition of a stable tube. For each $i \in \{1, \dots, r\}$, we choose some irreducible morphisms

$$v_{i,2} : X_i[1] \longrightarrow X_i[2] \quad \text{and} \quad q_{i,2} : X_i[2] \longrightarrow X_{i+1}[1]$$

in $\mod A$, and consider an almost split sequence

$$0 \longrightarrow X_i[1] \overset{u_{i,2}}{\longrightarrow} X_i[2] \overset{p_{i,2}}{\longrightarrow} X_{i+1}[1] \longrightarrow 0,$$

in $\mod A$. By our earlier observations, there exist scalars $\lambda_i, \mu_i \in K \setminus \{0\}$ and homomorphisms $w_{i,2} \in \operatorname{rad}_A^2(X_i[1], X_i[2])$ and $t_{i,2} \in \operatorname{rad}_A^2(X_i[2], X_{i+1}[1])$ such that $v_{i,2} = \lambda_i u_{i,2} + w_{i,2}$ and $q_{i,2} = \mu_i p_{i,2} + t_{i,2}$. Then we get the equalities

$$
\begin{aligned}
q_{i,2}v_{i,2} &= (\mu_i p_{i,2} + t_{i,2})(\lambda_i u_{i,2} + w_{i,2}) \\
&= \mu_i \lambda_i p_{i,2} u_{i,2} + (\mu_i p_{i,2} w_{i,2} + \lambda_i t_{i,2} u_{i,2} + t_{i,2} w_{i,2}) \\
&= \mu_i p_{i,2} w_{i,2} + \lambda_i t_{i,2} u_{i,2} + t_{i,2} w_{i,2} \in \operatorname{rad}_A^3(X_i[1], X_{i+1}[1]),
\end{aligned}
$$

and (c1) follows.

Assume that $s \geq 2$ and, for all $i \in \{1, \dots, r\}$ and $k \in \{2, \dots, s\}$, there exist irreducible morphisms

$$v_{i,k} : X_i[k{-}1] \longrightarrow X_i[k] \quad \text{and} \quad q_{i,k} : X_i[k] \longrightarrow X_{i+1}[k{-}1]$$

in $\mod A$ satisfying (c1) and (c2), for $k \in \{2, \dots, s-1\}$. Fix $i \in \{1, \dots, r\}$ and consider an almost split sequence

$$0 \longrightarrow X_i[s] \overset{\left[\begin{smallmatrix} u_{i,s+1} \\ p_{i,s} \end{smallmatrix}\right]}{\longrightarrow} X_i[s{+}1] \oplus X_{i+1}[s{-}1] \overset{[p_{i,s+1}\ u_{i+1,s}]}{\longrightarrow} X_{i+1}[s] \longrightarrow 0$$

in $\mod A$. Then there exist scalars $\lambda_{i+1}^{(s)}, \mu_i^{(s)} \in K \setminus \{0\}$ and homomorphisms $w_{i+1,s} \in \operatorname{rad}_A^2(X_{i+1}[s{-}1], X_{i+1}[s])$ and $t_{i,s} \in \operatorname{rad}_A^2(X_i[s], X_{i+1}[s{-}1])$ such that $v_{i+1,s} = \lambda_{i+1}^{(s)} u_{i+1,s} + w_{i+1,s}$ and $q_{i,s} = \mu_i^{(s)} p_{i,s} + t_{i,s}$.

Take the irreducible morphisms

- $v_{i,s+1} = u_{i,s+1} : X_i[s] \longrightarrow X_i[s+1]$, and
- $q_{i,s+1} = \lambda_{i+1}^{(s)} \mu_i^{(s)} p_{i,s+1} : X_i[s+1] \longrightarrow X_{i+1}[s]$

in mod A. Then we get the equalities

$$
\begin{aligned}
v_{i+1,s} q_{i,s} + q_{i,s+1} v_{i,s+1} &= (\lambda_{i+1}^{(s)} u_{i+1,s} + w_{i+1,s})(\mu_i^{(s)} p_{i,s} + t_{i,s}) \\
&\quad + (\lambda_{i+1}^{(s)} \mu_i^{(s)} p_{i,s+1}) u_{i,s+1} \\
&= \lambda_{i+1}^{(s)} \mu_i^{(s)} (u_{i+1,s} p_{i,s} + p_{i,s+1} u_{i,s+1}) \\
&\quad + (\lambda_{i+1}^{(s)} u_{i+1,s} t_{i,s} + \mu_i^{(s)} w_{i+1,s} p_{i,s} + w_{i+1,s} t_{i,s}) \\
&= \lambda_{i+1}^{(s)} u_{i+1,s} t_{i,s} + \mu_i^{(s)} w_{i+1,s} p_{i,s} + w_{i+1,s} t_{i,s}.
\end{aligned}
$$

Hence $v_{i+1,s} q_{i,s} + q_{i,s+1} v_{i,s+1} \in \mathrm{rad}_A^3(X_i[s], X_{i+1}[s])$ and (c2) follows. $\qquad\square$

Here, we would like to warn the reader that in general the relations (c1) and (c2) in (1.7) can not be replaced by the equalities

$$q_{i,2} v_{i,2} = 0 \quad \text{and} \quad v_{i+1,m} q_{i,m} + q_{i,m+1} v_{i,m+1} = 0,$$

for $i \in \{1, \dots, r\}$, $m \geq 2$, and suitably chosen irreducible morphisms $v_{i,m}$ and $q_{i,m}$. However, we show in the next section that this is possible, provided that the stable tube has special homological properties.

Now we establish an extraordinary property of irreducible morphisms between indecomposable modules lying in stable tubes of the quiver $\Gamma(\mathrm{mod}\,A)$ of an algebra A.

1.8. Proposition. *Let A be an algebra, $\mathcal{T}$ a stable tube of rank $r = r_{\mathcal{T}} \geq 1$ of $\Gamma(\mathrm{mod}\,A)$, and $(X_1, \dots, X_r)$ a τ_A-cycle of mouth modules of the tube $\mathcal{T}$. In the notation of (1.4), let $(\mathfrak{r}_i)$ and $(\mathfrak{c}_i)$ be the ray starting at X_i and the coray ending with X_i in $\mathcal{T}$, for each $i \in \{1, \dots, r\}$.*

(a) *Given $i \in \{1, \dots, r\}$ and $m \geq 2$, let $f_{i,m} : X_i[m-1] \longrightarrow X_i[m]$ be an irreducible morphism in mod A corresponding to the arrow $X_i[m-1] \longrightarrow X_i[m]$ in the ray $(\mathfrak{r}_i)$ of $\mathcal{T}$ starting at X_i. If $h : M \longrightarrow X_i[m-1]$ is a homomorphism in mod A such that $h \notin \mathrm{rad}_A^\ell(M, X_i[m-1])$, for some $\ell \geq 1$, then $f_{i,m} h \notin \mathrm{rad}_A^{\ell+1}(M, X_i[m])$.*

(b) *Given $s \in \{1, \dots, r\}$ and $m \geq 2$, let $g_{s,m} : [m]X_s \longrightarrow [m-1]X_s$ be an irreducible morphism in mod A corresponding to the arrow $[m]X_s \longrightarrow [m-1]X_s$ in the coray $(\mathfrak{c}_s)$ of $\mathcal{T}$ ending with X_s. If $h : [m-1]X_s \longrightarrow N$ is a homomorphism in mod A such that $h \notin \mathrm{rad}_A^\ell([m-1]X_s, N)$, for some $\ell \geq 1$, then $h g_{s,m} \notin \mathrm{rad}_A^{\ell+1}([m]X_s, N)$.*

Proof. We only prove the statement (a), because the proof of (b) is similar.

Assume, to the contrary, that $f : X_i[m{-}1] \longrightarrow X_i[m]$ is an irreducible morphism such that there are an integer $\ell \geq 1$ and a homomorphism $h :$ $M \longrightarrow X_i[m{-}1]$, with $h \notin \mathrm{rad}_A^\ell(M, X_i[m{-}1])$ and $fh \in \mathrm{rad}_A^{\ell+1}(M, X_i[m])$.

Without loss of generality, we may assume that f and h are chosen in such a way that $\ell \geq 1$ is minimal with respect to this property. Obviously, h is not a retraction and, hence, $\ell \geq 2$.

Note that the module $X_i[m{-}1]$ lies on the coray $(\mathfrak{c}_{i+m})$ and the module $X_i[m]$ lies on the coray $(\mathfrak{c}_{i+m+1})$, because there are isomorphisms $X_i[m{-}1] \cong [m{-}1]X_{i+m}$ and $X_i[m] \cong [m]X_{i+m+1}$, by (1.4). Because $f :$ $X_i[m{-}1] \longrightarrow X_i[m]$ is an irreducible morphism in $\mathrm{mod}\,A$ then, according to (IV.1.10), there exists an almost split sequence

$$0 \longrightarrow [m]X_{i+m} \xrightarrow{\left[\begin{smallmatrix} g_1 \\ f_1 \end{smallmatrix}\right]} [m{-}1]X_{i+m} \oplus [m{+}1]X_{i+m+1} \xrightarrow{[f,\,g]} [m]X_{i+m+1} \longrightarrow 0$$

in $\mathrm{mod}\,A$ and (1.6)(a) applies. It follows that there exists a homomorphism $h_1 : M \longrightarrow [m]X_{i+m}$ in $\mathrm{mod}\,A$ such that $h_1 \notin \mathrm{rad}_A^{\ell-1}(M, [m]X_{i+m})$ and $f_1 h_1 \in \mathrm{rad}_A^\ell(M, [m{+}1]X_{i+m+1})$. Moreover, $f_1 : [m]X_{i+m} \longrightarrow [m{+}1]X_{i+m+1}$ is an irreducible morphism corresponding to an arrow of the ray $(\mathfrak{r}_i)$, because it follows from (1.4) that $[m]X_{i+m} \cong X_i[m]$ and $[m{+}1]X_{i+m+1} \cong X_i[m{+}1]$. The homomorphism h_1 is not a retraction. Hence, $\ell - 1 \geq 2$ and we get a contradiction with the minimal choice of ℓ. $\qquad\square$

We finish the section with two important corollaries.

1.9. Corollary. *Let A be an algebra, $\mathcal{T}$ a stable tube of rank $r = r_{\mathcal{T}} \geq 1$ of $\Gamma(\mathrm{mod}\,A)$, and $(X_1, \dots, X_r)$ a τ_A-cycle of mouth modules of the tube $\mathcal{T}$. In the notation of (1.4), let $(\mathfrak{r}_i)$ and $(\mathfrak{c}_i)$ be the ray starting at X_i and the coray ending with X_i in $\mathcal{T}$, for each $i \in \{1, \dots, r\}$. Let $i \in \{1, \dots, r\}$ and $m \geq 2$ be fixed.*

    (a) *Let $f_{i,m} : X_i[m{-}1] \to X_i[m]$ be an irreducible morphism in $\mathrm{mod}\,A$ corresponding to the arrow $X_i[m{-}1] \to X_i[m]$ in the ray $(\mathfrak{r}_i)$ of $\mathcal{T}$ starting at X_i. If $h : M \to X_i[m{-}1]$ is a homomorphism in $\mathrm{mod}\,A$ such that $f_{i,m}h \in \mathrm{rad}_A^\infty(M, X_i[m])$, then $h \in \mathrm{rad}_A^\infty(M, X_i[m{-}1])$.*

    (b) *Let $g_{i,m} : [m]X_i \longrightarrow [m{-}1]X_i$ be an irreducible morphism in $\mathrm{mod}\,A$ corresponding to the arrow $[m]X_i \longrightarrow [m{-}1]X_i$ in the coray $(\mathfrak{c}_i)$ of $\mathcal{T}$ ending with X_i. If $h : [m{-}1]X_i \longrightarrow N$ is a homomorphism in $\mathrm{mod}\,A$ such that $hg_{i,m} \in \mathrm{rad}_A^\infty([m]X_i, N)$ then $h \in \mathrm{rad}_A^\infty([m{-}1]X_i, N)$.*

Proof. Apply (1.8). $\qquad\square$

In (X.5.8) we present an algebra A and a stable tube $\mathcal{T}$ of rank $r = r_{\mathcal{T}} = 1$ of $\Gamma(\mathrm{mod}\,A)$ with a mouth module E such that there exist an irreducible monomorphism $v : E[1] \longrightarrow E[2]$ and an irreducible epimorphism $p :$ $E[2] \longrightarrow E[1]$ satisfying $0 \neq pv \in \mathrm{rad}_A^\infty(E[1], E[1])$.

1.10. Corollary. *Let A be an algebra, $\mathcal{T}$ a stable tube of $\Gamma(\operatorname{mod} A)$, and*

$$M_0 \xrightarrow{h_1} M_1 \xrightarrow{h_2} M_2 \longrightarrow \ \ldots \ \longrightarrow M_{\ell-1} \xrightarrow{h_\ell} M_\ell$$

be a path of irreducible morphisms in $\operatorname{mod} A$ corresponding to a sectional path in $\mathcal{T}$. Then $h_\ell \cdot \ldots \cdot h_2 \cdot h_1 \in \operatorname{rad}_A^\ell(M_0, M_\ell) \setminus \operatorname{rad}_A^{\ell+1}(M_0, M_\ell)$.

Proof. Assume that $\mathcal{T}$ is a stable tube of rank $r = r_{\mathcal{T}} \geq 1$ of $\Gamma(\operatorname{mod} A)$ and $(X_1, \ldots, X_r)$ is a τ_A-cycle of mouth modules of the tube $\mathcal{T}$. If we use the notation of (1.4) then any sectional path in $\mathcal{T}$ of length $\ell \geq 2$ is a subpath

$$X_i[m] \longrightarrow X_i[m+1] \longrightarrow X_i[m+2] \longrightarrow \ \ldots \ \longrightarrow X_i[m+\ell-1] \longrightarrow X_i[m+\ell]$$

of a ray $(\mathfrak{r}_i)$ in $\mathcal{T}$ starting at X_i, for some $i \in \{1, \ldots, r\}$ and $m \geq 1$, or a subpath

$$[m+\ell]X_s \longrightarrow [m+\ell-1]X_s \longrightarrow \ \ldots \ \longrightarrow [m+2]X_s \longrightarrow [m+1]X_s \longrightarrow [m]X_s$$

of a coray $(\mathfrak{c}_s)$ in $\mathcal{T}$ ending with X_s, for some $s \in \{1, \ldots, r\}$ and $m \geq 1$. Because (IV.1.6) yields $h_i \in \operatorname{rad}_A(M_{i-1}, M_i) \setminus \operatorname{rad}_A^2(M_{i-1}, M_i)$, for any $i \in \{1, \ldots, r\}$ then the lemma follows by an iterated application of (1.8)(a) and (1.8)(b). $\qquad\square$

X.2. Standard stable tubes

In Chapter XI, we show that the regular components of the Auslander–Reiten quiver of a hereditary (or a concealed) algebra of Euclidean type are stable tubes. We start by giving a construction showing how stable tubes occur as components of the Auslander–Reiten quiver of an (arbitrary) algebra.

Let A be any algebra. We recall that a **brick** E in $\operatorname{mod} A$ is a (necessarily indecomposable) module E such that $\operatorname{End} E \cong K$. Two bricks E and E' in $\operatorname{mod} A$ are called **orthogonal** if $\operatorname{Hom}_A(E, E') = 0$, and $\operatorname{Hom}_A(E', E) = 0$.

Let $E_1, \ldots, E_r$ be a family of pairwise orthogonal bricks in $\operatorname{mod} A$, and let

$$\mathcal{E} = \mathcal{E}_A = \mathcal{EXT}_A(E_1, \ldots, E_r)$$

denote the full subcategory of $\operatorname{mod} A$ (called an extension category) whose non-zero objects are all the modules M such that there exists a chain of submodules $M = M_0 \supsetneq M_1 \supsetneq \ldots \supsetneq M_l = 0$, for some $l \geq 1$, with M_i/M_{i+1} isomorphic to one of the bricks $E_1, \ldots, E_r$ for all i such that $0 \leq i < l$. Thus $\mathcal{E}_A = \mathcal{EXT}_A(E_1, \ldots, E_r)$ is the smallest additive subcategory of $\operatorname{mod} A$ containing the bricks E_i and closed under extensions.

We say that $\mathcal{E}_A$ is an **exact subcategory** of $\operatorname{mod} A$ if the inclusion functor $\mathcal{E}_A \hookrightarrow \operatorname{mod} A$ is exact. An object S of $\mathcal{E}_A$ is said to be **simple**, if any non-zero subobject of S in $\mathcal{E}_A$ equals S.

We recall that an algebra A is hereditary if $\operatorname{Ext}_A^2(X, Y) = 0$, for each pair of A-modules X and Y in $\operatorname{mod} A$, or equivalently, if $\operatorname{pd} X \le 1$ and $\operatorname{id} X \le 1$, for any indecomposable (even simple) module X in $\operatorname{mod} A$. Following this definition we introduce some new concepts that are used throughout this book.

Let $\mathcal{H}$ be a family of modules in $\operatorname{mod} A$, or a full subcategory of $\operatorname{mod} A$.

- $\mathcal{H}$ is defined to be a **hereditary family** of $\operatorname{mod} A$, if $\operatorname{pd} X \le 1$ and $\operatorname{id} X \le 1$, for any module X in $\mathcal{H}$.
- $\mathcal{H}$ is defined to be a **self-hereditary family** of $\operatorname{mod} A$, if $\operatorname{Ext}_A^2(X, Y) = 0$, for each pair of A-modules X and Y in $\mathcal{H}$.

It is clear that $\mathcal{H} = \operatorname{mod} A$ is a hereditary family of $\operatorname{mod} A$ if and only if $\mathcal{H} = \operatorname{mod} A$ is self-hereditary. Note also that any hereditary family $\mathcal{H}$ of $\operatorname{mod} A$ is self-hereditary, but obviously the converse implication does not hold in general.

Basic properties of the extension category $\mathcal{EXT}_A(E_1, \ldots, E_r)$ are collected in the following lemma.

2.1. Lemma. *Let A be an algebra, and $\{E_1, \ldots, E_r\}$ a finite family of pairwise orthogonal bricks in $\operatorname{mod} A$.*

(a) *$\mathcal{E}_A = \mathcal{EXT}_A(E_1, \ldots, E_r)$ is an exact abelian subcategory of $\operatorname{mod} A$, $\mathcal{E}_A$ is closed under extensions, and $\{E_1, \ldots, E_r\}$ is a complete set of pairwise non-isomorphic simple objects in $\mathcal{E}_A$.*

(b) *The finite set $\{E_1, \ldots, E_r\}$ is a hereditary family of $\operatorname{mod} A$ if and only if $\mathcal{EXT}_A(E_1, \ldots, E_r)$ is a hereditary subcategory of $\operatorname{mod} A$.*

(c) *The finite set $\{E_1, \ldots, E_r\}$ is a self-hereditary family of $\operatorname{mod} A$ if and only if $\mathcal{EXT}_A(E_1, \ldots, E_r)$ is a self-hereditary subcategory of $\operatorname{mod} A$.*

Proof. (a) Let $\mathcal{E} = \mathcal{EXT}_A(E_1, \ldots, E_r)$ and let $f : M \to N$ be a homomorphism, with M and N in $\mathcal{E}$. We show that the modules $\operatorname{Ker} f$, $\operatorname{Im} f$ and $\operatorname{Coker} f$ lie in $\mathcal{E}$. Assume that

$$M = M_0 \supsetneq M_1 \supsetneq \ldots \supsetneq M_m = 0 \quad \text{and} \quad N = N_0 \supsetneq N_1 \supsetneq \ldots \supsetneq N_n = 0$$

are chains of submodules such that M_i/M_{i+1} and N_j/N_{j+1} belong to the set $\{E_1, \ldots, E_r\}$ for all i, j such that $0 \le i < m$ and $0 \le j < n$. We prove our statement by induction on $m + n$.

If $m + n \le 1$, then, clearly, $\operatorname{Ker} f$, $\operatorname{Im} f$ or $\operatorname{Coker} f$ are either zero or belong to $\{E_1, \ldots, E_r\}$. Therefore, assume that $m + n \ge 2$. Consider first the case where $f(M_{m-1}) = 0$. Then f induces a homomorphism $f' : M/M_{m-1} \longrightarrow N$ and we have

$$f(M) = f'(M/M_{m-1}), \quad \operatorname{Ker} f/M_{m-1} \cong \operatorname{Ker} f' \text{ and } \operatorname{Coker} f \cong \operatorname{Coker} f'.$$

Applying the induction hypothesis yields that $\operatorname{Ker} f$, $\operatorname{Im} f$ and $\operatorname{Coker} f$ lie in $\mathcal{E}$, as required. Assume now that $f(M_{m-1}) \ne 0$. Then there exists a largest j such that $f(M_{m-1}) \subseteq N_{j-1}$. Let $p : N_{j-1} \longrightarrow N_{j-1}/N_j$ denote the canonical projection and $f' : M_{m-1} \longrightarrow N_{j-1}$ the homomorphism induced by f. Then the composition $pf' : M_{m-1} \longrightarrow N_{j-1}/N_j$ is non-zero. Because $E_1, \dots, E_r$ are pairwise orthogonal bricks, this composition is an isomorphism. In particular, f' is a section and we have

$$N_{j-1} \cong N_j \oplus f'(M_{m-1}) = N_j \oplus f(M_{m-1}).$$

Setting $N'_s = N_{s+1} \oplus f(M_{m-1})$, for all s such that $j \le s \le n-1$, we thus obtain a chain of submodules of N

$$N = N_0 \supsetneq N_1 \supsetneq \dots \supsetneq N_{j-1} \supsetneq N'_j \supsetneq \dots \supsetneq N'_{n-1} \supsetneq N'_n = 0,$$

where the quotient of two consecutive terms belongs to $\{E_1, \dots, E_r\}$ and, moreover,

$$N'_{n-1} = N_n \oplus f(M_{m-1}) = f(M_{m-1}).$$

This shows that we may assume additionally that $f(M_{m-1}) = N_{n-1}$. Hence, in particular, f is an isomorphism between M_{m-1} and N_{n-1}, and thus it induces a homomorphism $\overline{f} : M/M_{m-1} \longrightarrow N/N_{n-1}$. It follows from the induction hypothesis that $\operatorname{Ker} \overline{f}$, $\operatorname{Im} \overline{f}$ and $\operatorname{Coker} \overline{f}$ lie in $\mathcal{E}$. Because $\operatorname{Im} \overline{f} = \operatorname{Im} f/N_{n-1}$, and N_{n-1} belongs to $\mathcal{E}$, which is closed under extensions, we deduce that $\operatorname{Im} f$ belongs to $\mathcal{E}$. Applying the Snake Lemma (I.5.1) to the commutative diagram

$$
\begin{array}{ccccccccc}
0 & \longrightarrow & M_{m-1} & \longrightarrow & M & \longrightarrow & M/M_{m-1} & \longrightarrow & 0 \\
& & \cong \downarrow & & \downarrow f & & \downarrow \overline{f} & & \\
0 & \longrightarrow & N_{n-1} & \longrightarrow & N & \longrightarrow & N/N_{n-1} & \longrightarrow & 0
\end{array}
$$

with exact rows we get isomorphisms $\operatorname{Ker} f \cong \operatorname{Ker} \overline{f}$ and $\operatorname{Coker} f \cong \operatorname{Coker} \overline{f}$ and, hence, the modules $\operatorname{Ker} f$ and $\operatorname{Coker} f$ lie in the category $\mathcal{E}$. This shows that $\mathcal{E}$ is an exact and abelian subcategory of $\operatorname{mod} A$.

Clearly, the modules $E_1, \dots, E_r$ are simple objects in the category $\mathcal{E}$ and, because they are pairwise orthogonal, they are also pairwise non-isomorphic.

(b) The sufficiency is obvious. To prove the necessity, we assume that the set $\{E_1, \dots, E_r\}$ is a hereditary family of $\operatorname{mod} A$, that is, $\operatorname{pd} E_j \le 1$ and $\operatorname{id} E_j \le 1$, for any $j \in \{1, \dots, r\}$, or equivalently, $\operatorname{Ext}_A^2(E_j, Y) = 0$ and $\operatorname{Ext}_A^2(X, E_j) = 0$, for any $j \in \{1, \dots, r\}$ and for all modules X and Y in $\operatorname{mod} A$. We fix a module Y in $\operatorname{mod} A$ and we prove that $\operatorname{Ext}_A^2(X, Y) = 0$, for any module X in $\mathcal{E}$. If X is any of the modules $E_1, \dots, E_r$, the assumption gives the result. Assume that X is an arbitrary non-zero object of $\mathcal{E}$. Then X contains a submodule X_0 isomorphic to one of the modules $E_1, \dots, E_r$, because they are all simple objects in the category $\mathcal{E}$, up to isomorphism, by (a). Then there exists an exact sequence $0 \to X_0 \to X \to \overline{X} \to 0$ in the category $\mathcal{E}$ and $\dim_K \overline{X} < \dim_K X$. We derive the induced exact sequence

$$\operatorname{Ext}_A^2(\overline{X}, Y) \longrightarrow \operatorname{Ext}_A^2(X, Y) \longrightarrow \operatorname{Ext}_A^2(X_0, Y).$$

By induction, the left hand term and the right hand term of the sequence are zero. Hence we get $\operatorname{Ext}_A^2(X, Y) = 0$, and the required result follows by induction on $\dim_K X$. The equality $\operatorname{Ext}_A^2(X, Y) = 0$, for any module X in $\operatorname{mod} A$ and any module Y in $\mathcal{E}_A$, follows in a similar way.

(c) Apply the arguments used in the proof of (b). The details are left to the reader.   □

An object U in the category $\mathcal{E}_A = \mathcal{EXT}_A(E_1, \dots, E_r)$ is defined to be **uniserial**, if $U_1 \subseteq U_2$ or $U_2 \subseteq U_1$, for each pair U_1, U_2 of subobjects of U in $\mathcal{E}_A$. In other words, an object U in $\mathcal{E}_A$ is uniserial if all subobjects of U in $\mathcal{E}_A$ form a chain with respect to the inclusion.

The length of the chain of subobjects of a uniserial object U of the category $\mathcal{E} = \mathcal{E}_A$ is called an $\mathcal{E}$-**length** of U, and is denoted by $\ell_{\mathcal{E}}(U)$.

Our next objective is to show that, if $r \geq 1$ and $E_1, \dots, E_r$ are pairwise orthogonal bricks satisfying the following two conditions:

(a) $\tau E_{i+1} = E_i$, for all $i \in \{1, \dots, r\}$, where we set $E_{r+1} = E_1$,
(b) $\operatorname{Ext}_A^2(E_i, E_j) = 0$, for all $i, j \in \{1, \dots, r\}$,

(the condition (b) is always satisfied if the algebra A is hereditary) then the indecomposable objects in the category $\mathcal{E}_A = \mathcal{EXT}_A(E_1, \dots, E_r)$ are uniserial in $\mathcal{E}_A$ and form an Auslander–Reiten component in $\Gamma(\operatorname{mod} A)$, that is a stable tube of rank r. We start by constructing indecomposable modules and almost split sequences in $\mathcal{E}_A$.

2.2. Theorem. *Let A be an algebra, and $(E_1, \dots, E_r)$, with $r \geq 1$, be a τ_A-cycle of pairwise orthogonal bricks in $\operatorname{mod} A$ such that $\{E_1, \dots, E_r\}$ is a self-hereditary family of $\operatorname{mod} A$. The abelian category*

$$\mathcal{E} = \mathcal{EXT}_A(E_1, \dots, E_r),$$

has the following properties.

(a) *For each pair (i, j), with $1 \leq i \leq r$ and $j \geq 1$, there exist a uniserial object $E_i[j]$ of $\mathcal{E}$-length $\ell_{\mathcal{E}}(E_i[j]) = j$ in the category $\mathcal{E}$, and homomorphisms*

$$u_{ij} : E_i[j-1] \longrightarrow E_i[j], \qquad p_{ij} : E_i[j] \longrightarrow E_{i+1}[j-1],$$

for $j \geq 2$, such that we have two short exact sequences in $\operatorname{mod} A$

$$0 \longrightarrow \quad E_i[j-1] \quad \xrightarrow{u_{ij}} \quad E_i[j] \quad \xrightarrow{p'_{ij}} \quad E_{i+j-1}[1] \quad \longrightarrow 0,$$

$$0 \longrightarrow \quad E_i[1] \quad \xrightarrow{u'_{ij}} \quad E_i[j] \quad \xrightarrow{p_{ij}} \quad E_{i+1}[j-1] \quad \longrightarrow 0,$$

where $p'_{ij} = p_{i+j-2,2} \dots p_{ij}$ and $u'_{ij} = u_{ij} \dots u_{i2}$. Moreover, for each $j \geq 2$, there exists an almost split sequence

$$0 \longrightarrow E_i[j-1] \xrightarrow{\begin{bmatrix} p_{i,j-1} \\ u_{ij} \end{bmatrix}} E_{i+1}[j-2] \oplus E_i[j] \xrightarrow{[u_{i+1,j-1} \; p_{ij}]} E_{i+1}[j-1] \longrightarrow 0,$$

in mod A, where we set $E_i[0] = 0$ and $E_{i+kr}[m] = E_i[m]$, for $m \geq 1$ and all $k \in \mathbb{Z}$.

(b) The indecomposable uniserial objects $E_i[j]$, with $i \in \{1,\ldots,r\}$ and $j \geq 1$, of the category $\mathcal{E}$, connected by the homomorphisms u_{ij} : $E_i[j-1] \longrightarrow E_i[j]$ and $p_{ij} : E_i[j] \longrightarrow E_{i+1}[j-1]$, form the infinite diagram **(2.3)** presented below.

(c) $\mathrm{Ext}_A^2(X,Y) = 0$, for each pair of objects X and Y of $\mathcal{E}$.

To understand (and visualise) the above statement, some comments may be useful. We first notice that the homomorphisms u_{ij} (and hence the u'_{ij}) are necessarily monomorphisms, while the p_{ij} (and hence p'_{ij}) are necessarily epimorphisms. We thus have the following diagram

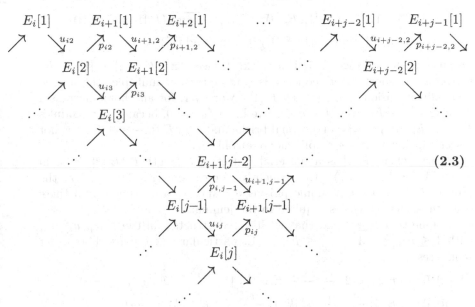

(2.3)

where we set $E_i[1] = E_{i+kr}[1]$, for all $k \in \mathbb{Z}$, all the arrows pointing down represent monomorphisms, all the arrows pointing up represent epimorphisms, and each of the squares represents an almost split sequence (as well as each of the triangles on the top).

If we set $E_i[1] = E_i$, for each i, then the first short exact sequence in (2.2) implies that $E_i[j-1]$ embeds into $E_i[j]$ as a maximal subject in the category $\mathcal{E} = \mathcal{EXT}_A(E_1,\ldots,E_r)$, because the quotient $E_{i+j-1}[1] = E_{i+j-1}$ is one of the simple objects $E_1, \ldots, E_r$ of the abelian category $\mathcal{E}$. Hence the uniseriality of $E_i[j]$ would imply that $E_i[j-1]$ is the unique maximal sub-object of $E_i[j]$ in $\mathcal{E}$, or, equivalently, that $f u_{ij} = 0$ for any homomorphism $f : E_i[j] \longrightarrow E_k$, for any k such that $1 \leq k \leq r$. The second short exact sequence in (2.2) entails similar consequences. We are now ready to proceed with the proof of the theorem.

Proof of the theorem. By our assumption, $r \geq 1$ and the modules $E_1, \ldots, E_r$ are pairwise orthogonal bricks such that $\tau_A E_1 \cong E_r, \tau_A E_2 = E_1, \ldots, \tau_A E_r = E_{r-1}$ and $\operatorname{Ext}_A^2(E_i, E_s) = 0$, for all $i, s \in \{1, \ldots, r\}$.

(a) We use induction on j. If $j = 1$, we set $E_i[1] = E_i$. Assume that $j = 2$. By hypothesis, we have $E_i = \tau E_{i+1}$ for each i, hence there is an almost split sequence

$$0 \longrightarrow E_i[1] \xrightarrow{u_{i2}} E_i[2] \xrightarrow{p_{i2}} E_{i+1}[1] \longrightarrow 0$$

in mod A. Applying the Auslander–Reiten formulae (IV.2.13) yields

$$1 \leq \dim_K \operatorname{Ext}_A^1(E_{i+1}[1], E_i[1]) = \dim_K D\overline{\operatorname{Hom}}_A(E_i[1], \tau E_{i+1}[1])$$

$$= \dim_K D\overline{\operatorname{Hom}}_A(E_i[1], E_i[1]) \leq \dim_K \operatorname{End} E_i[1] = 1,$$

because, by hypothesis, $E_i[1]$ is a brick. Hence $\operatorname{Ext}_A^1(E_{i+1}[1], E_i[1]) \cong K$, so the above short exact sequence is (up to a scalar multiple) the unique non-split extension of $E_{i+1}[1]$ by $E_i[1]$. Moreover, for any homomorphism $f : E_i[2] \longrightarrow E_k$, with $1 \leq k \leq r$, we have $f u_{i2} = 0$, because the modules $E_1, \ldots, E_r$ are pairwise orthogonal bricks and $u_{i2} : E_i[1] \longrightarrow E_i[2]$ is not a section, that is, is not a split monomorphism.

Clearly, then, $E_i[2]$ is a uniserial object of $\mathcal{E}$-length $\ell_\mathcal{E}(E_i[2]) = 2$ in $\mathcal{E} = \mathcal{EXT}_A(E_1, \ldots, E_r)$. Consequently, for each i with $1 \leq i \leq r$, the objects $E_i[2]$ and the homomorphisms u_{i2} and p_{i2} are defined, and there exist the exact sequences required in the lemma.

Assume that $k \geq 3$ and that we have constructed all the $E_i[j]$, u_{ij}, p_{ij} with $1 \leq i \leq r$ and $1 \leq j \leq k-1$. In particular, there exist short exact sequences

(i) $0 \longrightarrow E_{i+1}[k-2] \xrightarrow{u_{i+1,k-1}} E_{i+1}[k-1] \xrightarrow{p'_{i+1,k-1}} E_{i+k-1}[1] \longrightarrow 0,$

(ii) $0 \longrightarrow E_i[1] \xrightarrow{u'_{i,k-1}} E_i[k-1] \xrightarrow{p_{i,k-1}} E_{i+1}[k-2] \longrightarrow 0.$

Applying $\operatorname{Hom}_A(-, E_i[1])$ to (i) yields an exact sequence

$$\ldots \longrightarrow \operatorname{Ext}_A^1(E_{i+1}[k-1], E_i[1]) \xrightarrow{\operatorname{Ext}_A^1(u_{i+1,k-1}, E_i[1])} \operatorname{Ext}_A^1(E_{i+1}[k-2], E_i[1])$$

$$\longrightarrow \operatorname{Ext}_A^2(E_{i+k-1}[1], E_i[1]) \longrightarrow \ldots$$

Because, by hypothesis, $\operatorname{Ext}_A^2(E_{i+k-1}[1], E_i[1]) = 0$, $\operatorname{Ext}_A^1(u_{i+1,k-1}, E_i[1])$ is an epimorphism. Using (ii), there exist $E_i[k]$, u_{ik}, p_{ik}, $u'_{ik} = u_{ik}u'_{i,k-1}$

and $p'_{ik} = p'_{i+1,k-1}p_{ik}$ such that we have a commutative diagram with exact rows and columns

$$
\begin{array}{ccccccccc}
 & & & & 0 & & 0 & & \\
 & & & & \downarrow & & \downarrow & & \\
0 & \longrightarrow & E_i[1] & \xrightarrow{u'_{i,k-1}} & E_i[k-1] & \xrightarrow{p_{i,k-1}} & E_{i+1}[k-2] & \longrightarrow & 0 \\
 & & \| & & \downarrow{\scriptstyle u_{ik}} & & \downarrow{\scriptstyle u_{i+1,k-1}} & & \\
0 & \longrightarrow & E_i[1] & \xrightarrow{u'_{ik}} & E_i[k] & \xrightarrow{-p_{ik}} & E_{i+1}[k-1] & \longrightarrow & 0 \\
 & & & & \downarrow{\scriptstyle -p'_{ik}} & & \downarrow{\scriptstyle p'_{i+1,k-1}} & & \\
 & & & & E_{i+k-1}[1] & = & E_{i+k-1}[1] & & \\
 & & & & \downarrow & & \downarrow & & \\
 & & & & 0 & & 0 & &
\end{array}
$$

In particular, we get the first two short exact sequences required in the lemma for $j = k$. Further, the upper right corner of the above diagram gives the short exact sequence

(iii) $0 \longrightarrow E_i[k-1] \xrightarrow{\left[\begin{smallmatrix} p_{i,k-1} \\ u_{ik} \end{smallmatrix}\right]} E_{i+1}[k-2] \oplus E_i[k]$
$$\xrightarrow{[\,u_{i+1,k-1}\ p_{ik}\,]} E_{i+1}[k-1] \longrightarrow 0.$$

We now prove that $\operatorname{Im} u_{ik} \cong E_i[k-1]$ is the unique maximal subobject of $E_i[k]$ in $\mathcal{E}$. Because, by the induction hypothesis, $E_i[k-1]$ is uniserial, this would imply that so is $E_i[k]$.

Let $f : E_i[k] \longrightarrow E_l$ be a homomorphism in $\mathcal{E}$, with $1 \le l \le r$. We claim that $f u_{ik} = 0$. Because the module E_l is a simple object of $\mathcal{E}$, f is an epimorphism. We observe that $f u_{ik} u_{i,k-1} = 0$, because the module $\operatorname{Im} u_{i,k-1} \cong E_i[k-2]$ is the unique maximal subobject of the uniserial object $E_i[k-1]$ of $\mathcal{E}$, and the target of $f u_{ik} : E_i[k-1] \longrightarrow E_l$ is a simple object of $\mathcal{E}$. Hence, there exists a homomorphism $g : E_{i+k-2}[1] \longrightarrow E_l$ such that $f u_{ik} = g p'_{i,k-1} = g p_{i+k-2,2} \ldots p_{i,k-1}$. Letting

$$f' = g p_{i+k-2,2} \ldots p_{i+1,k-2} : E_{i+1}[k-2] \longrightarrow E_l,$$

the last equation becomes $f u_{ik} = f' p_{i,k-1}$. Thus $[-f'\ f]\left[\begin{smallmatrix} p_{i,k-1} \\ u_{ik} \end{smallmatrix}\right] = 0$. The exactness of (iii) yields $f'' : E_{i+1}[k-1] \longrightarrow E_l$ such that $[-f'\ f] = f''[u_{i+1,k-1}\ p_{ik}]$. But $-f' = f'' u_{i+1,k-1} = 0$ because, by the induction hypothesis, the image of $u_{i+1,k-1}$ is the unique maximal subobject of $E_{i+1}[k-1]$ in $\mathcal{E}$. Hence $f u_{ik} - f' p_{i,k-1} = 0$. This shows that $\operatorname{Im} u_{ik}$ is the unique maximal subobject of $E_i[k]$, and, consequently, the uniseriality of the latter.

We now prove that (iii) is an almost split sequence in $\operatorname{mod} A$. Because the middle term is the direct sum of two uniserial objects having $\mathcal{E}$-lengths

$k-2$ and k in $\mathcal{E}$, while the extreme terms are uniserial of $\mathcal{E}$-length $k-1$ in $\mathcal{E}$, the sequence (iii) does not split. In particular, $E_{i+1}[k-1]$ is not projective and hence $\tau E_{i+1}[k-1] \neq 0$. We claim that $\tau E_{i+1}[k-1] \cong E_i[k-1]$. By the induction hypothesis, we have an almost split sequence (corresponding to $j = k - 1$)

$$0 \longrightarrow E_i[k-2] \xrightarrow{\left[\begin{smallmatrix} p_{i,k-2} \\ u_{i,k-1} \end{smallmatrix}\right]} E_{i+1}[k-3] \oplus E_i[k-1] \xrightarrow{[\, u_{i+1,k-2} \; p_{i,k-1} \,]} E_{i+1}[k-2] \longrightarrow 0$$

for any i. Replacing i by $i + 1$, we get that the homomorphism $u_{i+1,k-1} : E_{i+1}[k - 2] \longrightarrow E_{i+1}[k - 1]$ is an irreducible morphism. It follows that there exists an irreducible morphism $\tau E_{i+1}[k - 1] \longrightarrow E_{i+1}[k - 2]$. The induction hypothesis yields that

- $E_{i+1}[k-3] = 0$, if $k = 3$, or
- $\tau^{-1} E_{i+1}[k-3] \cong E_{i+2}[k-3] \not\cong E_{i+1}[k-1]$, if $k \neq 3$.

Therefore, there is an isomorphism $\tau E_{i+1}[k-1] \cong E_i[k-1]$. Because, by (IV.3.2), an almost split sequence with the left term $E_i[k-1]$ and the right term $E_{i+1}[k-1]$ represents a non-zero element of the socle of $\mathrm{Ext}_A^1(E_{i+1}[k-1], E_i[k-1])$, then it remains to show that, for every non-invertible endomorphism $f : E_i[k-1] \longrightarrow E_i[k-1]$, the equality $\mathrm{Ext}_A^1(E_{i+1}[k-1], f) = 0$ holds.

For, let $f : E_i[k-1] \longrightarrow E_i[k-1]$ be such a homomorphism. Then $\mathrm{Ker}\, f \neq 0$ and, because the module $\mathrm{Im}\, u'_{i,k-1} \cong E_i[1]$ is a unique simple subobject of the uniserial object $E_i[k-1]$ in $\mathcal{E}$, we have $f u'_{i,k-1} = 0$. Hence there exists a homomorphism $f' : E_{i+1}[k-2] \longrightarrow E_i[k-1]$ such that $f = f' p_{i,k-1}$. On the other hand, the Auslander–Reiten formulae (IV.2.13) yields

$$\mathrm{Ext}_A^1(E_{i+1}[k-1], E_{i+1}[k-2]) \cong D\underline{\mathrm{Hom}}_A(\tau^{-1} E_{i+1}[k-2], E_{i+1}[k - 1])$$

$$\cong D\underline{\mathrm{Hom}}_A(E_{i+2}[k-2], E_{i+1}[k-1]).$$

Now $E_{i+1}[k-1]$ is a uniserial object of $\mathcal{E}$ and all its subobjects in $\mathcal{E}$ are of the form $E_{i+1}[j]$, with $j \leq k - 1$, while all epimorphic images of the object $E_{i+2}[k-2]$ in the category $\mathcal{E}$ are of the form $E_{i+l+1}[k-l]$, with $1 \leq l \leq k-1$. Hence $\mathrm{Hom}_A(E_{i+2}[k-2], E_{i+1}[k-1]) = 0$, and consequently

$$\mathrm{Ext}_A^1(E_{i+1}[k-1], E_{i+1}[k-2]) = 0.$$

Applying the functor $\mathrm{Ext}_A^1(E_{i+1}[k-1], -)$ to the short exact sequence

$$0 \longrightarrow E_i[1] \xrightarrow{u'_{i,k-1}} E_i[k-1] \xrightarrow{p_{i,k-1}} E_{i+1}[k-2] \longrightarrow 0$$

yields thus $\mathrm{Ext}_A^1(E_{i+1}[k-1], p_{i,k-1}) = 0$. Hence $\mathrm{Ext}_A^1(E_{i+1}[k-1], f) = 0$, because $f = f' p_{i,k-1}$, as required.

(b) The statement follows from (1.4), the discussion preceding the proof, and the arguments given in the proof of (a).

(c) This is a consequence of (2.1). $\square$

Before starting (and proving) the main result of this section, we need two further definitions, see Bongartz–Gabriel [34] and Ringel [215].

2.4. Definition. Let $\mathcal{C}$ be a component of the Auslander–Reiten quiver $\Gamma(\text{mod } A)$ of an algebra A, and assume, for simplicity, that $\mathcal{C}$ has no multiple arrows.

(a) The **path category** $K\mathcal{C}$ of $\mathcal{C}$ is the K-category defined as follows: the objects of $K\mathcal{C}$ are the points in $\mathcal{C}$, and the morphisms from $x \in \mathcal{C}_0$ to $y \in \mathcal{C}_0$ are the K-linear combinations of paths in $\mathcal{C}$ from x to y, with coefficients in K.

(b) We define an ideal $M_\mathcal{C}$ in the category $K\mathcal{C}$ as follows: to every non-projective point $x \in \mathcal{C}_0$ corresponds a mesh in $\mathcal{C}$ of the form

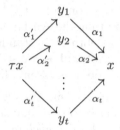

and to this mesh, we associate an element m_x of $\text{Hom}_{K\mathcal{C}}(\tau x, x)$, called the **mesh element**, and defined by the formula

$$m_x = \sum_{i=1}^{t} \alpha_i' \alpha_i.$$

We denote by $M_\mathcal{C}$ the ideal of $K\mathcal{C}$ generated by all the mesh elements m_x (where x ranges over all the non-projective points of $\mathcal{C}$).

(c) The **mesh category** is the quotient K-category $K(\mathcal{C}) = K\mathcal{C}/M_\mathcal{C}$.

Note that, an arbitrary element of the ideal $M_\mathcal{C}$ is a finite sum of the form $\sum_{i=1}^{m} u_i m_{x_i} v_i$, where $u_i m_{x_i} v_i$ is a morphism in $K\mathcal{C}$ from y_i to z_i (say) such that x_i is a non-projective point, u_i is a morphism from y_i to τx_i, and v_i is a morphism from x_i to z_i.

2.5. Definition. Let $\mathcal{C}$ be a component of the Auslander–Reiten quiver $\Gamma(\text{mod } A)$ of an algebra A.

(a) $\mathcal{C}$ is defined to be a **standard component** of $\Gamma(\text{mod } A)$ if there exists an equivalence of K-categories

$$K(\mathcal{C}) = K\mathcal{C}/M_\mathcal{C} \cong \text{ind}\,\mathcal{C},$$

where $\text{ind}\,\mathcal{C}$ is the full K-subcategory of $\text{mod } A$ whose objects are representatives of the isomorphism classes of the indecomposable modules in $\mathcal{C}$.

(b) $\mathcal{C}$ is defined to be a **self-hereditary component** of $\Gamma(\mathrm{mod}\,A)$ if, for each pair of indecomposable A-modules X and Y in $\mathcal{C}$, we have $\mathrm{Ext}_A^2(X,Y) = 0$.

It follows that the homomorphisms between two indecomposable modules lying in a standard component $\mathcal{C}$ of $\Gamma(\mathrm{mod}\,A)$ may be represented combinatorially.

As we shall see, this allows an easy computation of the homomorphism spaces between modules in $\mathcal{C}$.

One can show that every postprojective component $\mathcal{P}(A)$ without multiple arrows and every preinjective component $\mathcal{Q}(A)$ without multiple arrows is standard, and consequently that the Auslander–Reiten quiver $\Gamma(\mathrm{mod}\,A)$ of a representation-directed algebra A is standard. We do not need this fact, but we need that some stable tubes are standard.

The following important theorem is essentially due to Ringel [215].

2.6. Theorem. *Let A be an algebra, and $(E_1,\dots,E_r)$, with $r \geq 1$, be a τ_A-cycle of pairwise orthogonal bricks in $\mathrm{mod}\,A$ such that $\{E_1,\dots,E_r\}$ is a self-hereditary family of $\mathrm{mod}\,A$. The abelian category*

$$\mathcal{E} = \mathcal{EXT}_A(E_1,\dots,E_r),$$

has the following properties.

(a) *Every indecomposable object M of the category $\mathcal{E}$ is uniserial and is of the form $M \cong E_i[j]$, where $i \in \{1,\dots,r\}$ and $j \geq 1$.*

(b) *All indecomposable objects of the category $\mathcal{E} = \mathcal{EXT}(E_1,\dots,E_r)$ are uniserial in $\mathcal{E}$, and they form a self-hereditary component $\mathcal{T}_\mathcal{E}$ of $\Gamma(\mathrm{mod}\,A)$,*

(c) *The component $\mathcal{T}_\mathcal{E}$ is a standard stable tube of rank r.*

(d) *The modules $E_1,\dots,E_r$ form a complete set modules lying on the mouth of the tube $\mathcal{T}_\mathcal{E}$.*

Proof. By our assumption, $r \geq 1$ and the modules $E_1,\dots,E_r$ are pairwise orthogonal bricks such that $\tau_A E_1 \cong E_r, \tau_A E_2 \cong E_1, \dots, \tau_A E_r \cong E_{r-1}$ and $\mathrm{Ext}_A^2(E_i, E_s) = 0$, for all $i, s \in \{1,\dots,r\}$.

We first prove that every indecomposable object M in the category $\mathcal{E} = \mathcal{EXT}_A(E_1,\dots,E_r)$ is isomorphic to one of the objects $E_i[j]$ (for some $i \in \{1,\dots,r\}$, and some $j \geq 1$), as constructed in (2.2). Clearly, this implies that the indecomposable objects in $\mathcal{E}$ are uniserial in $\mathcal{E}$.

It follows from the definition of $\mathcal{E}$ that M contains a subobject of the form $E_i = E_i[1]$. Let $j \geq 1$ be maximal so that there exist $i \in \{1,\dots,r\}$, and a monomorphism $h : E_i[j] \longrightarrow M$. We claim that h is an isomorphism. Assume that this is not the case. The almost split sequence

$$0 \longrightarrow E_i[j] \xrightarrow{\left[\begin{smallmatrix} p_{ij} \\ u_{i,j+1} \end{smallmatrix}\right]} E_{i+1}[j-1] \oplus E_i[j+1] \xrightarrow{\left[\begin{smallmatrix} u_{i+1,j} & p_{i,j+1} \end{smallmatrix}\right]} E_{i+1}[j] \longrightarrow 0$$

yields a homomorphism

$$[h'' \; h'] : E_{i+1}[j-1] \oplus E_i[j+1] \longrightarrow M$$

such that $h = h''p_{ij} + h'u_{i,j+1}$. Because $\operatorname{Im} u'_{ij} \cong E_i[1]$ is the unique simple subobject of $E_i[j]$ in $\mathcal{E}$, and $h : E_i[j] \longrightarrow M$ is a monomorphism, we have $hu'_{ij} \neq 0$. On the other hand, the second short exact sequence of (2.2) yields $p_{ij}u'_{ij} = 0$. Hence

$$h'u'_{i,j+1} = h'u_{i,j+1}u'_{ij} + h''p_{ij}u'_{ij} = hu'_{ij} \neq 0.$$

But $E_i[j+1]$ is a uniserial object of $\mathcal{E}$ with unique simple subobject in $\mathcal{E}$ equal to $\operatorname{Im} u'_{i,j+1} \cong E_i[1]$. This implies that $h' : E_i[j+1] \longrightarrow M$ is a monomorphism, which contradicts our choice of j. Therefore, there is an isomorphism $M \cong E_i[j]$.

Let now $\mathcal{E}'$ denote the K-subcategory of $\mathcal{E}$ generated by the indecomposable modules $E_i[j]$, the identity homomorphisms, and the homomorphisms u_{ij}, p_{ij} between them. We claim that $\mathcal{E}'$ is a full subcategory of $\mathcal{E}$ and the equality $\mathcal{E}' = \mathcal{E}$ holds.

Let $f : E_i[j] \longrightarrow E_l[k]$ be a non-zero homomorphism in $\operatorname{mod} A$. We show by induction on $j + k$ that f belongs to $\mathcal{E}'$.

For $j + k = 2$, we have $j = k = 1$, and f is a homomorphism from E_i to E_l. Because $f \neq 0$, we have $i = l$ and, hence, f is an isomorphism of the form $\lambda \cdot 1_{E_i}$, for some $\lambda \in K \setminus \{0\}$, thus f belongs to $\mathcal{E}'$.

Assume that $j + k > 2$. Because $\operatorname{Im} f$ is a subobject of $E_l[k]$, which is uniserial in $\mathcal{E}$, then $\operatorname{Im} f$ is indecomposable in $\mathcal{E}$. Because any indecomposable object of $\mathcal{E}$ is isomorphic to an object of $\mathcal{E}'$, we may assume that f is an epimorphism or a monomorphism.

Suppose that f is an epimorphism. If f is not an isomorphism, then $\operatorname{Ker} f \neq 0$; hence $fu'_{ij} = 0$ and there exists a homomorphism $f' : E_{i+1}[j-1] \longrightarrow E_l[k]$ such that $f = f'p_{ij}$. By our induction hypothesis, f' lies in $\mathcal{E}'$, hence so does f. The proof is similar if f is a monomorphism, but not an isomorphism. Finally, if f is an isomorphism, then $i = l$, $j = k$ and $f = \lambda \cdot 1 + f'$, where $\lambda \in K$ is a non-zero scalar, and f' is not an isomorphism. As before, f' lies in $\mathcal{E}'$ and consequently so does f. This completes the proof of our claim that $\mathcal{E}' = \mathcal{E}$.

It follows easily from the description of the almost split sequences in (2.2) that the indecomposable objects in $\mathcal{E}$, namely the modules $E_i[j]$, form a component $\mathcal{T}_\mathcal{E}$ of $\Gamma(\operatorname{mod} A)$, and that this component is a stable tube of rank r. By (2.1), the component $\mathcal{T}_\mathcal{E}$ is self-hereditary, because we assume that $\{E_1, \ldots, E_r\}$ is a self-hereditary family of $\operatorname{mod} A$.

There remains to prove its standardness. We thus consider the mesh K-category $K(\mathcal{T}_\mathcal{E})$ of $\mathcal{T}_\mathcal{E}$ and the K-linear functor

$$F : K(\mathcal{T}_\mathcal{E}) \longrightarrow \operatorname{ind} \mathcal{T}_\mathcal{E}$$

that assigns to the point $E_i[j]$ in the tube $\mathcal{T}_\mathcal{E}$ the module $E_i[j]$, to the arrow $E_i[j-1]\longrightarrow E_i[j]$ in $\mathcal{T}_\mathcal{E}$ the homomorphism u_{ij}, and to the arrow $E_i[j]\longrightarrow E_{i+1}[j-1]$ the homomorphism p_{ij}. The above discussion shows that the functor F is full and dense. There remains to show that F is faithful.

We say that a homomorphism $E_i[j] \longrightarrow E_l[k]$ is standard provided it is a composition of homomorphisms of the form $p_{\alpha\beta}$ followed by homomorphisms of the form $u_{\gamma\delta}$.

Because the functor F is full, the standard homomorphisms from $E_i[j]$ to $E_l[k]$ generate the K-vector space $\operatorname{Hom}_A(E_i[j], E_l[k])$ (for any i, j, k, l).

Now we prove that the standard homomorphisms in $\operatorname{Hom}_A(E_i[j], E_l[k])$ are also linearly independent over K. Indeed, if f_t is a standard homomorphism in $\operatorname{Hom}_A(E_i[j], E_l[k])$ given as a composition of t homomorphisms of the form $p_{\alpha\beta}$ followed by homomorphisms of the form $u_{\gamma\delta}$, then it is easily seen that there is an isomorphism $\operatorname{Im} f_t \cong E_{i+t}[j-t]$.

Let now $f = \sum_t \lambda_t f_t$, with $\lambda_t \in K$. Assume that at least one of the scalars λ_t is non-zero, and let s be the smallest integer such that $\lambda_s \neq 0$. Then $\operatorname{Im}(\lambda_s f_s)$ is the unique subobject of $\mathcal{E}$-length $j - s$ in $E_l[k]$, and the module $\operatorname{Im}(\sum_{t \neq s} \lambda_t f_t)$ is properly contained in $\operatorname{Im}(\lambda_s f_s)$. Hence $f \neq 0$. This shows that the standard homomorphisms in $\operatorname{Hom}_A(E_i[j], E_l[k])$ form a K-basis of $\operatorname{Hom}_A(E_i[j], E_l[k])$. It follows that the functor F is faithful, and consequently, the tube $\mathcal{T}_\mathcal{E}$ is standard. $\qquad\square$

The preceding theorem provides us with a tool for computing homomorphisms between indecomposable modules in standard stable tubes.

2.7. Corollary. *Let A be an algebra, $\mathcal{T}$ a standard stable tube of rank $r \geq 1$ of $\Gamma(\operatorname{mod} A)$, and $\{E_1, \ldots, E_r\}$ a self-hereditary family of pairwise orthogonal bricks forming the mouth of the tube $\mathcal{T}$.*

(a) *In the notation of (2.2), the only homomorphisms between two indecomposable modules in $\mathcal{T}$ are K-linear combinations of compositions of the homomorphisms u_{ij}, p_{ij}, and the identity homomorphisms, and they are only subject to the relations arising from the almost split sequences in (2.2).*

(b) *In the notation of (2.2), given $i \in \{1, \ldots, r\}$ and $j \geq 1$, we have*
   - *$\operatorname{End} E_i[j] \cong K[t]/(t^m)$, for some $m \geq 1$,*
   - *$\operatorname{End} E_i[j] \cong K$ if and only if $j \leq r$, and*
   - *$\operatorname{Ext}_A^1(E_i[j], E_i[j]) \cong D\operatorname{Hom}_A(E_i[j], \tau E_i[j]) = 0$ if and only if $j \leq r - 1$.*

(c) *If the tube $\mathcal{T}$ is homogeneous then $\operatorname{Ext}_A^1(M, M) \neq 0$, for any indecomposable M in $\mathcal{C}$.*

Proof. Because the stable tube is standard and $\{E_1, \ldots, E_r\}$ a self-hereditary family of pairwise orthogonal bricks forming the mouth of the tube $\mathcal{T}$ then (2.2) and (2.6) applies to the extension category

$$\mathcal{E} = \mathcal{EXT}_A(E_1, \ldots, E_r)$$

and $\mathcal{T} = \mathcal{T}_\mathcal{E}$ is formed by the indecomposable uniserial modules of $\mathcal{E}$. Then (a) easily follows from (2.2) and (2.6), but (b) and (c) are an immediate consequence of (a). □

It follows from (2.7)(a) that a simple combinatorial argument allows to compute the Hom-spaces and the Ext-space between two indecomposable modules in a standard stable tube $\mathcal{T}$ satisfying the conditions of (2.7). The combinatorial technique is illustrated in the example (2.12) presented later.

We now prove two lemmata showing how modules in stable tubes and modules that lie outside stable tubes map to each other. The first is a new, more practical, version of (IV.5.1).

2.8. Lemma. *Let A be an algebra, $\mathcal{E} = \mathcal{EXT}_A(E_1, \ldots, E_r)$ be as in (2.6), $\mathcal{T}_\mathcal{E}$ be the tube of rank r formed by the indecomposable modules in $\mathcal{E}$, and let M be an indecomposable A-module that is not in $\mathcal{T}_\mathcal{E}$.*

(a) *If there exist $i \in \{1, \ldots, r\}$ and a non-zero homomorphism $f : E_i \longrightarrow M$ then, for any $j \geq 2$, there exists a homomorphism $g_j : E_i[j] \longrightarrow M$ such that $f = g_j u'_{ij}$.*

(b) *If there exist $i \in \{1, \ldots, r\}$ and a non-zero homomorphism $f : M \longrightarrow E_i$ then, for any $j \geq 2$, there exists a homomorphism $h_j : M \longrightarrow E_{i-j+1}[j]$ such that $f = p'_{i-1,2}h_j$.*

Proof. We only show (a), because the proof of (b) is similar. This is done by induction on j. Assume that $j = 2$. By hypothesis, there exists a non-isomorphism f from $E_i = E_i[1]$ to M, which must therefore factor through the left minimal almost split morphism $u'_{i2} = u_{i2} : E_i[1] \longrightarrow E_i[2]$. This gives g_2. Assume that $j > 2$, and that $g_j : E_i[j] \longrightarrow M$ such that $f = g_j u'_{ij}$ is given. Because M is not in $\mathcal{T}_\mathcal{E}$, g_j is not an isomorphism. Hence, it factors through the left minimal almost split morphism $\left[\begin{smallmatrix} u_{i,j+1} \\ p_{ij} \end{smallmatrix}\right]$: $E_i[j] \longrightarrow E_i[j+1] \oplus E_{i+1}[j-1]$, that is, there exists a homomorphism $[g_{j+1} \; g'_{j+1}] : E_i[j+1] \oplus E_{i+1}[j-1] \longrightarrow M$ such that

$$g_j = g_{j+1}u_{i,j+1} + g'_{j+1}p_{ij}.$$

Now it follows from the first exact sequence in (2.2) that $p_{ij}u'_{ij} = 0$. Thus $f = g_j u'_{ij} = g_{j+1}u_{i,j+1}u'_{ij} = g_{j+1}u'_{i,j+1}$. □

2.9. Lemma. *Let $\mathcal{E} = \mathcal{EXT}_A(E_1, \ldots, E_r)$ and $\mathcal{T}_\mathcal{E}$ be a standard stable tube as in (2.6), and M be an indecomposable A-module that is not in $\mathcal{T}_\mathcal{E}$.*

(a) *If there exists an indecomposable module L in the tube $\mathcal{T}_\mathcal{E}$ such that $\mathrm{Hom}_A(L, M) \neq 0$, then there exists an index $i \in \{1, \ldots, r\}$ such* *that* $\mathrm{Hom}_A(E_i, M) \neq 0$ *and, consequently,* $\mathrm{Hom}_A(E_i[j], M) \neq 0$, *for any* $j \geq 1$.

(b) *If there exists an indecomposable module* L *in the tube* $\mathcal{T}_\mathcal{E}$ *such that* $\mathrm{Hom}_A(M, L) \neq 0$, *then there exists an index* $i \in \{1, \ldots, r\}$ *such that* $\mathrm{Hom}_A(M, E_i) \neq 0$ *and, consequently,* $\mathrm{Hom}_A(M, E_{i-j+1}[j]) \neq 0$, *for any* $j \geq 1$.

Proof. We only prove (a), because the proof of (b) is similar. Let L be an indecomposable module in $\mathcal{T}_\mathcal{E}$, and $f : L \longrightarrow M$ be a non-zero homomorphism. There exist i and j such that $L \cong E_i[j]$, so that we have a monomorphism $u'_{ij} : E_i \longrightarrow L \cong E_i[j]$, with E_i lying on the mouth of $\mathcal{T}_\mathcal{E}$. If $fu'_{ij} \neq 0$, we are done. Assume that this is not the case, and consider the second exact sequence of (2.2)

$$0 \longrightarrow E_i[1] \xrightarrow{u'_{ij}} E_i[j] \xrightarrow{p_{ij}} E_{i+1}[j-1] \longrightarrow 0.$$

Because $fu'_{ij} = 0$, there exists $f' : E_{i+1}[j-1] \longrightarrow M$ such that $f = f'p_{ij}$. Our first claim thus follows by induction on j. The second claim is a direct consequence of the first, and of (2.8)(a). $\qquad\square$

In a characterisation of tilted algebras of Euclidean type given in Chapters XII and XVII, we essentially use a special class of stable tubes, namely the hereditary tubes in the following sense, see [238].

2.10. Definition. Let A be an algebra and $\mathcal{T}$ a stable tube of the Auslander–Reiten quiver $\Gamma(\mathrm{mod}\,A)$ of A.

(a) The tube $\mathcal{T}$ is defined to be **hereditary**, if $\mathrm{pd}\,X \leq 1$ and $\mathrm{id}\,X \leq 1$, for any indecomposable A-module X of $\mathcal{T}$.

(b) The tube $\mathcal{T}$ is defined to be **self-hereditary**, if $\mathrm{Ext}_A^2(X, Y) = 0$, for each pair of indecomposable A-modules X and Y of $\mathcal{T}$.

We remark that the standard stable tubes constructed in (2.2) and (2.6) are self-hereditary. Note also that, given a hereditary stable tube $\mathcal{T}$, the additive subcategory $\mathrm{add}\,\mathcal{T}$ of $\mathrm{mod}\,A$ is hereditary in the sense that $\mathrm{Ext}_A^2(M, N) = 0$, for each pair of the objects M and N of $\mathrm{add}\,\mathcal{T}$. Obviously, every stable tube $\mathcal{T}$ of the Auslander–Reiten quiver $\Gamma(\mathrm{mod}\,A)$ of a hereditary algebra A is hereditary and, clearly, every hereditary stable tube $\mathcal{T}$ is self-hereditary, but the converse implication does not hold in general. We show in Section 4 that faithful standard stable tubes are hereditary.

The following lemma shows that the hereditariness and the self-hereditariness of a stable tube $\mathcal{T}$ of rank $r \geq 1$ of $\Gamma(\mathrm{mod}\,A)$ is decided on the level of the set of all mouth modules of $\mathcal{T}$.

2.11. Lemma. *Let A be an algebra and $\mathcal{T}$ a stable in $\Gamma(\mathrm{mod}\,A)$.*

(a) *The tube $\mathcal{T}$ is self-hereditary if and only if the finite family of mouth modules of $\mathcal{T}$ is self-hereditary.*

(b) *The tube $\mathcal{T}$ is hereditary if and only if the finite family of mouth modules of $\mathcal{T}$ is hereditary.*

Proof. We only prove (a), because the proof of (b) is similar. The necessity is obvious. To prove the sufficiency, we assume that $\mathcal{T}$ is a stable tube of rank $r \geq 1$ in $\Gamma(\mathrm{mod}\,A)$ and $(X_1, \ldots, X_r)$ is a τ_A-cycle of mouth modules of $\mathcal{T}$ such that $\mathrm{Ext}_A^2(X_i, X_j) = 0$, for all $i, j \in \{1, \ldots, r\}$. We show that $\mathrm{Ext}_A^2(X, Y) = 0$, for each pair of indecomposable A-modules X and Y lying on the tube $\mathcal{T}$. Assume that X and Y are such modules. By (1.4), $X \cong X_i[m]$ and $Y \cong X_j[n]$, for some $i, j \in \{1, \ldots, r\}$ and some integers $m, n \geq 1$. First we assume that $n = 1$, that is $Y \cong X_j[1] = X_j$. We prove by induction on $m \geq 1$ that $\mathrm{Ext}_A^2(X_i[m], Y) = 0$. If $m = 1$ then $X_i[m] = X_i$ and we are done by our assumption. Assume that $m \geq 1$ is such that $\mathrm{Ext}_A^2(X_i[m], Y) = 0$ and $\mathrm{Ext}_A^2(X_i[m-1], Y) = 0$, for all $i \in \{1, \ldots, r\}$. By (1.4), there exists an almost split sequence

$$0 \longrightarrow X_i[m] \longrightarrow X_i[m+1] \oplus X_{i+1}[m-1] \longrightarrow X_{i+1}[m] \longrightarrow 0$$

in $\mathrm{mod}\,A$, where we set $X_j[0] = 0$ and $X_{i+kr}[t] = X_i[t]$, for all $i \in \{1, \ldots, r\}$, $t \geq 1$, and $k \in \mathbb{Z}$. Hence we derive the induced exact sequence

$$\mathrm{Ext}_A^2(X_{i+1}[m], Y) \longrightarrow \mathrm{Ext}_A^2(X_i[m+1] \oplus X_{i+1}[m-1], Y) \longrightarrow \mathrm{Ext}_A^2(X_i[m], Y).$$

The induction hypothesis yields $\mathrm{Ext}_A^2(X_{i+1}[m-1], Y) = 0$, the left hand term and the right hand term of the sequence is zero. It follows that $\mathrm{Ext}_A^2(X_{i+1}[m+1], Y) = 0$, for $i \in \{0, \ldots, r-1\}$. By the induction principle, we conclude that $\mathrm{Ext}_A^2(X, Y) = 0$, for Y and any indecomposable A-module X lying on the tube $\mathcal{T}$. Applying the same arguments, we prove by induction on $n \geq 1$ that $\mathrm{Ext}_A^2(X, X_j[n]) = 0$, for any indecomposable A-module X lying on the tube $\mathcal{T}$. This finishes the proof. $\square$

We end this section with a few examples of stable tubes. The first example shows that the vanishing of the second extension spaces of the brick is necessary for the validity of Theorem (2.6). The second one illustrates the use of Corollary (2.7), and the third one shows that there exist self-hereditary standard stable tubes $\mathcal{T}$ of $\Gamma(\mathrm{mod}\,A)$ that are not hereditary and gl.dim $A = 2$.

The reader is also referred to Section 4 for more examples of stable tubes. In particular, we construct there a standard stable tube $\mathcal{T}$ that is not self-hereditary, and consequently, theorems (2.2) and (2.6) do not apply to that tube $\mathcal{T}$.

2.11. Example. Let $A = K[t]/(t^2)$. Then A is a local self-injective Nakayama algebra, $S = K \cong K[t]/(t)$ is a unique non-projective indecomposable A-module, and, by (IV.4.1), the Auslander–Reiten quiver $\Gamma(\text{mod } A)$ of A is of the form

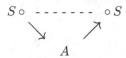

Observe that S is a brick and $\dim_K \text{Ext}^n_A(S, S) = 1$, for any $n \geq 1$.

We show later that there exist algebras A such that the Auslander–Reiten quiver $\Gamma(\text{mod } A)$ admits non-standard stable tubes, see (X.5.8).

2.12. Example. Let A be the path algebra of the quiver Q

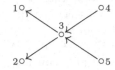

Because the underlying graph of Q is a Euclidean diagram (and thus is not a Dynkin diagram), the hereditary algebra $A = KQ$ is representation-infinite. By (VIII.2.1), $\Gamma(\text{mod } A)$ has a postprojective component and a preinjective component, and all its remaining components are regular (VIII.2.12). The straightforward calculation of the postprojective and the preinjective components shows that the simple module $S = S(3)$ is neither postprojective nor preinjective. Hence S is regular. We compute the component of $\Gamma(\text{mod } A)$ containing S. Applying the Nakayama functor $\nu = D\text{Hom}_A(-, A)$ to the minimal projective presentation

$$0 \longrightarrow P(1) \oplus P(2) \longrightarrow P(3) \longrightarrow S \longrightarrow 0$$

and using that $\nu P(a) = I(a)$, for any $a \in Q_0$, yields, by (IV.2.4), a left exact sequence

$$0 \longrightarrow \tau_A S \longrightarrow I(1) \oplus I(2) \longrightarrow I(3)$$

thus, $\tau_A S = E$ is the indecomposable module (viewed as a representation of Q)

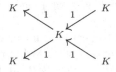

Computing again a minimal projective presentation for E, we get

$$0 \longrightarrow P(3) \longrightarrow P(5) \oplus P(4) \longrightarrow E \longrightarrow 0.$$

Applying the Nakayama functor yields a left exact sequence

$$0 \longrightarrow \tau_A E \longrightarrow I(3) \longrightarrow I(5) \oplus I(4)$$

so that $\tau_A E \cong S$. Because each of the modules S and E is clearly a brick, and the algebra A is hereditary, then $\mathrm{Ext}_A^2(-,-) = 0$ and we are in the situation of (2.6). Therefore the component $\mathcal{C}$ of $\Gamma(\mathrm{mod}\, A)$ containing S is a standard stable tube $\mathcal{T}$ of rank 2, which is of the form

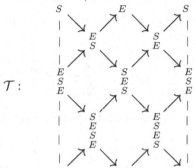

where the indecomposable modules are represented by their composition factors (see (I.3.10)) in the category $\mathcal{EXT}_A(E,S)$, and one identifies along the vertical dotted lines. The monomorphisms u_{ij} and the epimorphisms p_{ij} of (2.2) are particularly easy to understand in the above picture. We also show how to compute the indecomposable modules $\frac{E}{S}$ and $\frac{S}{E}$. The indecomposable module $\frac{S}{E}$ is the middle term of the almost split sequence

$$0 \longrightarrow E \longrightarrow \frac{S}{E} \longrightarrow S \longrightarrow 0$$

in $\mathrm{mod}\, A$, thus has a dimension vector equal to $\begin{smallmatrix}1\\1\end{smallmatrix}2\begin{smallmatrix}1\\1\end{smallmatrix}$, has S as a summand of its top, and E as a maximal submodule. It is then easy to see that $\frac{S}{E}$ and $\frac{E}{S}$ are given by the representations

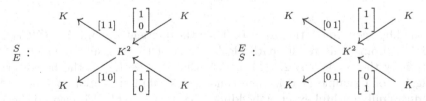

respectively. We notice that $\mathbf{dim}\,\frac{S}{E} = \mathbf{dim}\,\frac{E}{S}$, but clearly $\frac{S}{E} \not\cong \frac{E}{S}$.

It is particularly easy to compute the Hom-spaces between two modules of $\mathcal{C}$. This is done as for a Nakayama algebra (V.3), because the indecomposable objects of $\mathcal{C}$ are uniserial. Thus there exist two K-linearly independent homomorphisms from $\begin{smallmatrix}S\\E\\S\end{smallmatrix}$ to $\begin{smallmatrix}E\\S\\E\\S\end{smallmatrix}$ having as images the unique maximal subobject $\begin{smallmatrix}S\\E\\S\end{smallmatrix}$, and the simple object S in $\mathcal{EXT}_A(E,S)$. This also implies an easy computation of extension spaces, because, for M and N in $\mathcal{EXT}_A(E,S)$, we have $\mathrm{Ext}^1_A(M,N) \cong D\mathrm{Hom}_A(N,\tau_A M)$. For instance, taking $M = \begin{smallmatrix}S\\E\\S\\E\end{smallmatrix}$ and $N = \begin{smallmatrix}E\\S\\E\\S\end{smallmatrix}$, we have $\tau_A M \cong N$, and $\mathrm{End}\, N$ is two-dimensional. Hence there exist, up to scalars, two non-split extensions of M by N. One of them is the almost split sequence

$$0 \longrightarrow \begin{smallmatrix}E\\S\\E\\S\end{smallmatrix} \longrightarrow \begin{smallmatrix}E\\S\\E\end{smallmatrix} \oplus \begin{smallmatrix}S\\E\\S\\E\\S\end{smallmatrix} \longrightarrow \begin{smallmatrix}S\\E\\S\\E\end{smallmatrix} \longrightarrow 0.$$

The second non-split extension of M by N is the canonical exact sequence

$$0 \longrightarrow \begin{smallmatrix}E\\S\\E\\S\end{smallmatrix} \longrightarrow E \oplus \begin{smallmatrix}S\\E\\S\\E\\S\\E\\S\end{smallmatrix} \longrightarrow \begin{smallmatrix}S\\E\\S\\E\end{smallmatrix} \longrightarrow 0.$$

Now we give an example of an algebra C such that $\mathrm{gl.dim}\, C = 2$ and $\Gamma(\mathrm{mod}\, C)$ admits a self-hereditary standard stable tube that is not hereditary.

2.13. Example. Let C be the path algebra of the quiver

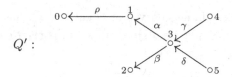

$$Q':$$

bound by one zero relation $\alpha\rho = 0$. Then the quotient algebra $A = C/Ce_0C$ of C is isomorphic to the path algebra KQ of the full subquiver Q of Ω given by the vertices 1, 2, 3, 4, and 5, that is $A \cong KQ$ is the hereditary algebra of Example (2.12). The canonical algebra surjection $C \longrightarrow A$ induces fully faithful exact embedding $\mathrm{mod}\, A \hookrightarrow \mathrm{mod}\, C$. It is easy to check that the indecomposable projective C-modules $P(0) = e_0C = S(0)$ and $P(1) = e_1C \cong I(0)$ are the unique indecomposable C-modules that do not lie in $\mathrm{mod}\, A \hookrightarrow \mathrm{mod}\, C$. The component $\mathcal{P}(C)$ of $\Gamma(\mathrm{mod}\, C)$ containing the module $P(0)$ is postprojective and has the form

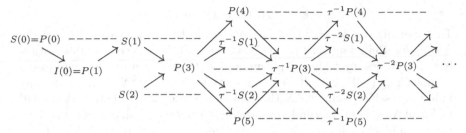

It is obtained from the unique postprojective component $\mathcal{P}(A)$ of $\Gamma(\mathrm{mod}\,A)$ by adding two points connected by one arrow corresponding to the irreducible embedding $P(0) = \mathrm{rad}\,P(1) \hookrightarrow P(1)$. The component $\mathcal{P}(C)$ contains all the indecomposable projective C-modules, up to isomorphism. Note that the indecomposable projective C-module $I(0) = P(1)$ is also injective.

It follows that the unique preinjective component $\mathcal{Q}(A)$ of $\Gamma(\mathrm{mod}\,A)$ is also the unique preinjective component $\mathcal{Q}(C)$ of $\Gamma(\mathrm{mod}\,C)$. It contains all the indecomposable injective C-modules, except the projective-injective C-module $I(0) = P(1)$, and is of the form

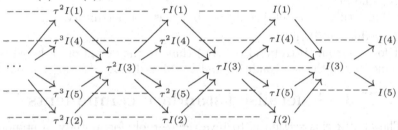

The standard stable tube $\mathcal{T}$ of rank $r = 2$ in $\Gamma(\mathrm{mod}\,A)$ with the mouth modules E and $S = S(3)$, constructed in Example (2.12), remains a standard stable tube in $\Gamma(\mathrm{mod}\,C)$, under the fully faithful exact embedding $\mathrm{mod}\,A \hookrightarrow \mathrm{mod}\,C$.

Now we show that
   (i) $\mathrm{pd}\,S(3) = 2$, $\mathrm{id}\,S(3) = 1$, $\mathrm{pd}\,E = 1$, and $\mathrm{id}\,E = 1$,
   (ii) $\mathrm{id}\,X = 1$, for any indecomposable module X lying in the tube $\mathcal{T}$,
   (iii) $\mathrm{gl.dim}\,C = 2$, and
   (iv) the tube $\mathcal{T}$ is self-hereditary, but it is not hereditary.
To prove (i), we note that A-modules $S(3)$ and E have minimal projective resolutions in $\mathrm{mod}\,C$ of the forms
$$0 \longrightarrow P(0) \longrightarrow P(1) \oplus P(2) \longrightarrow P(3) \longrightarrow S(3) \longrightarrow 0,$$
$$0 \longrightarrow P(3) \longrightarrow P(4) \oplus P(5) \longrightarrow E \longrightarrow 0,$$
and the minimal injective resolutions of $S(3)$ and of E are of the forms
$$0 \longrightarrow S(3) \longrightarrow I(3) \longrightarrow I(4) \oplus I(5) \longrightarrow 0,$$

$$0 \longrightarrow E \longrightarrow I(1) \oplus I(2) \longrightarrow I(3) \longrightarrow 0.$$

To prove (ii), we observe that the postprojective component $\mathcal{P}(C)$ of the Auslander–Reiten quiver $\Gamma(\operatorname{mod} C)$ contains all the indecomposable projective C-modules, up to isomorphism, and $\mathcal{P}(C)$ is closed under predecessors in $\operatorname{mod} C$, by (VIII. 2.5). It then follows that $\operatorname{Hom}_C(\tau_C^{-1}X, C_C) = 0$, for any indecomposable C-module X that is not postprojective. Thus, (IV.2.7)(b) yields $\operatorname{id} X \leq 1$, for any indecomposable C-module X that is not postprojective, and (ii) follows. Because one easily shows that $\operatorname{pd} S(j) \leq 1$, for all $j \neq 3$, then gl.dim $C = 2$. Because (ii) yields $\operatorname{Ext}_C^2(X, Y) = 0$, for any indecomposable modules X and Y in $\mathcal{T}$ then the tube $\mathcal{T}$ is self-hereditary.

Let $C' = C^{\mathrm{op}}$ be the algebra opposite to C. Because there is an isomorphism $A^{\mathrm{op}} \cong A$ of algebras then the preceding consideration implies that the Auslander–Reiten quiver $\Gamma(\operatorname{mod} C')$ dual to $\Gamma(\operatorname{mod} C)$ admits a standard stable tube $\mathcal{T}'$ of rank $r = 2$, with the mouth modules $E' = D(E)$ and $S'(3) = e_3 C' \cong D(S)$, that has the following properties

(i′) $\operatorname{id} S'(3) = 2$, $\operatorname{pd} S'(3) = 1$, $\operatorname{id} E' = 1$, and $\operatorname{pd} E' = 1$,

(ii′) $\operatorname{pd} Y = 1$, for any indecomposable module Y lying in the tube $\mathcal{T}'$,

(iii′) gl.dim $C' = 2$, and

(iv′) the tube $\mathcal{T}'$ is self-hereditary, but it is not hereditary.

that are dual to (i)-(iv).

It follows from (i)-(iv) and (i′)-(iv′) that, in the definition of hereditary tube, neither of the two conditions $\operatorname{pd} X \leq 1$ and $\operatorname{id} X = 1$ can be dropped.

X.3. Generalised standard components

The aim of this section is to investigate tools for a study of standard stable tubes of $\Gamma(\operatorname{mod} A)$ in terms of the infinite radical $\operatorname{rad}_A^\infty$ of the module category $\operatorname{mod} A$ of an algebra A. We do it by applying the following concept due to Skowroński [246].

3.1. Definition. A connected component $\mathcal{C}$ of the Auslander–Reiten quiver $\Gamma(\operatorname{mod} A)$ of an algebra A is defined to be **generalised standard** if $\operatorname{rad}_A^\infty(X, Y) = 0$, for each pair of indecomposable modules X and Y in $\mathcal{C}$.

The following proposition provides examples of generalised standard components.

3.2. Proposition. *Let A be an algebra and $\Gamma(\operatorname{mod} A)$ the Auslander–Reiten quiver of A.*

(a) *If A is representation-finite then the finite quiver $\Gamma(\operatorname{mod} A)$ is generalised standard.*

(b) *If $\mathcal{P}$ is a postprojective component of $\Gamma(\operatorname{mod} A)$ then $\mathcal{P}$ is generalised standard.*

(c) *If $\mathcal{Q}$ is a preinjective component of $\Gamma(\bmod A)$ then $\mathcal{Q}$ is generalised standard.*

(d) *Let $A = \operatorname{End} T_H$ be a tilted algebra, where T_H is a tilting module over a hereditary algebra H. If $\mathcal{C}_T$ is the connecting component of $\Gamma(\bmod A)$ determined by T then $\mathcal{C}_T$ is generalised standard.*

Proof. (a) Assume that A is a representation-finite algebra and let $d \geq 1$ be an integer such that $\dim_K X \leq d$, for every indecomposable A-module X. Then the Harada-Sai Lemma (IV.5.2) yields $\operatorname{rad}_A^m = 0$, for $m = 2^d - 1$. Then $\operatorname{rad}_A^\infty = 0$ and $\Gamma(\bmod A)$ is generalised standard.

(b) Assume that $\mathcal{P}$ is a postprojective component of $\Gamma(\bmod A)$ and let M be an indecomposable A-module in $\mathcal{P}$. It follows from (VIII.2.5)(a) that M has only finitely many predecessors in $\mathcal{P}$, and every indecomposable A-module L such that $\operatorname{Hom}_A(L, M) \neq 0$ is a predecessor of M in $\mathcal{P}$. Hence easily follows that $\operatorname{rad}_A^\infty(X, Y) = 0$, for each pair of indecomposable modules X and Y in $\mathcal{P}$, that is, $\mathcal{P}$ is a generalised standard component.

(c) Apply (VIII.2.5)(b) and dualise the arguments used in the proof of (b).

(d) Assume that $A = \operatorname{End} T_H$ is a tilted algebra, where T_H is a tilting module over a hereditary algebra H, and let $\mathcal{C}_T$ be the connecting component of $\Gamma(\bmod A)$ determined by T. It follows from (VIII.3.5) that the images $\operatorname{Hom}_A(T_A, I)$ of the indecomposable injective H-modules I, under the functor $\operatorname{Hom}_A(T_A, -) : \bmod H \longrightarrow \bmod A$, form a section Σ of $\mathcal{C}_T$ such that

- any predecessor of Σ in $\mathcal{C}_T$ lies in the torsion-free part $\mathcal{Y}(T)$, and
- any proper successor of Σ in $\mathcal{C}_T$ lies in the torsion part $\mathcal{X}(T)$

of the torsion pair $(\mathcal{X}(T), \mathcal{Y}(T))$ in $\bmod A$ induced by T_A.

Suppose, to the contrary, that there is a pair of indecomposable modules X and Y in $\mathcal{C}_T$ such that $\operatorname{rad}_A^\infty(X, Y) \neq 0$. Because $\operatorname{rad}_A^\infty(X, Y) = \bigcap_{m=0}^\infty \operatorname{rad}_A^m(X, Y)$ then, for each $t \geq 1$, there is a path of irreducible morphisms

$$X = Z_0 \xrightarrow{f_1} Z_1 \xrightarrow{f_2} Z_2 \longrightarrow \ \dots \ \longrightarrow Z_{t-1} \xrightarrow{f_t} Z_t,$$

between indecomposable modules in $\mathcal{C}_T$ and a homomorphism $g_t : Z_t \longrightarrow Y$ such that $g_t \cdot f_t \cdot \ldots \cdot f_2 \cdot f_1 \neq 0$. By applying the fact that the component $\mathcal{C}_T$ is acyclic and has only finitely many τ_A-orbits, we conclude that there exists an indecomposable module $Z = Z_t$ in $\mathcal{C}_T$ such that

- Z lies in $\mathcal{X}(T)$,
- Z is a proper successor of Y in $\mathcal{C}_T$, and
- $\operatorname{Hom}_A(Z, Y) \neq 0$.

Hence, in view of (IV.5.1)(b), we conclude that, for each $s \geq 1$, there is a path of irreducible morphisms

$$N_s \xrightarrow{h_s} N_{s-1} \longrightarrow \ldots \longrightarrow N_2 \xrightarrow{h_2} N_1 \xrightarrow{h_1} N_0 = Y,$$

between indecomposable modules in $\mathcal{C}_T$ and a homomorphism $u_s : Z \longrightarrow N_s$ such that $h_1 \cdot h_2 \cdot \ldots \cdot h_s \cdot u_s \neq 0$. Again, there exists an indecomposable module $N = N_t$ in $\mathcal{C}_T$ such that N lies in $\mathcal{Y}(T)$ and $\mathrm{Hom}_A(Z, N) \neq 0$. This is a contradiction, because $Z \in \mathcal{X}(T)$, $N \in \mathcal{Y}(T)$ and there are no non-zero homomorphisms from the torsion modules to torsion-free modules with respect to the torsion theory $(\mathcal{X}(T), \mathcal{Y}(T))$ in $\mathrm{mod}\,A$ and, hence, $\mathrm{Hom}_A(Z, N) = 0$. The contradiction proves that $\mathrm{rad}_A^\infty(X, Y) = 0$, for all indecomposable modules X and Y in $\mathcal{C}_T$. This finishes the proof. $\qquad\square$

The following result shows that the generalised standardness of a stable tube $\mathcal{T}$ is equivalent to the vanishing of the infinite radical rad_A^∞ on the mouth of $\mathcal{T}$.

3.3. Proposition. *Let A be an algebra and $\mathcal{T}$ a stable tube of rank $r \geq 1$ of the Auslander–Reiten quiver $\Gamma(\mathrm{mod}\,A)$ of A. Then $\mathcal{T}$ is generalised standard if and only if $\mathrm{rad}_A^\infty(E, E') = 0$, for each pair of mouth modules E and E' of $\mathcal{T}$.*

Proof. Assume that $\mathcal{T}$ is a stable tube of rank $r \geq 1$ of $\Gamma(\mathrm{mod}\,A)$. Then the mouth modules of $\mathcal{T}$ form a τ_B-cycle $(E_1, \ldots, E_r)$.

The necessity part of the proposition is obvious. To prove the sufficiency, we assume, to the contrary, that the stable tube $\mathcal{T}$ is not generalised standard, that is, there exist indecomposable modules X and Y in $\mathcal{T}$ such that $\mathrm{rad}_A^\infty(X, Y) \neq 0$. By (1.4), there exist $j, s \in \{1, \ldots, r\}$ and integers $n, m \geq 1$ such that $X \cong E_j[n]$ and $Y \cong E_s[m]$. Throughout, we freely use the notation of (1.4).

Choose two indices $j, s \in \{1, \ldots, r\}$ and two integers $n, m \geq 1$ such that $\mathrm{rad}_A^\infty(E_j[n], E_s[m]) \neq 0$ and $n + m \geq 2$ is minimal, with respect to this property.

We prove the proposition by showing that $n + m = 2$, that is, $n = 1$ and $m = 1$, and therefore $X \cong E_j[1] = E_j$ and $Y \cong E_s[1] = E_s$. Assume, to the contrary, that $n + m \geq 3$. Without loss of generality, we may suppose that $m \geq 2$. Then, according to (1.4), there exists an almost split sequence

$$0 \longrightarrow E_s[m{-}1] \xrightarrow{\left[\begin{smallmatrix} f \\ g \end{smallmatrix}\right]} E_s[m] \oplus E_{s+1}[m{-}2] \xrightarrow{[u,\,v]} E_{s+1}[m{-}1] \longrightarrow 0$$

in $\mathrm{mod}\,A$, where we set $E_s[m{-}2] = 0$, $g = 0$, and $v = 0$, if $m = 2$.

Let $h \in \mathrm{rad}_A^\infty(E_j[n], E_s[m])$ be a non-zero homomorphism. Because $uh \in \mathrm{rad}_A^\infty(E_j[n], E_{s+1}[m{-}1])$ then the minimality of $n + m$ yields $uh = 0$.

Consider the homomorphism $\left[\begin{smallmatrix} h \\ 0 \end{smallmatrix}\right] : E_j[n] \longrightarrow E_s[m] \oplus E_{s+1}[m-2]$ and note that $[u,\, v] \cdot \left[\begin{smallmatrix} h \\ 0 \end{smallmatrix}\right] = uh = 0$. It follows that there exists a homomorphism $h' : E_j[n] \longrightarrow E_s[m-1]$ such that $\left[\begin{smallmatrix} h \\ 0 \end{smallmatrix}\right] = \left[\begin{smallmatrix} f \\ g \end{smallmatrix}\right] \cdot h'$ and, hence, $h = fh'$. In the notation of (1.4), the homomorphism $f : E_s[m-1] \longrightarrow E_s[m]$ is an irreducible morphism in mod A corresponding to an arrow of the ray $(\mathfrak{r}_s)$ of $\mathcal{T}$ starting at E_s. Because the homomorphism $h \in \operatorname{rad}_A^\infty(E_j[n], E_s[m])$ is non-zero then (1.9) yields $0 \neq h \in \operatorname{rad}_A^\infty(E_j[n], E_s[m-1])$, and we get a contradiction with the minimality of $n + m$. It follows that $n = 1$, $m = 1$ and we get $\operatorname{rad}_A^\infty(X, Y) \cong \operatorname{rad}_A^\infty(E_j[1], E_s[1]) = \operatorname{rad}_A^\infty(E_j, E_s) = 0$. $\square$

X.4. Generalised standard stable tubes

The main objective of this section is to investigate standard stable tubes of $\Gamma(\operatorname{mod} A)$ in terms of the infinite radical $\operatorname{rad}_A^\infty$ of the module category mod A of an algebra A. We present a characterisation of standard stable tubes of $\Gamma(\operatorname{mod} A)$ in terms of $\operatorname{rad}_A^\infty$ and we prove that a stable tube $\mathcal{T}$ of $\Gamma(\operatorname{mod} A)$ is standard if the mouth of $\mathcal{T}$ consists of pairwise orthogonal bricks. We also show that any faithful generalised standard stable tube is hereditary. Our presentation is mainly based on [246], [247], and [254].

Let A be an algebra and M an A-module. We recall that the **right annihilator** of M is the two-sided ideal $\operatorname{Ann}_A M = \{a \in A; \ Ma = 0\}$ of A. Recall also that the module M is said to be **faithful** if the ideal $\operatorname{Ann}_A M$ is zero.

4.1. Definition. Let A be an algebra and $\mathcal{C}$ a component of the Auslander–Reiten quiver $\Gamma(\operatorname{mod} A)$ of A.

(a) The **annihilator** of $\mathcal{C}$ is the intersection $\operatorname{Ann}_A \mathcal{C} = \bigcap\limits_{X \in \mathcal{C}} \operatorname{Ann}_A X$ of the annihilators of all indecomposable A-modules X lying in $\mathcal{C}$.

(b) The component $\mathcal{C}$ is said to be **faithful** if $\operatorname{Ann}_A \mathcal{C} = 0$.

We remark that if $\mathcal{C}$ is a component of $\Gamma(\operatorname{mod} A)$ and $B = A/\operatorname{Ann}_A \mathcal{C}$ then $\mathcal{C}$ is a faithful component of $\Gamma(\operatorname{mod} B)$ under the fully faithful exact embedding mod $B \hookrightarrow$ mod A induced by the canonical algebra surjection $A \longrightarrow B$.

The following simple lemma is very useful.

4.2. Lemma. *Let A be an algebra, and $\mathcal{C}$ be a component of $\Gamma(\operatorname{mod} A)$.*

(a) *There exists a module M in add $\mathcal{C}$ such that $\operatorname{Ann}_A \mathcal{C} = \operatorname{Ann}_A M$.*

(b) *The component $\mathcal{C}$ is faithful if and only if the category add $\mathcal{C}$ admits a faithful A-module.*

Proof. (a) Given a module X in add $\mathcal{C}$ there is an isomorphism of A-modules $X \cong X_1 \oplus X_2 \oplus \ldots \oplus X_s$, where $X_1, X_2, \ldots, X_s$ are indecomposable A-modules in $\mathcal{C}$. Then

$$\mathrm{Ann}_A X = \mathrm{Ann}_A(X_1 \oplus X_2 \oplus \ldots \oplus X_s) = \mathrm{Ann}_A X_1 \cap \mathrm{Ann}_A X_2 \cap \ldots \cap \mathrm{Ann}_A X_s.$$

Moreover, $\mathrm{Ann}_A Y \supseteq \mathrm{Ann}_A Z$ if Y is a submodule of Z. Then the ideals of the form $\mathrm{Ann}_A(X_1 \oplus X_2 \oplus \ldots \oplus X_s)$, where $X_1, X_2, \ldots, X_s$ are indecomposable A-modules in $\mathcal{C}$, form a partially ordered set, with respect to the inclusion. Because the algebra A is finite dimensional then the family contains a minimal element $\mathrm{Ann}_A(M_1 \oplus \ldots \oplus M_\ell)$, for some indecomposable A-modules $M_1, \ldots, M_\ell$ in $\mathcal{C}$. It follows that $\mathrm{Ann}_A \mathcal{C} = \mathrm{Ann}_A M$, where $M = M_1 \oplus \ldots \oplus M_\ell$ is a module in add $\mathcal{C}$.

The statement (b) follows immediately from (a). $\qquad\square$

4.3. Example. Let $A = KQ$ be the path algebra of the Euclidean quiver

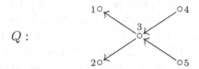

$$Q:$$

Then A has the lower triangular matrix form

$$A = \begin{bmatrix} K & 0 & 0 & 0 & 0 \\ 0 & K & 0 & 0 & 0 \\ K & K & K & 0 & 0 \\ K & K & K & K & 0 \\ K & K & K & 0 & K \end{bmatrix}.$$

Let $\mathcal{C}$ be the standard stable tube of rank 2 of $\Gamma(\mathrm{mod}\, A)$ constructed in Example (2.12). Then the mouth of $\mathcal{C}$ consists of two modules S and E, where $S = S(3)$ is the simple module at the vertex 3 and E is the regular module

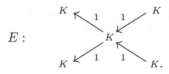

$$E:$$

It is clear that E is faithful, that is, $\mathrm{Ann}_A E = 0$. It follows that $\mathrm{Ann}_A \mathcal{C} = 0$, that is, the tube $\mathcal{C}$ is faithful. Observe also that the annihilator $\mathrm{Ann}_A S(3)$ of the simple module $S = S(3)$ is the two-sided ideal

$$\mathrm{Ann}_A S(3) = \begin{bmatrix} K & 0 & 0 & 0 & 0 \\ 0 & K & 0 & 0 & 0 \\ K & K & 0 & 0 & 0 \\ K & K & K & K & 0 \\ K & K & K & 0 & K \end{bmatrix} = AeA,$$

where $e = e_1+e_2+e_4+e_5$ and e_1, e_2, e_4, and e_5 are the primitive idempotents of A corresponding to the vertices 1, 2, 4, and 5 of the quiver Q.

Now we prove that any faithful generalised standard stable tube is hereditary.

4.4. Theorem. *Let A be an algebra and $\mathcal{T}$ a faithful generalised standard stable tube of $\Gamma(\operatorname{mod} A)$. Then $\operatorname{pd} X \leq 1$ and $\operatorname{id} X \leq 1$, for any indecomposable module X of $\mathcal{T}$, that is, the tube $\mathcal{T}$ is hereditary.*

Proof. Assume that $\mathcal{T}$ is a faithful generalised standard stable tube of $\Gamma(\operatorname{mod} A)$ and let X be an indecomposable A-module in $\mathcal{T}$. We only prove that $\operatorname{pd} X \leq 1$, because the proof of the inequality $\operatorname{id} X \leq 1$ is similar.

Assume, to the contrary, that $\operatorname{pd} X \geq 2$. Then, it follows from (IV.2.7) that $\operatorname{Hom}_A(D(_AA), \tau_A X) \neq 0$. Let $f : D(_AA) \longrightarrow \tau_A X$ be a non-zero homomorphism in $\operatorname{mod} A$. Because the stable tube $\mathcal{T}$ is faithful then the category $\operatorname{add} \mathcal{T}$ admits a faithful A-module M, by (4.2), and it follows from (VI.2.2) that the A-module $D(_AA)$ is cogenerated by M, that is, there exist an integer $t \geq 1$ and an epimorphism $h : M^t \longrightarrow D(_AA)$ of A-modules. Hence, there exists an indecomposable direct summand Z of M^t such that the composite homomorphism $Z \overset{g}{\longrightarrow} D(_AA) \overset{f}{\longrightarrow} \tau_A X$ is non-zero, where g is the restriction of h to the summand Z of M^t. Note that the A-module $D(_AA)$ is injective and the tube $\mathcal{T}$ contains no indecomposable injective A-modules. Then the indecomposable modules Z and $\tau_A X$ are not injective, because they lie in the tube $\mathcal{T}$. This yields

- $\operatorname{rad}_A(Z, D(_AA)) = \operatorname{rad}_A^\infty(Z, D(_AA))$, and
- $\operatorname{rad}_A(D(_AA), \tau_A X) = \operatorname{rad}_A^\infty(D(_AA), \tau_A X)$.

Consequently, we get $0 \neq fg \in \operatorname{rad}_A^\infty(Z, \tau_A X)$. This contradicts the assumption that the tube $\mathcal{T}$ is generalised standard, and finishes the proof of the theorem. $\square$

Now we present a characterisation of (generalised) standard stable tubes.

4.5. Theorem. *Let A be an algebra and let $\mathcal{T}$ be a stable tube of $\Gamma(\operatorname{mod} A)$. The following three statements are equivalent.*

(a) $\mathcal{T}$ *is standard.*
(b) *The mouth of $\mathcal{T}$ consists of pairwise orthogonal bricks.*
(c) $\mathcal{T}$ *is generalised standard.*

Proof. Assume that $\mathcal{T}$ is a stable tube of rand $r \geq 1$ in $\Gamma(\operatorname{mod} A)$. Then the mouth A-modules of $\mathcal{T}$ form a τ_A-cycle $(E_1, \ldots, E_r)$.

(a)$\Rightarrow$(b) Assume that the stable tube $\mathcal{T}$ is standard, that is, there is an equivalence of K-categories $K(\mathcal{T}) = K\mathcal{T}/M_{\mathcal{T}} \cong \operatorname{ind} \mathcal{T}$, where $\operatorname{ind} \mathcal{T}$ is the full K-subcategory of $\operatorname{mod} A$ whose objects are representatives of the isomorphism classes of the indecomposable modules in $\mathcal{T}$ and $K(\mathcal{T}) = K\mathcal{T}/M_{\mathcal{T}}$ is the mesh category of $\mathcal{T}$. Therefore, it follows that, for each $i \in \{1, \ldots, r\}$ and $j \geq 2$, there exist irreducible morphisms

$$u_{i,j} : E_i[j-1] \longrightarrow E_i[j] \quad \text{and} \quad p_{i,j} : E_i[j] \longrightarrow E_{i+1}[j-1]$$

such that the following mesh relations are satisfied:

(i) $p_{i2}u_{i2} = 0$, for all $i \in \{1, \ldots, r\}$, and
(ii) $u_{i+1,j}p_{ij} + p_{i,j+1}u_{i,j+1} = 0$, for all $i \in \{1, \ldots, r\}$ and $j \geq 2$.

Moreover, for each pair of indecomposable modules X and Y in the tube $\mathcal{T}$, any homomorphism $f : X \longrightarrow Y$ is a K-linear combination of compositions of the irreducible morphisms $u_{i,j}$, $p_{i,j}$, and the identity homomorphisms; and subject to the relations (i) and (ii). It follows that

- End $E_i[1] \cong K$, for all $i \in \{1, \ldots, r\}$, and
- $\mathrm{Hom}_A(E_i, E_s) = \mathrm{Hom}_A(E_i[1], E_s[1]) = 0$, for all $i, s \in \{1, \ldots, r\}$ with $i \neq s$,

that is, the mouth modules $E_1, \ldots, E_r$ are pairwise orthogonal bricks.

(b)$\Rightarrow$(c) Assume that the mouth modules $E_1, \ldots, E_r$ are pairwise orthogonal bricks. Then, for all $i, s \in \{1, \ldots, r\}$, we have $\mathrm{rad}_A(E_i, E_s) = \mathrm{Hom}_A(E_i, E_s) = 0$ and, hence, $\mathrm{rad}_A^\infty(E_i, E_s) = 0$. Then, it follows from (3.3) that the stable tube $\mathcal{T}$ is generalised standard.

(c)$\Rightarrow$(a) Assume that the stable tube $\mathcal{T}$ is generalised standard. Let $I = \mathrm{Ann}_A \mathcal{T}$ be the annihilator of $\mathcal{T}$ and we set $B = A/I$. Then $\mathcal{T}$ is a stable tube in $\Gamma(\mathrm{mod}\, B)$, under the fully faithful exact embedding $\mathrm{mod}\, B \hookrightarrow \mathrm{mod}\, A$ induced by the canonical algebra surjection $A \longrightarrow B$. Note also that then $\mathcal{T}$ is a generalised standard stable tube in $\Gamma(\mathrm{mod}\, B)$. Moreover, by (3.2), $\mathcal{T}$ is a faithful tube of $\Gamma(\mathrm{mod}\, B)$. Hence, by (4.4), $\mathcal{T}$ is hereditary, that is, $\mathrm{pd}_B X \leq 1$ and $\mathrm{id}_B X \leq 1$, for any indecomposable module X in $\mathcal{T}$. It follows that $\mathrm{Ext}_B^2(E_i, E_s) = 0$, for all $i, s \in \{1, \ldots, r\}$.

Because the subcategory $\mathrm{ind}\,\mathcal{T}$ of $\mathrm{mod}\, A$ lies in the subcategory $\mathrm{mod}\, B \hookrightarrow \mathrm{mod}\, A$ of $\mathrm{mod}\, A$ then, in view of (2.6), to show that $\mathcal{T}$ is a standard stable tube of $\mathrm{mod}\, A$ it is sufficient to prove that the mouth modules $E_1, \ldots, E_r$ of $\mathcal{T}$ are pairwise orthogonal bricks.

To prove the later statement, we apply (1.7). For each $i \in \{1, \ldots, r\}$ and $j \geq 2$, we choose irreducible morphisms

$$v_{i,j} : X_i[j-1] \longrightarrow X_i[j] \quad \text{and} \quad q_{i,j} : X_i[j] \longrightarrow X_{i+1}[j-1]$$

in $\mathrm{mod}\, A$ such that

- $q_{i,2}v_{i,2} \in \mathrm{rad}_A^3(X_i[1], X_{i+1}[1])$, for all $i \in \{1, \ldots, r\}$,
- $v_{i+1,j}q_{i,j} + q_{i,j+1}v_{i,j+1} \in \mathrm{rad}_A^3(X_i[j], X_{i+1}[j])$, for all $i \in \{1, \ldots, r\}$ and $j \geq 2$.

Now we fix $i, k \in \{1, \ldots, r\}$. Without loss of generality, we may assume that $i \leq k$. Then $E_i \cong \tau_A^s E_k$, where $s = k - i \geq 0$. To prove that

$\text{Hom}_A(E_i, E_k) = 0$, for $i \neq k$, and $\text{Hom}_A(E_i, E_i) \cong K$, we show that $\text{rad}_A(E_i, E_k) = 0$.

It is easy to see that

- any non-trivial path in $\mathcal{T}$ from $E_i = E_i[1]$ to $E_k = E_k[1]$ is of length $2s + 2\ell r$, where $\ell \geq 0$, $s = k - i \geq 0$ and $r \geq 1$ is the rank of the tube $\mathcal{T}$,
- $\text{rad}_A(E_i, E_k) = \text{rad}_A^{2s}(E_i, E_k)$, if $i \neq k$,
- $\text{rad}_A(E_i, E_k) = \text{rad}_A^{2r}(E_i, E_k)$, if $i = k$, and
- $\text{rad}_A^{2s+2\ell r+1}(E_i, E_k) = \text{rad}_A^{2s+2(\ell+1)r}(E_i, E_k)$, for any $\ell \geq 0$, and
- $\text{rad}_A^{\infty}(E_i, E_k) = \text{rad}_A^{m}(E_i, E_k) = 0$, for some $m \geq 0$.

The final statement follows from (1.5) and the assumption that the tube $\mathcal{T}$ is generalised standard.

Then, to prove the equality $\text{rad}_A(E_i, E_k) = 0$, it is sufficient to show that the inclusion $\text{rad}_A^p(E_i, E_k) \subseteq \text{rad}_A^{p+1}(E_i, E_k)$ holds, for any $p \in \{1, \dots, m-1\}$.

Fix $p \in \{1, \dots, m-1\}$ and choose a homomorphism $h \in \text{rad}_A^p(E_i, E_k)$. We may assume that h is non-zero, $p \geq 2s$ (for $i \neq k$), and $p \geq 2r$ (for $i = k$). Observe that rad_A^p, viewed as a left ideal of the category $\text{mod}\,A$, is generated by the compositions of p irreducible morphisms between indecomposable modules in $\text{mod}\,A$, apply (IV.5.1). Because $\text{rad}_A^m(M, N) = 0$ then, according to (IV.5.1), the non-zero homomorphism h has the form $h = h_1 + h_2 + \dots + h_n$, where the summand h_t is the composite homomorphism

$$E_i = X_{t1} \xrightarrow{h_{t1}} X_{t2} \longrightarrow \dots \longrightarrow X_{t p_t} \xrightarrow{h_{t p_t}} X_{t,p_t+1} = E_k,$$

$p_t \geq p$ and $h_{t1}, \dots, h_{t p_t}$ are irreducible morphisms between indecomposable modules in $\text{mod}\,A$, for $t \in \{1, \dots, n\}$. It follows that, for each $t \in \{1, \dots, n\}$, there exists $j_t \in \{2, \dots, p_t - 1\}$ such that $X_{t,j_t+1} \cong E_{i+1}[j_t-1]$ and $X_{tj} \cong E_i[j]$, for $j \in \{1, \dots, j_t\}$. Because we have

- $\dim_K \left[\text{rad}_A(E_i[j_t], E_{i+1}[j_t-1]) / \text{rad}_A^2(E_i[j_t], E_{i+1}[j_t-1]) \right] = 1$, and
- $\dim_K \left[\text{rad}_A(E_i[j], E_i[j+1]) / \text{rad}_A^2(E_i[j], E_i[j+1]) \right] = 1$, for all $j \in \{1, \dots, j_t - 1\}$,

then there exist scalars $\lambda_1^{(t)}, \dots, \lambda_{j_t}^{(t)} \in K \setminus \{0\}$ such that

- $h_{t j_t} - \lambda_{j_t}^{(t)} q_{i,j_t} \in \text{rad}_A^2(E_i[j_t], E_{i+1}[j_t-1])$, and
- $h_{tj} - \lambda_j^{(t)} v_{i,j} \in \text{rad}_A^2(E_i[j], E_i[j+1]) = 1$, for all $j \subset \{1, \dots, j_t - 1\}$.

Therefore, for each $t \in \{1, \dots, n\}$, we get the equalities

$$h_t + \mathrm{rad}_A^{p+1}(E_i, E_k) = h_{tp_t} \cdot \ldots \cdot h_{t2} \cdot h_{t1} + \mathrm{rad}_A^{p+1}(E_i, E_k)$$

$$= \lambda_1^{(t)} \cdot \ldots \cdot \lambda_{j_t}^{(t)} h_{tp_t} \cdot \ldots \cdot h_{tj_t+1} \cdot q_{i,j_t} \cdot v_{i,j_t} \cdot \ldots \cdot v_{i,2} + \mathrm{rad}_A^{p+1}(E_i, E_k)$$

$$= \lambda_1^{(t)} \cdot \ldots \cdot \lambda_{j_t}^{(t)} h_{tp_t} \cdot \ldots \cdot h_{tj_t+1} \cdot v_{i+1,j_t-1} \cdot \ldots \cdot v_{i+1,2} \cdot q_{i,2} \cdot v_{i,2}$$

$$+ \mathrm{rad}_A^{p+1}(E_i, E_k)$$

$$= 0 + \mathrm{rad}_A^{p+1}(E_i, E_k),$$

because $q_{i,2} \cdot v_{i,2} \in \mathrm{rad}_A^3$.

It follows that $h_1, \ldots, h_n \in \mathrm{rad}_A^{p+1}(E_i, E_k)$ and, hence, $h = h_1 + \ldots + h_n \in \mathrm{rad}_A^{p+1}(E_i, E_k)$. This shows that the inclusion

$$\mathrm{rad}_A^p(E_i, E_k) \subseteq \mathrm{rad}_A^{p+1}(E_i, E_k)$$

holds, for any $p \in \{1, \ldots, m-1\}$, and consequently, we get $\mathrm{rad}_A(E_i, E_k) = \mathrm{rad}_A^m(E_i, E_k) = \mathrm{rad}_A^\infty(E_i, E_k) = 0$, because the component $\mathcal{T}$ is generalised standard. Then the proof of the theorem is complete.   □

4.6. Corollary. *Let A be an algebra, $\mathcal{T}$ a standard stable tube of the Auslander–Reiten quiver $\Gamma(\mathrm{mod}\, A)$ of A, and $B = A/\mathrm{Ann}_A \mathcal{T}$. Then $\mathcal{T}$ is a hereditary standard stable tube of $\Gamma(\mathrm{mod}\, B)$, under the fully faithful exact embedding $\mathrm{mod}\, B \hookrightarrow \mathrm{mod}\, A$ induced by the canonical algebra surjection $A \longrightarrow B$.*

Proof. Apply (4.4) and (4.5)   □

4.7. Corollary. *Let A be an algebra and $\mathcal{T}$ a faithful stable tube of $\Gamma(\mathrm{mod}\, A)$. The following three conditions are equivalent.*

  (a) *$\mathcal{T}$ is standard.*
  (b) *$\mathcal{T}$ is hereditary and the mouth modules of $\mathcal{T}$ are pairwise orthogonal bricks.*
  (c) *$\mathcal{T}$ is self-hereditary and the mouth modules of $\mathcal{T}$ are pairwise orthogonal bricks.*

Proof. The implication (a)⇒(b) is a consequence of (4.5), and the implication (b)⇒(c) is obvious.

To prove the implication (c)⇒(a), we assume that $\mathcal{T}$ is a faithful stable tube of rank $r = r_{\mathcal{T}} \geq 1$ of $\Gamma(\mathrm{mod}\, A)$, and $(E_1, \ldots, E_r)$ is a τ_A-cycle of mouth modules of $\mathcal{T}$. By our assumption, $\mathcal{T}$ is self-hereditary and the modules $E_1, \ldots, E_r$ are pairwise orthogonal bricks. Then, by applying (2.6) to the extension category $\mathcal{E} = \mathcal{EXT}_A(E_1, \ldots, E_r)$, we conclude that the stable tube $\mathcal{T}$ is standard.   □

We end this section with two examples of (generalised) standard stable tubes. In the first one we construct an algebra B such that $\mathrm{gl.dim}\, B = 3$ and

$\Gamma(\mathrm{mod}\,B)$ admits a (generalised) standard stable tube that is self-hereditary, but is neither faithful nor hereditary. In the second one we construct an algebra R, with gl.dim $R = \infty$, such that $\Gamma(\mathrm{mod}\,R)$ admits a standard stable tube $\mathcal{T}$ that is not self-hereditary.

4.8. Example. Let B be the path algebra of the quiver

Δ :

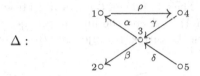

bound by two zero relations $\alpha\rho = 0$ and $\rho\gamma = 0$. It is easy to see that the quotient algebra $A = B/\mathcal{I}$, where $\mathcal{I}$ is the two sided ideal of B generated by the arrow ρ, is isomorphic to the path algebra KQ of the quiver

Q :

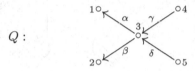

considered in Example (2.12). Then we have a fully faithful exact embedding $\mathrm{mod}\,A \hookrightarrow \mathrm{mod}\,B$ induced by the canonical algebra surjection $B \longrightarrow A$. It is easy to see that there is precisely one indecomposable A-module X, up to isomorphism, such that X does not lie in the subcategory $\mathrm{mod}\,A$ of $\mathrm{mod}\,B$. The module X is isomorphic with the unique projective-injective B-module

$P(1) = I(4)$:

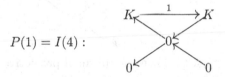

It follows that the standard stable tube $\mathcal{T}$ of rank 2 of $\Gamma(\mathrm{mod}\,A)$ constructed in (2.12) remains a standard stable tube $\mathcal{T}$ of $\Gamma(\mathrm{mod}\,B)$ and the annihilator $\mathrm{Ann}_B\mathcal{T}$ of $\mathcal{T}$ is just the ideal $\mathcal{I}$ of B generated by the arrow ρ.

The simple B-module $S = S(3)$ at the vertex 3 lying on the mouth of the tube $\mathcal{T}$ has a minimal projective resolution in $\mathrm{mod}\,B$ of the form

$$0 \longrightarrow P(3) \longrightarrow P(4) \longrightarrow P(1) \oplus P(2) \longrightarrow P(3) \longrightarrow S(3) \longrightarrow 0,$$

a minimal injective resolution in $\mathrm{mod}\,B$ of the form

$$0 \longrightarrow S(3) \longrightarrow I(3) \longrightarrow I(4) \oplus I(5) \longrightarrow I(1) \longrightarrow I(3) \longrightarrow 0.$$

Hence, pd $S(3) = 3$ and id $S(3) = 3$. It follows that gl.dim $B = 3$, because the simple B-module $S(2)$ is projective and the remaining simple B-modules $S(1)$, $S(4)$, and $S(5)$ have minimal projective resolutions in $\mathrm{mod}\,B$ of the forms

$$0 \longrightarrow P(3) \longrightarrow P(4) \longrightarrow P(1) \longrightarrow S(1) \longrightarrow 0,$$
$$0 \longrightarrow P(3) \longrightarrow P(4) \longrightarrow S(4) \longrightarrow 0,$$
$$0 \longrightarrow P(3) \longrightarrow P(5) \longrightarrow S(5) \longrightarrow 0.$$

Observe also that the quiver $\Gamma(\mathrm{mod}\,B)$ admits a component $\mathcal{C}$ containing all the indecomposable projective B-modules and all the indecomposable injective B-modules. The component $\mathcal{C}$ is obtained from the unique postprojective component $\mathcal{P}(A)$ of A and the unique preinjective component $\mathcal{P}(A)$ of A by a glueing with the projective-injective B-module $P(1)_B = I(4)_B$ as follows

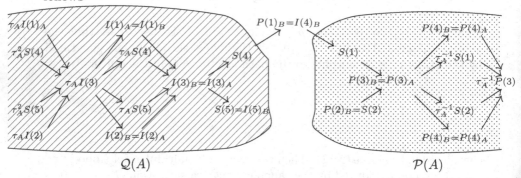

$$\mathcal{Q}(A) \qquad\qquad\qquad\qquad \mathcal{P}(A)$$

Now we show that $\mathcal{T}$ is a self-hereditary stable tube of $\Gamma(\mathrm{mod}\,B)$. Recall from (2.12) that the mouth module E of $\mathcal{T}$ has the form

$$E:$$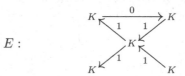

and $\tau_A S(3) \cong E \cong \tau_B S(3)$. Then the minimal projective resolution of E in $\mathrm{mod}\,B$ is of the form

$$0 \longrightarrow P(3) \longrightarrow P(4) \oplus P(5) \longrightarrow E \longrightarrow 0,$$

and the minimal injective resolution of E in $\mathrm{mod}\,B$ is of the form

$$0 \longrightarrow E \longrightarrow I(1) \oplus I(2) \longrightarrow I(3) \longrightarrow 0.$$

Hence, $\mathrm{pd}\,E = 1$ and $\mathrm{id}\,E = 1$. It follows that $\mathrm{Ext}_B^2(E, S(3) \oplus E) = 0$ and $\mathrm{Ext}_B^2(S(3) \oplus E, E) = 0$. Moreover, the short exact sequence

$$0 \longrightarrow S(1) \oplus S(2) \longrightarrow P(3) \longrightarrow S(3) \longrightarrow 0$$

yields isomorphisms of vector spaces

$$\mathrm{Ext}_B^2(S(3), S(3)) \cong \mathrm{Ext}_B^1(S(1) \oplus S(2), S(3)) \cong \mathrm{Ext}_B^1(S(1), S(3)),$$

because the B-module $S(2)$ is projective. By applying the Auslander–Reiten formula and the shape of the component of $\Gamma(\mathrm{mod}\,B)$ containing the module $S(1)$ (see the figure presented above), we get

$\mathrm{Ext}^1_B(S(1), S(3)) \cong D\overline{\mathrm{Hom}}_B(S(3), \tau_B S(1)) \cong D\overline{\mathrm{Hom}}_B(S(3), S(4)) = 0,$
and, consequently, $\mathrm{Ext}^2_B(S(3), S(3)) = 0.$ It follows that the two element family $\{S(3), E\}$ of mouth B-modules of the tube $\mathcal{T}$ is self-hereditary and consists of pairwise orthogonal bricks. Then, by (2.1) and (2.2), the tube $\mathcal{T}$ is self-hereditary. Because pd $S(3) = 3$ and $S(3)$ lies on $\mathcal{T}$ then the tube $\mathcal{T}$ is not hereditary. This finishes the example.

Now we give an example of an algebra R, with gl.dim $R = \infty$, such that $\Gamma(\mathrm{mod}\, R)$ admits a standard stable tube $\mathcal{T}$ that is not self-hereditary.

4.9. Example. Let $R = K\Omega/\mathcal{I}$ be the bound quiver algebra, where

$\Omega :$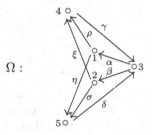

and $\mathcal{I}$ is the two-sided ideal of the path algebra $K\Omega$ generated by the elements

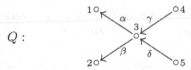

$$\rho\gamma - \eta\delta, \quad \xi\gamma - \sigma\delta, \quad \alpha\rho - \beta\xi, \quad \alpha\eta - \beta\sigma, \quad \rho\gamma\beta, \quad \gamma\beta\sigma, \quad \sigma\delta\alpha, \text{ and } \delta\alpha\rho.$$

Denote by J the two-sided ideal of R generated by the cosets $\rho + \mathcal{I}$, $\sigma + \mathcal{I}$, $\xi + \mathcal{I}$, and $\eta + \mathcal{I}$ of the arrows ρ, σ, ξ, and η of Ω. Then the quotient algebra $A = R/J$ is isomorphic to the path algebra KQ of the quiver

$Q :$

considered in Example (2.12). The canonical algebra surjection $R \longrightarrow A$ induces a fully faithful exact embedding $\mathrm{mod}\, A \hookrightarrow \mathrm{mod}\, R$.

Let $S = S(3)$ be the simple R-module at the vertex 3 of Ω, and let E be the indecomposable R-module

$E :$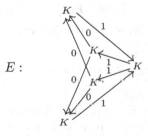

Because $SJ = 0$ and $EJ = 0$ then the modules S and E lie in the subcategory $\operatorname{mod} A \hookrightarrow \operatorname{mod} R$ of $\operatorname{mod} R$. We recall from Example (2.12) that the Auslander–Reiten quiver $\Gamma(\operatorname{mod} A)$ of A admits a standard stable tube $\mathcal{T}$ of rank 2 of the form

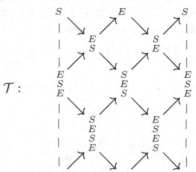

where the indecomposable modules are represented by their composition factors (see (I.3.10)) in the extension subcategory $\mathcal{EXT}_A(E, S)$ of $\operatorname{mod} A$, and one identifies them along the vertical dotted lines. In particular, the R-modules S and E are orthogonal bricks in $\operatorname{mod} A$ and (E, S) is a τ_A-cycle, that is, $\tau_A E \cong S$ and $\tau_A S \cong E$.

Now we show that (E, S) is a τ_R-cycle, that is, $\tau_R E \cong S$ and $\tau_R S \cong E$. Observe that the module $S = S(3)$ admits the minimal projective presentation

$$P(1) \oplus P(2) \longrightarrow P(3) \longrightarrow S(3) \longrightarrow 0$$

in $\operatorname{mod} R$. Applying the Nakayama functor $\nu_R = D\operatorname{Hom}_R(-, R)$ and using the isomorphism $\nu_R P(a) = I(a)$, for any vertex $a \in \Omega_0$, we get, by (IV.2.4), an exact sequence

$$0 \longrightarrow \tau_R S(3) \longrightarrow I(1) \oplus I(2) \longrightarrow I(3)$$

in $\operatorname{mod} R$. Hence, $\tau_R S(3) \cong E \cong \tau_A S(3)$.

Note also that the minimal projective presentation of E in $\operatorname{mod} R$ has the form

$$P(3) \longrightarrow P(5) \oplus P(4) \overset{\pi}{\longrightarrow} E \longrightarrow 0$$

and $\operatorname{Ker} \pi \cong P(3)/S(3)$. Applying the Nakayama functor ν_R yields a left exact sequence

$$0 \longrightarrow \tau_R E \longrightarrow I(3) \longrightarrow I(5) \oplus I(4)$$

and hence $\tau_R E \cong S(3) \cong \tau_A E$.

Because the R-modules $S = S(3)$ and E are orthogonal bricks in $\operatorname{mod} A$ they are also orthogonal bricks in $\operatorname{mod} R$ and we can form the extension subcategory $\mathcal{EXT}_R(E, S)$ of mod R that is abelian, exact and closed under extensions in mod C, by (2.1). It follows that

$$\mathcal{E}_R = \mathcal{EXT}_R(E, S) = \mathcal{EXT}_A(E, S),$$

because the subcategory mod $A \hookrightarrow$ mod R of mod R is closed under extensions. In particular, $\mathcal{E}_R$ consists entirely of A-modules and every simple object in $\mathcal{E}_R$ is isomorphic to E or to S. It follows that almost split sequences in mod A starting from indecomposable modules lying in $\mathcal{T}$ remain almost split in mod R and all indecomposable summands of their terms lie also in $\mathcal{T}$. Hence we conclude that the standard stable tube $\mathcal{T}$ of $\Gamma(\text{mod } A)$ remains a standard stable tube $\mathcal{T}$ of $\Gamma(\text{mod } R)$.

Now we show that the tube $\mathcal{T}$ of $\Gamma(\text{mod } R)$ is not self-hereditary, by proving that $\text{Ext}_R^2(E, S) \neq 0$. By applying the functor $\text{Hom}_R(-, S)$ to the short exact sequence

$$0 \longrightarrow P(3)/S(3) \longrightarrow P(5) \oplus P(4) \overset{\pi}{\longrightarrow} E \longrightarrow 0$$

we derive an isomorphism

$$\text{Ext}_R^2(E, S) \cong \text{Ext}_R^1(E, P(3)/S(3)).$$

Because the canonical exact sequence

$$0 \longrightarrow S(3) \longrightarrow P(3) \overset{\pi}{\longrightarrow} P(3)/S(3) \longrightarrow 0$$

in mod R does not split then $\text{Ext}_R^1(E, P(3)/S(3)) \neq 0$ and, consequently, $\text{Ext}_R^2(E, S) \neq 0$.

The preceding two short exact sequences give a minimal projective presentation

$$0 \longrightarrow S(3) \longrightarrow P(3) \longrightarrow P(5) \oplus P(4) \overset{\pi}{\longrightarrow} E \longrightarrow 0$$

of length 3 of the module E. Hence, using the minimal projective presentation of $S(3)$, we get a non-split exact sequence

$$0 \longrightarrow E \longrightarrow P(1) \oplus P(2) \longrightarrow P(3) \longrightarrow S(3) \longrightarrow 0$$

in mod R. By combining these two exact sequences, we get a periodic infinite minimal projective resolution of the module E. This shows that gl.dim $R = \infty$. One can also show that the algebra R is self-injective, see Section V.3. This finishes the example.

X.5. Exercises

1. Under the notation and assumption of (2.2), show that:

(a) if $f : E_i[j] \longrightarrow E_k[l]$ is a homomorphism, which is a composition of t homomorphisms of the form $p_{\alpha\beta}$ followed by the homomorphisms of the form $u_{\gamma\delta}$, then there is an isomorphism $\operatorname{Im} f \cong E_{i+t}[j{-}t]$, and

(b) if $\operatorname{Hom}(E_i[j], E_k[l]) \neq 0$, then $i \leq k \leq i+j-1$ and $i+j-k \leq l$.

2. Let M be an indecomposable regular module over an arbitrary algebra A lying in a standard stable tube of $\Gamma(\operatorname{mod} A)$. Show that there exist an integer $m \geq 1$ and an isomorphism of K-algebras $\operatorname{End}_A M \cong K[t]/(t^m)$.
Hint: Apply (2.6) and (2.7).

3. Let $A = KQ$ be the path K-algebra of the following quiver

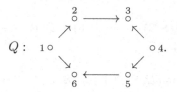

(a) Show that the following four indecomposable A-modules

$$E_1 = \begin{matrix} & K \longrightarrow 0 & \\ \nearrow & & \nwarrow \\ 0 & & 0, \\ \searrow & & \swarrow \\ & 0 \longleftarrow 0 & \end{matrix} \qquad E_2 = \begin{matrix} & 0 \longrightarrow 0 & \\ \nearrow & & \nwarrow \\ K & & 0, \\ {\scriptstyle 1}\searrow & & \swarrow \\ & K \longleftarrow 0 & \end{matrix}$$

$$E_3 = \begin{matrix} & 0 \longrightarrow 0 & \\ \nearrow & & \nwarrow \\ 0 & & 0, \\ \searrow & & \swarrow \\ & 0 \longleftarrow K & \end{matrix} \qquad E_4 = \begin{matrix} & 0 \longrightarrow K & \\ \nearrow & & \nwarrow{\scriptstyle 1} \\ 0 & & K, \\ \searrow & & \swarrow \\ & 0 \longleftarrow 0 & \end{matrix}$$

(viewed as the representation of the quiver Q) form the mouth of a standard stable tube $\mathcal{T}_0$ of $\Gamma(\operatorname{mod} A)$ of rank 4 such that $\tau E_4 \cong E_3$, $\tau E_3 \cong E_2$, $\tau E_2 \cong E_1$, and $\tau E_1 \cong E_4$.

(b) Show that the following two indecomposable A-modules

$$F_1 = \begin{matrix} & K \overset{1}{\longrightarrow} K & \\ {\scriptstyle 1}\nearrow & & \nwarrow \\ K & & 0, \\ \searrow & & \swarrow \\ & 0 \longleftarrow 0 & \end{matrix} \qquad F_2 = \begin{matrix} & 0 \longrightarrow 0 & \\ \nearrow & & \nwarrow \\ 0 & & K, \\ {\scriptstyle 1}\searrow & & \swarrow \\ & K \underset{1}{\longleftarrow} K & \end{matrix}$$

form the mouth of a standard stable tube $\mathcal{T}_1$ of $\Gamma(\mathrm{mod}\,A)$ of rank 2 such that $\tau F_2 \cong F_1$, and $\tau F_1 \cong F_2$.

(c) Show that the following indecomposable A-module

$$
R = \quad K \quad
\begin{array}{c}
K \xrightarrow{\;1\;} K \\
\end{array}
\quad K,
$$

forms the mouth of a standard homogeneous tube $\mathcal{T}$ of $\Gamma(\mathrm{mod}\,A)$.

4. Let $n \geq 1$ be an integer and let $C = C(n)$ be the path algebra of the following acyclic quiver

$$
\Delta(n):
$$

of the Euclidean type $\widetilde{\mathbb{A}}_n$. Note that the algebra $C(1)$ is isomorphic to the the Kronecker algebra $\begin{bmatrix} K & 0 \\ K^2 & K \end{bmatrix}$.

(a) Show that the simple C-modules

$$
F_1 = S(1),\ E_2 = S(2),\ \ldots,\ E_{n-1} = S(n-1),
$$

together with the indecomposable C-module

$$
E_n =
$$

(viewed as the representation of the quiver $\Delta(n)$) form the mouth of a standard stable tube $\mathcal{T}_0$ of $\Gamma(\mathrm{mod}\,C)$ of rank n.

(b) Show that the indecomposable C-module

$$
F =
$$

forms the mouth of a standard homogeneous tube $\mathcal{T}_1$ of $\Gamma(\mathrm{mod}\,C)$.

(c) Show that the indecomposable C-module

$$
R =
$$

forms the mouth of a standard homogeneous tube $\mathcal{T}_2$ of $\Gamma(\mathrm{mod}\,C)$.

5. Let $A = KQ$ be the path K-algebra of the following quiver

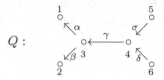

of the Euclidean type $\widetilde{\mathbb{D}}_5$.

(a) Show that the simple A-modules $S(3)$, $S(4)$, and the indecomposable A-module

$$
E = \begin{array}{ccc}
K & & K \\
\nwarrow^{1} & & \nearrow^{1} \\
K \xleftarrow{\ 1\ } K & \\
\nearrow^{1} & & \nwarrow^{1} \\
K & & K
\end{array}
$$

(viewed as the representation of the quiver Q) form the mouth of a standard stable tube $\mathcal{T}_0$ of $\Gamma(\mathrm{mod}\,A)$ of rank 3 such that $\tau S(3) \cong E$, $\tau S(4) \cong S(3)$, and $\tau E \cong S(4)$.

(b) Show that the following two indecomposable A-modules

$$
F_1 = \begin{array}{ccc}
K & & K \\
\nwarrow^{1} & & \nearrow^{1} \\
K \xleftarrow{\ 1\ } K & \\
\swarrow & & \nwarrow \\
0 & & 0
\end{array}
\qquad
F_2 = \begin{array}{ccc}
0 & & 0 \\
\nwarrow & & \nearrow \\
K \xleftarrow{\ \ } K & \\
\nearrow^{1} & & \nwarrow^{1} \\
K & & K
\end{array}
$$

form the mouth of a standard stable tube $\mathcal{T}_1$ of $\Gamma(\mathrm{mod}\,A)$ of rank 2 such that $\tau F_1 \cong F_2$ and $\tau F_2 \cong F_1$.

(c) Show that the following two indecomposable A-modules

$$
R_1 = \begin{array}{ccc}
K & & 0 \\
\nwarrow^{1} & & \nearrow \\
K \xleftarrow{\ 1\ } K & \\
\swarrow & & \nwarrow^{1} \\
0 & & K
\end{array}
\qquad
R_2 = \begin{array}{ccc}
0 & & K \\
\nwarrow & & \nearrow^{1} \\
K \xleftarrow{\ 1\ } K & \\
\nearrow^{1} & & \nwarrow \\
K & & 0
\end{array}
$$

form the mouth of a standard stable tube $\mathcal{T}_2$ of $\Gamma(\mathrm{mod}\,A)$ of rank 2 such that $\tau R_1 \cong R_2$ and $\tau R_2 \cong R_1$.

6. Let B be the path K-algebra of the following quiver

bound by the commutativity relation $\alpha\gamma = \beta\delta$.

(a) Show that the following three indecomposable B-modules

$$E_1 = \begin{array}{ccccccc} 0 & & K & & & & \\ & \searrow & \swarrow & \searrow & & & \\ & K & & \searrow & \swarrow & & 0, \\ & \swarrow & & \searrow & \swarrow & & \\ & K & & & 0 & & \end{array} \quad E_2 = \begin{array}{ccccccc} K & & & 0 & & \\ & \searrow & \swarrow & \searrow & & \\ & K & & & \searrow & \swarrow & 0, \\ & \swarrow & & \searrow & \swarrow & & \\ & 0 & & & K & & \end{array}$$

$$E_3 = \begin{array}{ccccc} 0 & & 0 & & \\ & \searrow & \swarrow & \searrow & \\ & 0 & & \searrow & K \\ & \swarrow & & \searrow & \swarrow \\ & 0 & & 0 & \end{array}$$

form the mouth of a standard stable tube $\mathcal{T}_0$ of $\Gamma(\mathrm{mod}\,B)$ of rank 3 such that $\tau E_1 \cong E_3$, $\tau E_2 \cong E_1$, and $\tau E_3 \cong E_2$.

(b) Show that the following two indecomposable B-modules

$$F_1 = \begin{array}{ccccccc} 0 & & K & & & \\ & \searrow & \swarrow & \searrow^1 & & \\ & K & & & K, \\ & \swarrow & \searrow_1 & \swarrow_1 & \\ & 0 & & K & \end{array} \quad F_2 = \begin{array}{ccccccc} K & & 0 & & \\ & \searrow & \swarrow & \searrow & \\ & K & & & 0 \\ & \swarrow_1 & \searrow & \swarrow & \\ & K & & 0 & \end{array}$$

form the mouth of a standard stable tube $\mathcal{T}_1$ of $\Gamma(\mathrm{mod}\,B)$ of rank 2 such that $\tau F_1 \cong F_2$ and $\tau F_2 \cong F_1$.

(c) Show that the following two indecomposable B-modules

$$R_1 = \begin{array}{ccccccc} K & & K & & \\ & \searrow^1 & \swarrow^1 & \searrow^1 & \\ & K & & & K, \\ & \swarrow_1 & \searrow_1 & \swarrow_1 & \\ & K & & K & \end{array} \quad R_2 = \begin{array}{ccccccc} 0 & & 0 & & \\ & \searrow & \swarrow & \searrow & \\ & K & & & 0 \\ & \swarrow & \searrow & \swarrow & \\ & 0 & & 0 & \end{array}$$

form the mouth of a standard stable tube $\mathcal{T}_2$ of $\Gamma(\mathrm{mod}\,B)$ of rank 2 such that $\tau R_1 \cong R_2$ and $\tau R_2 \cong R_1$.

7. Prove that the algebra $C = K\Delta/\mathcal{I}$ of Example (4.9) is self-injective.

8. Let Λ be the path K-algebra of the following quiver

bound by the three relations $\alpha^2 = 0$, $\beta^2 = 0$, and $\alpha\beta = \beta\alpha$. For each integer $d \geq 1$, denote by $E[d]$ the Λ-module

$$f_\alpha^{(d)} \; \bigcirc \!\!\! \longrightarrow K^{2d} \longleftarrow \!\!\! \bigcirc \; f_\beta^{(d)},$$

where the K-linear endomorphisms $f_\alpha^{(d)}, f_\beta^{(d)} : K^{2d} \longrightarrow K^{2d}$ are given, in the canonical basis of $K^{2d} = K^d \oplus K^d$, by the $2d \times 2d$ square matrices

$$f_\alpha^{(d)} = \left[\begin{array}{c|c} 0 & 0 \\ \hline E & 0 \end{array} \right] \quad \text{and} \quad f_\beta^{(d)} = \left[\begin{array}{c|c} 0 & 0 \\ \hline J_{d,0} & 0 \end{array} \right].$$

Here $E = \begin{bmatrix} 1 & 0 & 0 & \ldots & 0 \\ 0 & 1 & 0 & \ldots & 0 \\ 0 & 0 & 1 & \ldots & 0 \\ \vdots & \vdots & \vdots & \ddots & \vdots \\ 0 & 0 & 0 & \ldots & 1 \end{bmatrix} \in \mathbb{M}_d(K)$ is the $d \times d$ identity matrix and

$$J_{d,0} = \begin{bmatrix} 0 & 1 & 0 & \ldots & 0 & 0 \\ 0 & 0 & 1 & \ldots & 0 & 0 \\ \vdots & \ddots & \ddots & \ddots & & \vdots \\ 0 & \ldots & 0 & 0 & & 1 \\ 0 & \ldots & 0 & 0 & & 0 \end{bmatrix} \in \mathbb{M}_d(K)$$

is the the $d \times d$ Jordan block corresponding to the eigenvalue $\lambda = 0$.

Prove that:

(a) there is a K-algebra isomorphism $\Lambda \cong K[t_1, t_2]/(t_1^2, t_2^2)$,

(b) $\dim_K \Lambda = 4$,

(c) the Λ-module $E[d]$ is indecomposable, for each $d \geq 1$,

(d) the Auslander–Reiten quiver $\Gamma(\mathrm{mod}\,\Lambda)$ of Λ admits a homogeneous stable tube $\mathcal{T}$, with the unique ray of the form

$$E[1] \longrightarrow E[2] \longrightarrow E[3] \longrightarrow \ldots \longrightarrow E[d] \longrightarrow E[d+1] \longrightarrow \ldots \,,$$

(e) the exact sequence $0 \longrightarrow E[1] \overset{u}{\longrightarrow} E[2] \overset{p}{\longrightarrow} E[1] \longrightarrow 0$ is almost split in $\mathrm{mod}\,\Lambda$, where $u : K^2 \longrightarrow K^4$ and $p : K^4 \longrightarrow K^2$ are the linear maps given by the matrices $u = \begin{bmatrix} 1 & 0 \\ 0 & 0 \\ 0 & 1 \\ 0 & 0 \end{bmatrix}$ and $p = \begin{bmatrix} 0 & 1 & 0 & 0 \\ 0 & 0 & 0 & 1 \end{bmatrix}$,

(f) the endomorphism $h : E[2] \longrightarrow E[2]$ defined by the linear map $h : K^4 \longrightarrow K^4$, $(x_1, x_2, x_3, x_4) \mapsto (0, 0, 0, x_1)$, has a factorisation through the unique simple Λ-module S and belongs to $\mathrm{rad}_\Lambda^\infty(E[2], E[2])$,

(g) the homomorphism $v = u + hu : E[1] \longrightarrow E[2]$ is an irreducible morphism in $\mathrm{mod}\,\Lambda$, the composition $pv = phu$ of p and v is non-zero, and $pv \in \mathrm{rad}_\Lambda^\infty(E[1], E[1])$,

(h) the tube $\mathcal{T}$ is not standard.

Chapter XI

Module categories over concealed algebras of Euclidean type

The main aim of this chapter is to describe the structure of the category add $\mathcal{R}(B)$ of regular modules over a (representation-infinite) concealed algebra B of Euclidean type, and hence the structure of the whole category mod B of finite dimensional right B-modules over such an algebra B, because of the disjoint union decomposition

$$\Gamma(\operatorname{mod} B) = \mathcal{P}(B) \cup \mathcal{R}(B) \cup \mathcal{Q}(B)$$

of the Auslander–Reiten quiver $\Gamma(\operatorname{mod} B)$ of B, where $\mathcal{P}(B)$ is a unique postprojective component containing all the indecomposable projective B-modules, $\mathcal{Q}(B)$ is a unique preinjective component containing all the indecomposable injective B-modules, and $\mathcal{R}(B)$ is the family of the remaining components being called regular (see (VIII.2.12), (VIII.4.5) and Chapter X).

In particular, we show that the category add $\mathcal{R}(B)$ over a concealed algebra B of Euclidean type is abelian and serial, and the family $\mathcal{R}(B)$ of regular components of the Auslander–Reiten quiver $\Gamma(\operatorname{mod} B)$ of B is a disjoint union of a family

$$\mathcal{T}^B = \{\mathcal{T}_\lambda^B\}_{\lambda \in \Lambda}$$

of pairwise orthogonal standard stable tubes $\mathcal{T}_\lambda$, all but a finite number of them being homogeneous.

We recall that B is a concealed algebra of Euclidean type if B is the endomorphism algebra of a postprojective tilting module over a hereditary algebra $A = KQ$ of Euclidean type.

In Section 1 we collect the main properties of the Coxeter transformation $\Phi_A : K_0(A) \longrightarrow K_0(A)$ of a hereditary algebra $A = KQ$ of Euclidean type Q, where $Q = (Q_0, Q_1)$ is an acyclic quiver, whose underlying graph $\overline{Q}$ is one of the Euclidean diagrams $\widetilde{\mathbb{A}}_m$, $\widetilde{\mathbb{D}}_m$, $\widetilde{\mathbb{E}}_6$, $\widetilde{\mathbb{E}}_7$, $\widetilde{\mathbb{E}}_8$, and $K_0(A)$ is the Grothendieck group of A. In particular, we prove the periodicity of the action $\overline{\Phi}_A : \overline{K_0(A)} \longrightarrow \overline{K_0(A)}$ induced by Φ_A on the quotient group $\overline{K_0(A)} = K_0(A)/\operatorname{rad} q_A$ of $K_0(A)$ modulo the radical $\operatorname{rad} q_A$

of the Euler quadratic form $q_A : K_0(A) \longrightarrow \mathbb{Z}$ of the algebra A. This allows us to introduce the concept of a defect

$$\partial_A : K_0(A) \longrightarrow \mathbb{Z}$$

of A, being an important tool for the study of the module category $\operatorname{mod} A$ of a hereditary algebra A of Euclidean type.

In Sections 2 and 3 we collect main properties of the category $\operatorname{add} \mathcal{R}(A)$ of regular modules over a hereditary algebra A of Euclidean type, and then of the category $\operatorname{add} \mathcal{R}(B)$ over a concealed algebra B of Euclidean type. In particular, we prove that, given an indecomposable A-module M, the following holds:

- $M \in \mathcal{P}(A)$ if and only if $\partial_A(\operatorname{\mathbf{dim}} M) < 0$.
- $M \in \mathcal{R}(A)$ if and only if $\partial_A(\operatorname{\mathbf{dim}} M) = 0$.
- $M \in \mathcal{Q}(A)$ if and only if $\partial_A(\operatorname{\mathbf{dim}} M) > 0$.

In Section 4 we study in detail the case when $A = KQ$ and Q is the **Kronecker quiver**

$$1 \; \circ \; \Longleftarrow \; \circ \; 2,$$

of the Euclidean type $\widetilde{\mathbb{A}}_1$ and we give a complete description of the category of finite dimensional indecomposable modules over the **Kronecker K-algebra**

$$A = \begin{bmatrix} K & 0 \\ K^2 & K \end{bmatrix}.$$

In other words, we give a solution of the well-known **Kronecker problem**, that is, the problem of classifying the indecomposable K-linear representations of the Kronecker quiver $\circ \; \Longleftarrow \; \circ$.

In Section 5 we prove a useful characterisation of concealed algebras of Euclidean type, that we frequently use in Chapter XII.

XI.1. The Coxeter matrix and the defect of a hereditary algebra of Euclidean type

Throughout this section, we denote by

$$A = KQ$$

the path K-algebra of an acyclic quiver Q, with n points, whose underlying graph $\overline{Q}$ is one of the **Euclidean graphs** listed below. They are also called **Euclidean diagrams**.

The Euclidean graphs

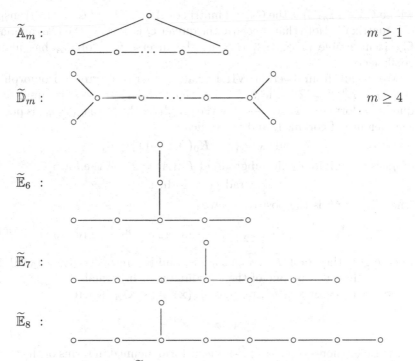

$\widetilde{\mathbb{A}}_m$: $m \geq 1$

$\widetilde{\mathbb{D}}_m$: $m \geq 4$

$\widetilde{\mathbb{E}}_6$:

$\widetilde{\mathbb{E}}_7$:

$\widetilde{\mathbb{E}}_8$:

The Euclidean diagram $\widetilde{\mathbb{A}}_1$: $\circ\!=\!=\!=\!=\!=\!=\!\circ$ is called the **Kronecker graph**, or **Kronecker diagram**.

We recall from (VIII.2.2) that an indecomposable A-module M is called postprojective (or preinjective) if M belongs to the postprojective component $\mathcal{P}(A)$ (or to the preinjective component $\mathcal{Q}(A)$, respectively). Our objective is to describe the additive full subcategory $\operatorname{add}\mathcal{R}(A)$ of $\operatorname{mod} A$ whose objects are the regular modules, that is, finite direct sums of indecomposable modules that are neither postprojective nor preinjective (see (VIII.2.12)). In particular, we describe the shape of the regular components of $\Gamma(\operatorname{mod} A)$.

We recall from (III.3.5) of Volume 1, that the Grothendieck group $K_0(A)$ of the algebra $A = KQ$ is isomorphic to the free abelian group $\mathbb{Z}^n$, where $n = |Q_0|$. As usual, we denote by $\mathbf{e}_1, \dots, \mathbf{e}_n$ the canonical $\mathbb{Z}$-basis of $\mathbb{Z}^n$. We know from (III.3.11) that under the isomorphisms $K_0(A) \cong \mathbb{Z}^{|Q_0|} = \mathbb{Z}^n$, the **Euler quadratic form**

$$q_A : \mathbb{Z}^n \longrightarrow \mathbb{Z}$$

of the algebra A is defined by the formula

$$q_A(\mathbf{x}) = \mathbf{x}^t (\mathbf{C}_A^{-1})^t \mathbf{x}, \quad \text{with } \mathbf{x} = \begin{bmatrix} x_1 \\ \vdots \\ x_n \end{bmatrix} = [x_1, \dots, x_n]^t \in \mathbb{Z}^n = K_0(A),$$

where $\mathbf{C}_A \in \mathbb{M}_n(\mathbb{Z})$ is the Cartan matrix of $A = KQ$ and C^t is the transpose of a matrix C. Note that, because the quiver Q is acyclic, the Cartan matrix $\mathbf{C}_A$ is invertible in $\mathbb{M}_n(\mathbb{Z})$, that is, the inverse $\mathbf{C}_A^{-1}$ of $\mathbf{C}_A$ has integral coefficients.

We recall from Section VII.4 that, under the group isomorphisms $K_0(A) \cong \mathbb{Z}^{|Q_0|} = \mathbb{Z}^n$, the Euler form $q_A : K_0(A) \longrightarrow \mathbb{Z}$ is equal to the quadratic form $q_Q : \mathbb{Z}^n \longrightarrow \mathbb{Z}$ of the graph Q, the form $q_A = q_Q$ is positive semidefinite of corank 1, and the radical

$$\operatorname{rad} q_A = \{x \in K_0(A); \; q_A(x) = 0\}$$

of q_A is an infinite cyclic subgroup of $K_0(A) = \mathbb{Z}^n$ of the form

$$\operatorname{rad} q_A = \mathbb{Z} \cdot \mathbf{h}_Q,$$

where $\mathbf{h}_Q \in \mathbb{Z}^n$ is the positive vector

$$\begin{smallmatrix} & 1 & \cdots & 1 \\ 1 & 1 & \cdots & 1 & 1 \end{smallmatrix}, \quad \begin{smallmatrix} 1 \\ 1 \end{smallmatrix} 2 \cdots 2 \begin{smallmatrix} 1 \\ 1 \end{smallmatrix}, \quad 1\,2\begin{smallmatrix}2\\3\end{smallmatrix}2\,1, \quad 1\,2\,3\begin{smallmatrix}2\\4\end{smallmatrix}3\,2\,1, \quad \text{and} \quad 2\,4\begin{smallmatrix}3\\6\end{smallmatrix}5\,4\,3\,2\,1 \qquad (1.1)$$

in case $\overline{Q}$ is the graph $\widetilde{\mathbb{A}}_m$, $\widetilde{\mathbb{D}}_m$, $\widetilde{\mathbb{E}}_6$, $\widetilde{\mathbb{E}}_7$, and $\widetilde{\mathbb{E}}_8$, respectively, see (VII.4.2). We note that at least one of the coordinates of $\mathbf{h}_Q$ equals 1.

For each vector $\mathbf{x} \in \mathbb{Z}^n$, we have $q_A(\mathbf{x}) = \langle \mathbf{x}, \mathbf{x} \rangle_A$, where

$$\langle -, - \rangle_A : \mathbb{Z}^n \times \mathbb{Z}^n \longrightarrow \mathbb{Z}$$

is the Euler (non-symmetric) $\mathbb{Z}$-bilinear form defined in terms of the Cartan matrix $\mathbf{C}_A$ of A by $\langle \mathbf{x}, \mathbf{y} \rangle_A = \mathbf{x}^t (\mathbf{C}_A^{-1})^t \mathbf{y}$, for $\mathbf{x}, \mathbf{y} \in \mathbb{Z}^n$. We also recall from (III.3.14) that the Coxeter matrix $\mathbf{\Phi}_A$ of A is the matrix

$$\mathbf{\Phi}_A = -\mathbf{C}_A^t \mathbf{C}_A^{-1} \in \mathbb{M}_n(\mathbb{Z}).$$

Because, by (III.3.10), $\mathbf{C}_A \in \mathrm{Gl}(n, \mathbb{Z})$, then we have $\mathbf{\Phi}_A \in \mathrm{Gl}(n, \mathbb{Z})$ as well. Here $\mathrm{Gl}(n, \mathbb{Z}) = \{A \in \mathbb{M}_n(\mathbb{Z}); \; \det A \in \{-1, 1\}\}$, is the general linear group over $\mathbb{Z}$. The group homomorphism

$$\mathbf{\Phi}_A : \mathbb{Z}^n \longrightarrow \mathbb{Z}^n$$

defined by the formula $\mathbf{\Phi}_A(\mathbf{x}) = \mathbf{\Phi}_A \cdot \mathbf{x}$, for all $\mathbf{x} = [x_1 \dots x_n]^t \in \mathbb{Z}^n$, is called the Coxeter transformation of A.

It follows from (VII.4.7) that the Coxeter matrix $\mathbf{\Phi}_A$ of $A = KQ$ is the matrix of the Coxeter transformation $c : \mathbb{Z}^n \longrightarrow \mathbb{Z}^n$ of the quiver Q in the canonical basis of $\mathbb{Z}^n$, where

$$c = s_{a_n} \dots s_{a_2} s_{a_1} : \mathbb{Z}^n \longrightarrow \mathbb{Z}^n$$

is defined with respect to a fixed admissible numbering $a_1, \ldots, a_n$ of the points of the quiver Q, and $s_{a_j} : \mathbb{Z}^n \longrightarrow \mathbb{Z}^n$ is the reflection homomorphism with respect to the point a_j of Q. We recall from (VII.4) that the Coxeter transformation c does not depend on the choice of the admissible sequence $a_1, \ldots, a_n$, because the quiver Q is acyclic. However, as shown in Examples (1.6) and (1.7) below, the Coxeter matrix Φ_A of $A = KQ$ depends on the orientation of the quiver Q.

The Coxeter matrices Φ_A of the path algebras $A = K\Delta$ of canonically oriented Euclidean quivers Δ are presented in Section XIII.1.

Throughout we identify the group homomorphism $c : \mathbb{Z}^n \longrightarrow \mathbb{Z}^n$ with the Coxeter transformation $\Phi_A : \mathbb{Z}^n \longrightarrow \mathbb{Z}^n$.

We start with a result for an arbitrary acyclic quiver.

1.2. Proposition. *Let Q be an arbitrary acyclic quiver, $A = KQ$ be the path K-algebra of Q and let $\Phi_A : \mathbb{Z}^n \longrightarrow \mathbb{Z}^n$ be the Coxeter transformation of A. If M is an indecomposable A-module which is not postprojective then, for each $m \geq 0$, we have*

$$\dim \tau^m M = \Phi_A^m(\dim M).$$

Proof. Assume that M is an indecomposable A-module which is not postprojective. By (VIII.2.1), for each $m \geq 0$, the module $\tau^m M$ is non-zero and non-projective. It follows that $\operatorname{Hom}_A(M, A) = 0$, because the algebra A is hereditary. Then the equality $\dim \tau^m M = \Phi_A^m(\dim M)$ is a consequence of (IV.2.9). $\square$

1.3. Proposition. *Let Q be an acyclic quiver whose underlying graph is Euclidean, let $A = KQ$ and let $\Phi_A : \mathbb{Z}^n \longrightarrow \mathbb{Z}^n$ be the Coxeter transformation of A, where $n = |Q_0|$ and $\mathbb{Z}^n = \mathbb{Z}^{|Q_0|} = K_0(A)$.*

(a) *If $q_A : \mathbb{Z}^n \longrightarrow \mathbb{Z}$ is the Euler form of A and $\mathbf{h}_Q \in \mathbb{Z}^n$ is a positive generator of the group $\operatorname{rad} q_A$, then $\Phi_A(\mathbf{h}_Q) = \mathbf{h}_Q$ and $\Phi_A(\operatorname{rad} q_A) = \operatorname{rad} q_A$.*

(b) *There exists a least $d = d_Q \geq 0$ such that, for each $\mathbf{x} \in \mathbb{Z}^n$, we have $\Phi_A^d(\mathbf{x}) - \mathbf{x} \in \operatorname{rad} q_A$.*

Proof. Let $\mathbf{h} = \mathbf{h}_Q \in \mathbb{Z}^n$ be the positive generator of the group $\operatorname{rad} q_A$, see (1.1).

(a) Because the Coxeter matrix Φ_A is the matrix of the Coxeter transformation $c : \mathbb{Z}^n \longrightarrow \mathbb{Z}^n$ and (VII.4.11) yields $c(\mathbf{h}) = \mathbf{h}$, then $\Phi_A(\mathbf{h}) = \mathbf{h}$ and (a) follows.

(b) We consider the quotient group $\overline{K_0(A)} = K_0(A)/\mathbb{Z} \cdot \mathbf{h}$ of the Grothendieck group $K_0(A) = \mathbb{Z}^n$ of A modulo the subgroup $\mathbb{Z} \cdot \mathbf{h} = \operatorname{rad} q_A$, where $\mathbf{h} = \mathbf{h}_Q$ is the positive generator of $\operatorname{rad} q_A$, see (1.1). Let $\pi : K_0(A) \longrightarrow \overline{K_0(A)}$

be the canonical group epimorphism defined by the formula $\pi(v) = \overline{v} :=$ $v + \mathbb{Z} \cdot \mathbf{h}$, for each $v \in K_0(A)$. Because the canonical group embedding $K_0(A) = \mathbb{Z}^n \hookrightarrow \mathbb{Q}^n$ induces an embedding $\overline{K_0(A)} \longrightarrow \mathbb{Q}^n/\mathbb{Q} \cdot \mathbf{h} \cong \mathbb{Q}^{n-1}$ of abelian groups and the group $\overline{K_0(A)}$ is generated by the cosets $\overline{\mathbf{e}}_1, \dots, \overline{\mathbf{e}}_n$ of the canonical $\mathbb{Z}$-basis vectors $\mathbf{e}_1, \dots, \mathbf{e}_n$ of the group $K_0(A) = \mathbb{Z}^n$, then the group $\overline{K_0(A)}$ is torsion-free of rank $n - 1$; that is, there is a group isomorphism $\overline{K_0(A)} \cong \mathbb{Z}^{n-1}$.

The Coxeter transformation $\boldsymbol{\Phi}_A : K_0(A) \longrightarrow K_0(A)$ of A induces the group automorphism

$$\overline{\boldsymbol{\Phi}}_A : \overline{K_0(A)} \longrightarrow \overline{K_0(A)}$$

defined by the formula $\overline{\boldsymbol{\Phi}}_A(\overline{v}) = \overline{\boldsymbol{\Phi}_A(v)}$, for each $v \in K_0(A)$, because $\boldsymbol{\Phi}_A(\mathbf{h}) = \mathbf{h}$ and $\operatorname{rad} q_A = \mathbb{Z} \cdot \mathbf{h}$ is an $\boldsymbol{\Phi}_A$-invariant subgroup of $K_0(A)$. Moreover, the Euler (non-symmetric) $\mathbb{Z}$-bilinear form $b_A(-,-) = \langle -, - \rangle_A :$ $K_0(A) \times K_0(A) \longrightarrow \mathbb{Z}$ of the algebra A induces the $\mathbb{Z}$-bilinear form

$$\overline{b}_A(-,-) : \overline{K_0(A)} \times \overline{K_0(A)} \longrightarrow \mathbb{Z}$$

defined by the formula $\overline{b}_A(\overline{v}, \overline{w}) = b_A(v, w)$, for each pair of vectors $v, w \in K_0(A) = \mathbb{Z}^n$, because $\langle \mathbf{h}, \mathbf{h} \rangle_A = q_A(\mathbf{h}) = 0$. Consequently, the map

$$\overline{q}_A : \overline{K_0(A)} \longrightarrow \mathbb{Z}$$

defined by the formula $\overline{q}_A(\overline{v}) = \overline{b}_A(\overline{v}, \overline{v}) = b_A(v, v) = q_A(v)$, for each $v \in K_0(A) = \mathbb{Z}^n$, is a quadratic form on the group $\overline{K_0(A)} \cong \mathbb{Z}^{n-1}$.

By our assumption, the quiver Q is acyclic and the underlying graph $\overline{Q}$ of Q is Euclidean. It follows that the quadratic form $\overline{q}_A : \mathbb{Z}^{n-1} \longrightarrow \mathbb{Z}$ is positive definite, because $\overline{q}_A(\overline{v}) = q_A(v) \geq 0$, for each $v \in K_0(A) = \mathbb{Z}^n$, and the equality $q_A(\overline{v}) = q_A(v) = 0$ yields $v \in \operatorname{rad} q_A = \mathbb{Z} \cdot \mathbf{h}$, that is, $\overline{v} = 0$. Hence we conclude that the set

$$\mathcal{R} = \{\overline{v} \in \overline{K_0(A)}; \ \overline{q}_A(\overline{v}) = 1\} \subseteq \overline{K_0(A)} \cong \mathbb{Z}^{n-1}$$

of all roots of the quadratic form $\overline{q}_A$ is finite, because one shows that the Euclidean norm $||\overline{v}||$ of any vector $\overline{v} \in \mathcal{R}$ (viewed as a vector of the Euclidean space $\mathbb{R}^{n-1}$) is bounded by $\frac{1}{\sqrt{\mu}}$, where $\mu = \inf\{\overline{q}_A(u); \ u \in \mathbb{R}^{n-1}, ||u|| = 1\}$ is the minimum of the values $\overline{q}_A(u)$ of the quadratic function $\overline{q}_A : \mathbb{R}^{n-1} \longrightarrow \mathbb{R}$ restricted to the unit sphere $\mathcal{S}^{n-2} = \{u \in \mathbb{R}^{n-1}, ||u|| = 1\}$; compare with (VII.3.4) and its proof, see also Exercise (XI.6.1).

Now we show that the finite subset $\mathcal{R}$ of $\overline{K_0(A)} \cong \mathbb{Z}^{n-1}$ is $\overline{\boldsymbol{\Phi}}_A$-invariant, that is, $\overline{\boldsymbol{\Phi}}_A(\mathcal{R}) = \mathcal{R}$. For, if $\overline{v} \in \mathcal{R}$ then

$$\overline{q}_A(\overline{\boldsymbol{\Phi}}_A(\overline{v})) = \overline{q_A(\boldsymbol{\Phi}_A(v))} = \overline{q_A(v)} = \overline{q}_A(\overline{v}) = 1,$$

because the definitions of q_A and of the Coxeter matrix yield

$$q_A(\mathbf{\Phi}_A(\mathbf{x})) = \mathbf{\Phi}_A(\mathbf{x})^t \cdot (\mathbf{C}_A^{-1})^t \cdot \mathbf{\Phi}_A(\mathbf{x})$$
$$= (\mathbf{\Phi}_A \cdot \mathbf{x})^t \cdot (\mathbf{C}_A^{-1})^t \cdot \mathbf{\Phi}_A \cdot \mathbf{x}$$
$$= \mathbf{x}^t \cdot \mathbf{\Phi}_A^t \cdot (\mathbf{C}_A^{-1})^t \cdot \mathbf{\Phi}_A \cdot \mathbf{x}$$
$$= \mathbf{x}^t \cdot (-\mathbf{C}_A^t \cdot \mathbf{C}_A^{-1})^t \cdot (\mathbf{C}_A^{-1})^t \cdot (-\mathbf{C}_A^t \mathbf{C}_A^{-1}) \cdot \mathbf{x}$$
$$= \mathbf{x}^t \cdot (\mathbf{C}_A^{-1})^t \cdot \mathbf{x} = q_A(\mathbf{x}),$$

for each vector $\mathbf{x} \in \mathbb{Z}^n = K_0(A)$, see (III.3.16). It follows that $\overline{q}_A(\overline{\mathbf{\Phi}}_A(\overline{\mathbf{x}})) = \overline{q}_A(\overline{\mathbf{x}})$, for any vector $\overline{\mathbf{x}} \in \mathbb{Z}^{n-1} = \overline{K_0(A)}$.

The cosets $\overline{\mathbf{e}}_1, \ldots, \overline{\mathbf{e}}_n \in \overline{K_0(A)}$ of the canonical $\mathbb{Z}$-basis vectors $\mathbf{e}_1, \ldots, \mathbf{e}_n$ of the group $K_0(A) = \mathbb{Z}^n$ are roots of the quadratic form $\overline{q}_A$, because $\overline{q}_A(\overline{\mathbf{e}}_j) = q_A(\mathbf{e}_j) = 1$, for $j = 1, \ldots, n$. In other words, the vectors $\overline{\mathbf{e}}_1, \ldots, \overline{\mathbf{e}}_n$ belong to the set $\overline{\mathcal{R}}$.

It follows that the group automorphism $\overline{\mathbf{\Phi}}_A : \overline{K_0(A)} \longrightarrow \overline{K_0(A)}$ is of finite order $d_Q \geq 2$, because the finite set $\overline{\mathcal{R}}$ is $\overline{\mathbf{\Phi}}_A$-invariant and contains the set $\{\overline{\mathbf{e}}_1, \ldots, \overline{\mathbf{e}}_n\}$ of generators of the group $\overline{K_0(A)}$. If we set $d = d_Q$ then $\overline{\mathbf{\Phi}}_A^d(\overline{v}) = \overline{v}$, for each $v \in K_0(A)$, and it follows that, for any $\mathbf{x} \in \mathbb{Z}^n = K_0(A)$, the vector $\mathbf{\Phi}_A^d(\mathbf{x}) - \mathbf{x}$ belongs to $\mathbb{Z} \cdot \mathbf{h} = \mathrm{rad}\, A$. This finishes the proof. $\square$

From now on, we denote by $d_Q = d_A$ the least integer $d \geq 1$ satisfying the condition of (1.3)(b). This allows us to introduce an important concept of defect of A.

For this purpose, we note that by (1.3)(b), for each $\mathbf{x} \in \mathbb{Z}^n = K_0(A)$, there exists an integer $\partial_A(\mathbf{x}) \in \mathbb{Z}$ such that

$$\mathbf{\Phi}_A^{d_Q}(\mathbf{x}) = \mathbf{x} + \partial_A(\mathbf{x}) \cdot \mathbf{h}_Q,$$

where $A = KQ$ is the path algebra of a Euclidean quiver Q and $\mathbf{h}_Q \in \mathbb{Z}^n$ is the positive generator of the subgroup $\mathrm{rad}\, q_A$ of $\mathbb{Z}^n$, see (1.1).

A straightforward computation shows that the function $\partial_A : \mathbb{Z}^n \longrightarrow \mathbb{Z}$ thus defined is an abelian group homomorphism ($\mathbb{Z}$-linear). We are thus led to the following definition introduced by Gelfand and Ponomarev in [97] for the four subspace type quiver $\widetilde{\mathbb{D}}_4$, and extended by Dlab and Ringel in [63] to arbitrary path algebras of quivers whose underlying graph is Euclidean.

1.4. Definition. Let Q be an acyclic quiver whose underlying graph is Euclidean. Let $A = KQ$ be the path algebra of Q, $n = |Q_0|$, let $K_0(A) = \mathbb{Z}^n$ be the Grothendieck group of A,

$$\mathbf{\Phi}_A : K_0(A) \longrightarrow K_0(A)$$

be the Coxeter transformation of A, $\mathbf{h}_Q \in \mathbb{Z}^n$ be the positive generator of $\mathrm{rad}\, q_A$ (see (1.1)).

(a) The **defect number** d_Q of Q is the least integer $d \geq 1$ satisfying the condition of (1.3)(b). The **defect number** d_A of the Euclidean algebra $A = KQ$ is defined to be the defect number $d_A = d_Q$ of the quiver Q.

(b) The **defect** of the algebra A is the abelian group homomorphism

$$\partial_A : \mathbb{Z}^n \longrightarrow \mathbb{Z}$$

such that $\Phi_A^{d_A}(\mathbf{x}) = \mathbf{x} + \partial_A(\mathbf{x}) \cdot \mathbf{h}_Q$, for any $\mathbf{x} \in \mathbb{Z}^n$, where $d_A = d_Q$ is the defect number of A.

Now we show that the defect $\partial_A : \mathbb{Z}^n \longrightarrow \mathbb{Z}$ is a Φ_A-invariant homomorphism.

1.5. Lemma. *Let Q be an acyclic quiver whose underlying graph is Euclidean. If $\partial_A : \mathbb{Z}^n \longrightarrow \mathbb{Z}$ is the defect of the path algebra $A = KQ$ and $K_0(A) = \mathbb{Z}^n$ is the Grothendieck group of A, then $\partial_A(\mathbf{x}) = \partial_A(\Phi_A(\mathbf{x}))$, for any $\mathbf{x} \in \mathbb{Z}^n$.*

Proof. Let $\mathbf{h} = \mathbf{h}_Q \in \mathbb{Z}^n$ be the positive generator of the group $\operatorname{rad} q_A$ and $d = d_Q$ the defect number of Q. By (1.3)(a), we have $\Phi_A(\mathbf{h}) = \mathbf{h}$, and, hence, we get the equalities

$$\Phi_A(\mathbf{x}) + \partial_A(\Phi_A(\mathbf{x})) \cdot \mathbf{h} = \Phi_A^d(\Phi_A(\mathbf{x}))$$
$$= \Phi_A \Phi_A^d(\mathbf{x})$$
$$= \Phi_A(\mathbf{x}) + \partial_A(\mathbf{x})\Phi_A(\mathbf{h})$$
$$= \Phi_A(\mathbf{x}) + \partial_A(\mathbf{x}) \cdot \mathbf{h}.$$

The statement follows. $\qquad\square$

The following two examples show that the defect ∂_A and the Coxeter transformation $\Phi_A : K_0(A) \longrightarrow K_0(A)$ of the path algebra $A = KQ$ of a Euclidean quiver Q depend on the orientation of Q.

1.6. Example. Let A be the path algebra of the quiver Q

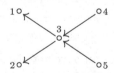

as in (X.2.12). We show that the defect $\partial_A : \mathbb{Z}^5 \longrightarrow \mathbb{Z}$ of A is defined by the formula $\partial_A(\mathbf{x}) = x_5 + x_4 - x_2 - x_1$.

First we note that $(1, 2, 3, 4, 5)$ is an admissible ordering of the vertices of Q and, by (VII.4.7), the Coxeter matrix Φ_A of $A = KQ$ is the matrix of the Coxeter transformation $c = s_5 s_4 s_3 s_2 s_1 : \mathbb{Z}^5 \longrightarrow \mathbb{Z}^5$ of the quiver Q in the canonical basis of $\mathbb{Z}^5$.

We easily compute, as in (VII.4.7), the matrices of the reflections $s_i : \mathbb{Z}^5 \longrightarrow \mathbb{Z}^5$, with $1 \le i \le 5$. It follows that the matrices of s_1, s_2, s_3, s_4, s_5, in the canonical basis $\mathbf{e}_1, \dots, \mathbf{e}_5$ of $\mathbb{Z}^5$, are of the form

$$
s_1 = \begin{bmatrix} -1 & 0 & 1 & 0 & 0 \\ 0 & 1 & 0 & 0 & 0 \\ 0 & 0 & 1 & 0 & 0 \\ 0 & 0 & 0 & 1 & 0 \\ 0 & 0 & 0 & 0 & 1 \end{bmatrix}, \quad
s_2 = \begin{bmatrix} 1 & 0 & 0 & 0 & 0 \\ 0 & -1 & 1 & 0 & 0 \\ 0 & 0 & 1 & 0 & 0 \\ 0 & 0 & 0 & 1 & 0 \\ 0 & 0 & 0 & 0 & 1 \end{bmatrix}, \quad
s_3 = \begin{bmatrix} 1 & 0 & 0 & 0 & 0 \\ 0 & 1 & 0 & 0 & 0 \\ 1 & 1 & -1 & 1 & 1 \\ 0 & 0 & 0 & 1 & 0 \\ 0 & 0 & 0 & 0 & 1 \end{bmatrix},
$$

$$
s_4 = \begin{bmatrix} 1 & 0 & 0 & 0 & 0 \\ 0 & 1 & 0 & 0 & 0 \\ 0 & 0 & 1 & 0 & 0 \\ 0 & 0 & 1 & -1 & 0 \\ 0 & 0 & 0 & 0 & 1 \end{bmatrix}, \quad
s_5 = \begin{bmatrix} 1 & 0 & 0 & 0 & 0 \\ 0 & 1 & 0 & 0 & 0 \\ 0 & 0 & 1 & 0 & 0 \\ 0 & 0 & 0 & 1 & 0 \\ 0 & 0 & 1 & 0 & -1 \end{bmatrix}.
$$

Hence the Coxeter matrix $\boldsymbol{\Phi}_A = s_5 s_4 s_3 s_2 s_1$ is given by

$$
\boldsymbol{\Phi}_A = \begin{bmatrix} -1 & 0 & 1 & 0 & 0 \\ 0 & -1 & 1 & 0 & 0 \\ -1 & -1 & 1 & 1 & 1 \\ -1 & -1 & 1 & 0 & 1 \\ -1 & -1 & 1 & 1 & 0 \end{bmatrix}.
$$

It follows from (VII.4.2) that $\operatorname{rad} q_Q = \mathbb{Z} \cdot \mathbf{h}$, where $\mathbf{h} = \mathbf{h}_Q = \begin{smallmatrix} & 1 \\ 1 & 2 & 1 \\ & 1 \end{smallmatrix}$ is the positive generator of the group $\operatorname{rad} q_A$, see (1.1). By applying $\boldsymbol{\Phi}_A$ to the basis vectors, we obtain

$$\boldsymbol{\Phi}_A^2(\mathbf{e}_1) = \mathbf{e}_1 - \mathbf{h}, \qquad \boldsymbol{\Phi}_A^2(\mathbf{e}_2) = \mathbf{e}_2 - \mathbf{h}, \qquad \boldsymbol{\Phi}_A^2(\mathbf{e}_3) = \mathbf{e}_3,$$
$$\boldsymbol{\Phi}_A^2(\mathbf{e}_4) = \mathbf{e}_4 + \mathbf{h}, \qquad \boldsymbol{\Phi}_A^2(\mathbf{e}_5) = \mathbf{e}_5 + \mathbf{h}, \qquad \boldsymbol{\Phi}_B(\mathbf{e}_5) - \mathbf{e}_5 \notin \mathbb{Z} \cdot \mathbf{h}.$$

We can also compute the Coxeter matrix $\boldsymbol{\Phi}_A = -\mathbf{C}_A^t \cdot \mathbf{C}_A^{-1}$ of A by using the Cartan matrix of A and its inverse:

$$
\mathbf{C}_A = \begin{bmatrix} 1 & 0 & 1 & 1 & 1 \\ 0 & 1 & 1 & 1 & 1 \\ 0 & 0 & 1 & 1 & 1 \\ 0 & 0 & 0 & 1 & 0 \\ 0 & 0 & 0 & 0 & 1 \end{bmatrix}, \quad
\mathbf{C}_A^{-1} = \begin{bmatrix} 1 & 0 & -1 & 0 & 0 \\ 0 & 1 & -1 & 0 & 0 \\ 0 & 0 & 1 & -1 & -1 \\ 0 & 0 & 0 & 1 & 0 \\ 0 & 0 & 0 & 0 & 1 \end{bmatrix}.
$$

Thus, if $\mathbf{x} = x_1 \mathbf{e}_1 + x_2 \mathbf{e}_2 + x_3 \mathbf{e}_3 + x_4 \mathbf{e}_4 + x_5 \mathbf{e}_5$, we get

$$\boldsymbol{\Phi}_A^2(\mathbf{x}) = \mathbf{x} + (x_5 + x_4 - x_2 - x_1) \cdot \mathbf{h}.$$

It follows that $d_Q = 2$ and the defect $\partial_A : \mathbb{Z}^5 \longrightarrow \mathbb{Z}$ of the algebra A is given by the formula $\partial_A(\mathbf{x}) = x_5 + x_4 - x_2 - x_1$, for any vector $\mathbf{x} = [x_1, x_2, x_3, x_4, x_5]^t \in \mathbb{Z}^5$.

Now we show that the Coxeter transformation $\boldsymbol{\Phi}_A$ and the defect ∂_A of the path algebra $A = KQ$ depend on the orientation of the quiver Q.

1.7. Example. Let Q' be the quiver

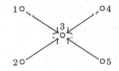

obtained from the quiver Q of Example (1.6) by reverting the orientation of the two arrows ending at the points 1 and 2. It follows that the path algebra $B = KQ'$ of Q' is isomorphic to the algebra

$$B = \begin{bmatrix} K & 0 & K & 0 & 0 \\ 0 & K & K & 0 & 0 \\ 0 & 0 & K & 0 & 0 \\ 0 & 0 & K & K & 0 \\ 0 & 0 & K & 0 & K \end{bmatrix}.$$

Then the Cartan matrix $\mathbf{C}_B$ of B and its inverse $\mathbf{C}_B^{-1}$ have the forms

$$\mathbf{C}_B = \begin{bmatrix} 1 & 0 & 0 & 0 & 0 \\ 0 & 1 & 0 & 0 & 0 \\ 1 & 1 & 1 & 1 & 1 \\ 0 & 0 & 0 & 1 & 0 \\ 0 & 0 & 0 & 0 & 1 \end{bmatrix} \quad \text{and} \quad \mathbf{C}_B^{-1} = \begin{bmatrix} 1 & 0 & 0 & 0 & 0 \\ 0 & 1 & 0 & 0 & 0 \\ -1 & -1 & -1 & -1 & -1 \\ 0 & 0 & 0 & 1 & 0 \\ 0 & 0 & 0 & 0 & 1 \end{bmatrix}.$$

Hence the Coxeter matrix $\Phi_B = -\mathbf{C}_B^t \cdot \mathbf{C}_B^{-1}$ of B is of the form

$$\Phi_B = \begin{bmatrix} 0 & 1 & -1 & 1 & 1 \\ 1 & 0 & -1 & 1 & 1 \\ 1 & 1 & -1 & 1 & 1 \\ 1 & 1 & -1 & 0 & 1 \\ 1 & 1 & -1 & 1 & 0 \end{bmatrix}.$$

Note also that $(3, 4, 5, 2, 1)$ is an admissible ordering of the vertices of Q' and, by (VII.4.7), $\Phi_B = s_1 s_2 s_5 s_4 s_3$, for the reflections s_1, s_2, s_3, s_4, s_5. Recall also that $\operatorname{rad} q_{Q'} = \mathbb{Z} \cdot \mathbf{h}$, where $\mathbf{h} = \mathbf{h}_Q = \begin{smallmatrix} & 1 & \\ 1 & 2 & 1 \\ & 1 & \end{smallmatrix}$ is the positive generator of the group $\operatorname{rad} q_B$, see (1.1). By applying the matrix Φ_B to the canonical basis vectors, we obtain

$$\Phi_B^2(\mathbf{e}_1) = \mathbf{e}_1 + \mathbf{h}, \qquad \Phi_B^2(\mathbf{e}_2) = \mathbf{e}_2 + \mathbf{h}, \qquad \Phi_B^2(\mathbf{e}_3) = \mathbf{e}_3 - 2\mathbf{h},$$
$$\Phi_B^2(\mathbf{e}_4) = \mathbf{e}_4 + \mathbf{h}, \qquad \Phi_B(\mathbf{e}_5) = \mathbf{e}_5 + \mathbf{h}, \qquad \Phi_B(\mathbf{e}_5) - \mathbf{e}_5 \notin \mathbb{Z} \cdot \mathbf{h}.$$

Then, for any $\mathbf{x} = [x_1, x_2, x_3, x_4, x_5]^t \in \mathbb{Z}^5$, we get

$$\Phi_B^2(\mathbf{x}) = \mathbf{x} + (x_1 + x_2 + x_4 + x_5 - 2x_3) \cdot \mathbf{h}.$$

It follows that $d_{Q'} = 2$ and the defect $\partial_B : \mathbb{Z}^5 \longrightarrow \mathbb{Z}$ of B is given by the formula $\partial_B(\mathbf{x}) = x_1 + x_2 + x_4 + x_5 - 2x_3$, for any vector $\mathbf{x} = [x_1, x_2, x_3, x_4, x_5]^t \in \mathbb{Z}^5$. In particular, we get $\Phi_A \neq \Phi_B$ and $\partial_A \neq \partial_B$, where $A = KQ$ is the algebra of Example (1.6).

XI.2. The category of regular modules over a hereditary algebra of Euclidean type

We now recall from (VIII.2.12) that, for a representation-infinite hereditary algebra $A = KQ$ as in (1.3), the Auslander–Reiten quiver $\Gamma(\operatorname{mod} A)$ of A is the union of the postprojective component $\mathcal{P}(A)$, the preinjective component $\mathcal{Q}(A)$, and the union $\mathcal{R}(A)$ of regular components. We may visualise the shape of $\Gamma(\operatorname{mod} A)$ as follows.

2.1. Figure

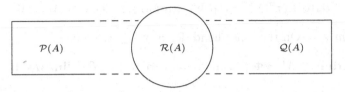

By (VIII.2.9), the regular part $\mathcal{R}(A)$ of $\Gamma(\mathrm{mod}\,A)$ is not empty, and it follows from (VIII.2.13) that, in the above picture, the non-zero homomorphisms can only go from the left to the right, that is, we have

$$\mathrm{Hom}_A(\mathcal{R}(A),\mathcal{P}(A)) = 0, \mathrm{Hom}_A(\mathcal{Q}(A),\mathcal{P}(A)) = 0 \text{ and } \mathrm{Hom}_A(\mathcal{Q}(A),\mathcal{R}(A)) = 0.$$

We also recall from (VIII.2.14) that the mutually inverse equivalences

$$\underline{\mathrm{mod}}\,A \underset{\tau^{-1}}{\overset{\tau}{\rightleftarrows}} \overline{\mathrm{mod}}\,A,$$

(IV.2.11) induced by the Auslander–Reiten translations τ and τ^{-1}, induce mutually inverse equivalences of categories

$$\mathrm{add}\,\mathcal{R}(A) \underset{\tau^{-1}}{\overset{\tau}{\rightleftarrows}} \mathrm{add}\,\mathcal{R}(A), \qquad (2.2)$$

where $\mathrm{add}\,\mathcal{R}(A)$ is the full subcategory of $\mathrm{mod}\,A$ whose objects are all the regular A-modules.

Now we characterise the components of $\Gamma(\mathrm{mod}\,A)$ in terms of the defect ∂_A.

2.3. Proposition. *Assume that Q is an acyclic quiver whose underlying graph is Euclidean. Let $A = KQ$ be the path algebra of Q, let $\partial_A : \mathbb{Z}^n \longrightarrow \mathbb{Z}$ be the defect of A, and let M be an indecomposable A-module.*

(a) *$M \in \mathcal{P}(A)$ if and only if $\partial_A(\mathbf{dim}\,M) < 0$.*
(b) *$M \in \mathcal{R}(A)$ if and only if $\partial_A(\mathbf{dim}\,M) = 0$.*
(c) *$M \in \mathcal{Q}(A)$ if and only if $\partial_A(\mathbf{dim}\,M) > 0$.*

Proof. Let $\Phi_A : \mathbb{Z}^n \longrightarrow \mathbb{Z}^n$ be the Coxeter transformation of A, where $n = |Q_0|$, let $\mathbf{h} = \mathbf{h}_Q \in \mathbb{Z}^n$ be the positive generator of the group $\mathrm{rad}\,q_A$, see (1.1).

Assume that the indecomposable module M is not postprojective. It follows from (1.2) that, for any $m \geq 0$, we have

$$\mathbf{dim}\,\tau^m M = \Phi_A^m(\mathbf{dim}\,M).$$

Suppose now $\partial_A(\dim M) < 0$. Applying (1.5), we get

$$\partial_A(\dim \tau^m M) = \partial_A(\Phi_A^m(\dim M)) = \partial_A(\dim M) < 0$$

for any $m \geq 0$. On the other hand, for any $s \geq 1$, we have

$$\dim \tau^{sd} M = \Phi_A^{sd}(\dim M) = \dim M + s \cdot \partial_A(\dim M) \cdot \mathbf{h},$$

where $d = d_Q$, see (1.4). Hence we obtain the inequalities

$$\dim M > \dim \tau^d M > \ldots > \dim \tau^{sd} M > \dim \tau^{(s+1)d} M > \ldots .$$

and we get a contradiction with the inequality $\dim \tau^{sd} M > 0$, for any $s \geq 1$. Therefore, $\partial_A(\dim M) \geq 0$. On the other hand, if M belongs to $\mathcal{Q}(A)$, then $\partial_A(\dim M) > 0$. Indeed, if we assume to the contrary that $\partial_A(\dim M) = 0$, then

$$\dim \tau^d M = \Phi_A^d(\dim M) = \dim M + \partial_A(\dim M) \cdot \mathbf{h} = \dim M,$$

and we get a contradiction to the fact that all indecomposable modules in preinjective components are directing (by (IX.1.1)) and hence are uniquely determined, up to isomorphism, by their dimension vectors (by (IX.3.1)).

Dually, we have $\partial_A(\dim M) \leq 0$, for any M in $\mathcal{R}(A)$, and $\partial_A(\dim M) < 0$, for any M in $\mathcal{P}(A)$. In particular, $\partial_A(\dim M) = 0$, for any M in $\mathcal{R}(A)$. $\square$

2.4. Proposition. *Let Q be an acyclic quiver whose underlying graph is Euclidean, and let $A = KQ$. The full subcategory* add $\mathcal{R}(A)$ *of* mod A *is abelian and closed under extensions.*

Proof. Let U, V be two objects from add $\mathcal{R}(A)$, and $f : U \longrightarrow V$ be a non-zero homomorphism. We first claim that Im $f \in$ add $\mathcal{R}(A)$. Indeed, let M be an indecomposable summand of Im f. There exist indecomposable summands L of U and N of V, respectively, such that $\mathrm{Hom}_A(L, M) \neq 0$ and $\mathrm{Hom}_A(M, N) \neq 0$. By (VIII.2.13), this implies that M belongs to $\mathcal{R}(A)$. Therefore, Im f, being a direct sum of indecomposable regular modules, is itself regular. In particular, $\partial_A(\dim \mathrm{Im}\, f) = 0$, where $\partial_A : \mathbb{Z}^n \longrightarrow \mathbb{Z}$ is the defect of A, $n = |Q_0|$, and we identify $\mathbb{Z}^n$ with the Grothendieck group $K_0(A)$ of A.

The short exact sequence

$$0 \longrightarrow \mathrm{Ker}\, f \longrightarrow U \longrightarrow \mathrm{Im}\, f \longrightarrow 0$$

gives that $\dim \mathrm{Ker}\, f = \dim U - \dim \mathrm{Im}\, f$. Hence we get

$$\partial_A(\dim \mathrm{Ker}\, f) = \partial_A(\dim U) - \partial_A(\dim \mathrm{Im}\, f) = 0.$$

Let M be an indecomposable summand of $\operatorname{Ker} f$, then $\operatorname{Hom}_A(M, U) \neq 0$. By (VIII.2.13), this implies that the module M is postprojective or regular. By (2.3), $\partial_A(\operatorname{\mathbf{dim}} M) \leq 0$. Because this holds for any indecomposable summand M of $\operatorname{Ker} f$, the vanishing of $\partial_A(\operatorname{\mathbf{dim}} \operatorname{Ker} f)$ implies that $\partial_A(\operatorname{\mathbf{dim}} M) = 0$ for any such direct summand, which is therefore regular; hence, so is $\operatorname{Ker} f$. Similarly, the short exact sequence

$$0 \longrightarrow \operatorname{Im} f \longrightarrow V \longrightarrow \operatorname{Coker} f \longrightarrow 0$$

gives that $\operatorname{Coker} f$ is regular. $\square$

The above proposition allows us to introduce an important numerical invariant of regular A-modules.

2.5. Definition. Let Q be an acyclic quiver whose underlying graph is Euclidean, and let $A = KQ$. A non-zero regular A-module E having no proper regular submodules is said to be **simple regular**. Then, for any regular module M, there exists a chain

$$M = M_0 \supsetneq M_1 \supsetneq \cdots \supsetneq M_l = 0$$

of regular submodules of M such that M_{i-1}/M_i is simple regular for any i with $1 \leq i \leq l$, and l is called the **regular length** of M, which we denote by $r\ell(M)$.

We need two more definitions.

2.6. Definition. Two components $\mathcal{C}$ and $\mathcal{C}'$ of the Auslander–Reiten quiver of an algebra A are said to be **orthogonal** if $\operatorname{Hom}_A(\mathcal{C}, \mathcal{C}') = 0$ and $\operatorname{Hom}_A(\mathcal{C}', \mathcal{C}) = 0$, that is, $\operatorname{Hom}_A(C, C') = 0$ and $\operatorname{Hom}_A(C', C) = 0$, for any module C in $\mathcal{C}$ and any module C' in $\mathcal{C}'$.

2.7. Definition. The category $\operatorname{add} \mathcal{R}(A)$ is said to be **serial** if every indecomposable object of $\operatorname{add} \mathcal{R}(A)$ is uniserial.

We are now able to prove the main result of this section.

2.8. Theorem. *Let Q be an acyclic quiver whose underlying graph is Euclidean, and let $A = KQ$.*

(a) *The components of $\mathcal{R}(A)$ form a family $\boldsymbol{\mathcal{T}}^A = \{\mathcal{T}_\lambda^A\}_{\lambda \in \Lambda}$ of pairwise orthogonal standard stable tubes $\mathcal{T}_\lambda^A$.*

(b) *The category $\operatorname{add} \mathcal{R}(A)$ is serial.*

Proof. By (VIII.2.14), the Auslander–Reiten translations τ and τ^{-1} induce mutually inverse self-equivalences of the abelian subcategory add $\mathcal{R}(A)$ of mod A. Thus, they preserve the regular length of objects in add $\mathcal{R}(A)$.

Let E be a simple regular module. The τ-orbit

$$\mathcal{O}(E) = \{\tau^m E \mid m \in \mathbb{Z}\}$$

of E consists of simple regular modules. We claim that $\mathcal{O}(E)$ is a finite set. Indeed, suppose that the modules $\tau^m E$, with $m \in \mathbb{Z}$, are pairwise non-isomorphic. Then $\operatorname{Hom}_A(\tau^s E, \tau^t E) = 0$ whenever $s \neq t$, because $\tau^s E$ and $\tau^t E$ are non-isomorphic simple objects of add $\mathcal{R}(A)$. Consider the regular module

$$M = E \oplus \tau^2 E \oplus \tau^4 E \oplus \ldots \oplus \tau^{2n} E,$$

where $n = |Q_0|$. Then $\operatorname{Hom}_A(M, \tau M) = 0$, and M is the direct sum of $n+1$ pairwise non-isomorphic indecomposable A-modules. Because $n \geq 2$ is the rank of the Grothendieck group $K_0(A)$ of A, this contradicts (VIII.5.3). Thus $\mathcal{O}(E)$ is finite, that is, there exists a positive integer r such that $\mathcal{O}(E)$ consists of the modules

$$E_r = E, \ E_{r-1} = \tau E, \ \ldots, \ E_1 = \tau^{r-1} E$$

(so that $\tau E_1 = \tau^r E \cong E = E_r$). Because $E_1, \ldots, E_r$ are simple regular modules, it follows from Schur's Lemma (I.3.1) that their endomorphism algebras are division K-algebras, and consequently are equal to K, because K is an algebraically closed field. Thus $E_1, \ldots, E_r$ are bricks. Moreover, $\operatorname{Ext}_A^2(-,-) = 0$, because A is hereditary. It follows from (X.2.1) that $\mathcal{E} = \mathcal{EXT}_A(E_1, \ldots, E_r)$ is an abelian subcategory of mod A, and clearly also of add $\mathcal{R}(A)$, whose indecomposable objects are uniserial and form a component $\mathcal{T}_{\mathcal{E}}$ of mod A, which is a standard stable tube of rank r.

We have proved that $\Gamma(\operatorname{mod} A)$ contains a family $\boldsymbol{T}^A = \{\mathcal{T}_\lambda^A\}_{\lambda \in \Lambda}$ of standard stable tubes such that the modules lying on the mouth of the tubes $\mathcal{T}_\lambda^A$, with $\lambda \in \Lambda$, form a complete family of the simple objects in add $\mathcal{R}(A)$. Observe also that, if M and N are indecomposable modules lying in two different tubes, then M and N have pairwise distinct simple regular composition factors. Hence $\operatorname{Hom}_A(M, N) = 0$. Thus the tubes in the family $\boldsymbol{T}^A = \{\mathcal{T}_\lambda^A\}_{\lambda \in \Lambda}$ are pairwise orthogonal.

Let now L be an arbitrary indecomposable regular A-module. Then there exists a chain of regular submodules $L = L_0 \supset L_1 \supset \ldots \supset L_l = 0$, with L_{i-1}/L_i simple regular, for each i such that $1 \leq i \leq l$. Consequently, L belongs to one of the tubes $\mathcal{T}_\lambda^A$. Hence the stable tubes in the family $\boldsymbol{T}^A = \{\mathcal{T}_\lambda^A\}_{\lambda \in \Lambda}$ are all the regular components of $\Gamma(\operatorname{mod} A)$. $\qquad\square$

We notice that, if M is an indecomposable regular module lying in a tube $\mathcal{T}_{\mathcal{E}}$, with $\mathcal{E} = \mathcal{E}\mathcal{X}\mathcal{T}_A(E_1, \ldots, E_r)$, then there exist i and j such that $1 \leq i \leq r$, $j \geq 1$, and $M \cong E_i[j]$. In this situation, $r\ell(M) = j$, and the modules $E_1, \ldots, E_r$ are simple regular.

The preceding theorem gives no information on the ranks r_λ of the standard stable tubes $\mathcal{T}_\lambda^A$, with $\lambda \in \Lambda$, in $\Gamma(\mathrm{mod}\,A)$. We now show that almost all of them are homogeneous (that is, of rank one).

2.9. Proposition. *Let Q be an acyclic quiver whose underlying graph is Euclidean, let $A = KQ$, and let*

$$\boldsymbol{\mathcal{T}}^A = \{\mathcal{T}_\lambda^A\}_{\lambda \in \Lambda}$$

be the family of the standard stable tubes in $\Gamma(\mathrm{mod}\,A)$ as in (2.8).

(a) *If, for each $\lambda \in \Lambda$, $r_\lambda^A \geq 1$ is the rank of the tube $\mathcal{T}_\lambda^A$ in $\mathcal{R}(A)$, then*

$$\sum_{\lambda \in \Lambda} (r_\lambda^A - 1) \leq n - 2,$$

where $n = |Q_0|$.

(b) *All but finitely many of the tubes $\mathcal{T}_\lambda^A$ in $\mathcal{R}(A)$ are homogeneous, and there are at most $n - 2$ non-homogeneous tubes $\mathcal{T}_\lambda$.*

Proof. Consider the subset $\Lambda_0 = \{\lambda \in \Lambda \mid r_\lambda^A \geq 2\}$ of Λ. We claim that Λ_0 has at most n elements. Indeed, assume to the contrary that $|\Lambda_0| \geq n+1$, and let $\lambda_1, \ldots, \lambda_n, \lambda_{n+1}$ be pairwise distinct elements of Λ_0. For each $i \in \{1, \ldots, n+1\}$, let $E_i \in \mathcal{T}_{\lambda_i}^A$ be a simple regular module, and let

$$E = E_1 \oplus \ldots \oplus E_{n+1}.$$

Because each $\mathcal{T}_{\lambda_i}^A$ is a standard stable tube of rank at least two, then we have $\mathrm{Hom}_A(E_i, \tau E_i) = 0$. Moreover, $\mathrm{Hom}_A(E_i, \tau E_j) = 0$, for $i \neq j$, because the tubes $\mathcal{T}_{\lambda_1}^A, \ldots, \mathcal{T}_{\lambda_{n+1}}^A$ are pairwise orthogonal. Therefore, $\mathrm{Hom}_A(E, \tau E) = 0$, and we get a contradiction with (VIII.5.3). Hence $|\Lambda_0| \leq n$ and, without loss of generality, we may assume that $\Lambda_0 = \{1, \ldots, m\}$, where $m \leq n$.

We next prove that $\sum_{\lambda \in \Lambda} (r_\lambda^A - 1) \leq n - 1$. For each $i \in \{1, \ldots, m\}$, we choose a ray

$$E_i[1] \xrightarrow{u_{i2}} E_i[2] \xrightarrow{u_{i3}} E_i[3] \longrightarrow \cdots \longrightarrow E_i[r_i - 2] \xrightarrow{u_{i,r_i-1}} E_i[r_i-1]$$

in $\mathcal{T}_i^A$, with $E_i[1]$ lying on the mouth of $\mathcal{T}_i^A$, and we set

$$M = \bigoplus_{i=1}^{m} \bigoplus_{j=1}^{r_i^A - 1} E_i[j].$$

For each $i \in \Lambda_0$, we have

$$\mathrm{Hom}_A\left(\bigoplus_{j=1}^{r_i^A - 1} E_i[j], \bigoplus_{j=1}^{r_i - 1} \tau E_i[j] \right) = 0,$$

because each of the tubes $\mathcal{T}_i^A$ is standard. Moreover, $\mathrm{Hom}_A(M, \tau M) = 0$, because the tubes $\mathcal{T}_i^A$ are pairwise orthogonal. Hence, (VIII.5.3) yields $\sum_{i=1}^{m}(r_i^A - 1) \leq n$. If the equality holds, then M is the direct sum of n pairwise non-isomorphic indecomposable A-modules, $\mathrm{pd}_A M \leq 1$ (because A is hereditary), and $\mathrm{Ext}_A^1(M, M) \cong D\mathrm{Hom}_A(M, \tau M) = 0$. By (VI.4.4), M is a tilting A-module. Moreover, $B = \mathrm{End}\, M$ is a product of hereditary algebras of Dynkin type $\mathbb{A}_{r_i^A - 1}$, with $1 \leq i \leq m$. Then, by (VI.5.6), M is a separating tilting module. Because B is representation-finite, so is A. But this contradicts the fact that $A = KQ$ and the quiver Q is Euclidean. Therefore, we get $\sum_{i=1}^{m}(r_i^A - 1) \leq n - 1$.

Finally, suppose, to the contrary, that $\sum_{i=1}^{m}(r_i^A - 1) = n - 1$, and show that this leads to a contradiction. Because M is a partial tilting module, and M is the sum of $n - 1$ pairwise non-isomorphic indecomposable A-modules, there exists, by (VI.2.4) and (VI.4.4), an indecomposable A-module N such that $T = M \oplus N$ is a tilting A-module. In particular, $\mathrm{Hom}_A(T, \tau T) \cong D\mathrm{Ext}_A^1(T, T) = 0$.

We claim that the module N is not regular. Because $\mathrm{Ext}_A^1(N, N) = 0$, the module N does not belong to a homogeneous tube. Assume now that N lies in $\mathcal{T}_i^A$, for some $i \in \{1, \ldots, m\}$. Because the tube $\mathcal{T}_i^A$ is standard and of rank r_i^A, then N must be of regular length at most $r_i^A - 1$. But this implies the existence of some $j \in \{1, \ldots, r_i^A - 1\}$ such that $\mathrm{Hom}_A(N, \tau E_i[j]) \neq 0$; a contradiction. This establishes our claim.

Let $B = \mathrm{End}\, T_A$. We claim that the algebra B is hereditary. We already know that $\mathrm{End}\, M$ is the direct product of hereditary algebras of Dynkin types $\mathbb{A}_{r_i^A - 1}$, with $1 \leq i \leq m$. Because N is postprojective or preinjective, it is directing, hence $\mathrm{End}\, N \cong K$ (by (IX.1.4)). Assume that N is preinjective. Then $\mathrm{Hom}_A(N, M) = 0$ and $\mathrm{Hom}_A(N, \tau M) = 0$. On the other hand, by (1.3), we have a short exact sequence

$$0 \longrightarrow E_i[j-1] \xrightarrow{u_{ij}} E_i[j] \xrightarrow{p'_{ij}} E \longrightarrow 0,$$

for each $j \in \{2, \ldots, r_i^A - 1\}$, where $E = E_{i+j-1}[1]$. Applying the functor $\mathrm{Hom}_A(-, N)$, we get an exact sequence

$$0 \to \mathrm{Hom}_A(E, N) \to \mathrm{Hom}_A(E_i[j], N) \to \mathrm{Hom}_A(E_i[j-1], N) \to \mathrm{Ext}_A^1(E, N).$$

Because $\mathrm{Hom}_A(\tau^{-1}E_i[j-1], N) = 0$, and E is the simple regular top of the module $\tau^{-1}E_i[j-1]$, we have $\mathrm{Hom}_A(E, N) = 0$. Also $\mathrm{Ext}^1_A(E, N) \cong$ $D\mathrm{Hom}_A(N, \tau E) = 0$. Thus we have an isomorphism

$$\mathrm{Hom}_A(u_{ij}, N) : \mathrm{Hom}_A(E_i[j], N) \xrightarrow{\;\simeq\;} \mathrm{Hom}_A(E_i[j-1]), N),$$

for each $j \in \{2, \dots, r^A_i - 1\}$. It follows that the algebra $B = \mathrm{End}\,T_A$ is hereditary and isomorphic to the path algebra $K\Delta$ of the following quiver

$$
\Delta: \quad
\begin{array}{ccccccccc}
(1,1) & \longleftarrow & (1,2) & \longleftarrow & \cdots & \longleftarrow & (1, r^A_1 - 2) & \longleftarrow & (1, r^A_1 - 1) \\
\vdots & & \vdots & & & & \vdots & & \vdots \\
(j,1) & \longleftarrow & (j,2) & \longleftarrow & \cdots & \longleftarrow & (j, r^A_j - 2) & \longleftarrow & (j, r^A_j - 1) \quad \omega, \\
\vdots & & \vdots & & & & \vdots & & \vdots \\
(m,1) & \longleftarrow & (m,2) & \longleftarrow & \cdots & \longleftarrow & (m, r^A_m - 2) & \longleftarrow & (m, r^A_m - 1)
\end{array}
$$

where the number of arrows from the source vertex ω to the vertex $(j, r^A_j - 1)$ equals $n_j = \dim_K \mathrm{Hom}_A(E_j[r^A_j - 1], N)$, for $j \in \{1, \dots, m\}$. In fact, we have $n_j \geq 1$, for any $j \in \{1, \dots, m\}$, because the algebra $B = \mathrm{End}\,T_A$ is connected, see (VI.3.5).

Assume now that the module N is postprojective. Then $\mathrm{Hom}_A(M, N) = 0$ and $\mathrm{Hom}_A(N, \tau M) \cong \mathrm{Ext}^1_A(M, N) = 0$. Now, if $\mathrm{Hom}_A(N, E_i[j]) \neq 0$, for some j, then a non-zero homomorphism from N to $E_i[j]$ must, by (IV.5.1), factor through τM, or else $j = r^A_i - 1$. It follows that $j = r^A_i - 1$ and the algebra $B = \mathrm{End}\,T_A$ is hereditary. More precisely, $B = \mathrm{End}\,T_A$ is isomorphic to the path algebra $K\Delta'$ of the following quiver

$$
\Delta': \quad
\begin{array}{ccccccccc}
(1,1) & \longleftarrow & (1,2) & \longleftarrow & \cdots & \longleftarrow & (1, r^A_1 - 2) & \longleftarrow & (1, r^A_1 - 1) \\
\vdots & & \vdots & & & & \vdots & & \vdots \\
(j,1) & \longleftarrow & (j,2) & \longleftarrow & \cdots & \longleftarrow & (j, r^A_j - 2) & \longleftarrow & (j, r^A_j - 1) \quad \omega', \\
\vdots & & \vdots & & & & \vdots & & \vdots \\
(m,1) & \longleftarrow & (m,2) & \longleftarrow & \cdots & \longleftarrow & (m, r^A_m - 2) & \longleftarrow & (m, r^A_m - 1)
\end{array}
$$

where the number of arrows from the vertex $(j, r^A_j - 1)$ to the sink vertex ω' equals $n'_j = \dim_K \mathrm{Hom}_A(N, E_j[r^A_j - 1])$, for $j \in \{1, \dots, m\}$, and again $n'_j \geq 1$, for any $j \in \{1, \dots, m\}$, because the algebra $B = \mathrm{End}\,T_A$ is connected.

Applying again (VI.5.6), we infer that T is a separating tilting module. Therefore, any indecomposable A-module belongs either to $\mathcal{F}(T)$ or to $\mathcal{T}(T)$. Because $E_i[1] \in \mathcal{T}(T)$, and any module in $\mathcal{T}_i^A$ is a successor of $E_i[1]$, we deduce that $\mathcal{T}_i^A \subseteq \mathcal{T}(T)$. On the other hand, $\operatorname{Hom}_A(T, \tau M) = 0$ implies that $\tau E_i[1] \in \mathcal{F}(T)$, a contradiction. This completes the proof that $\sum_{i=1}^{m} (r_i^A - 1) \leq n - 2$. This finishes the proof of the proposition, because the statement (b) is an immediate consequence of (a). $\qquad\square$

2.10. Example. Let A be a K-algebra given by the quiver

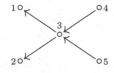

as in Examples (X.2.12) and (1.6). We have constructed in (X.2.12) a stable tube $\mathcal{T}_1^A$ of rank $r_1^A = 2$ containing the simple regular modules $S = S(3)$ and E given by $\mathbf{dim}\, E = \begin{smallmatrix} 1 & & 1 \\ & 1 & \\ 1 & & 1 \end{smallmatrix}$. On the other hand, the indecomposable module

$$E_1 = \quad$$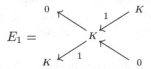

is regular. Indeed, using the defect ∂_A computed in Example (1.6), we have $\partial_A(\mathbf{dim}\, E_1) = 0$. An easy computation shows that

$$E_2 \cong \tau E_1 = \quad$$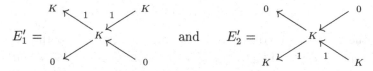

Moreover, $E_1 \cong \tau E_2$ and each of the modules E_1 and E_2 is a brick. Hence, by (X.2.2) and (X.2.6), the indecomposable modules in $\mathcal{EXT}_A(E_1, E_2)$ are uniserial and form a standard stable tube $\mathcal{T}_2^A$ of rank $r_2^A = 2$ in $\Gamma(\operatorname{mod} A)$. Similarly,

$$E_1' = \qquad \text{and} \qquad E_2' =$$

are regular bricks such that $\tau E_1' \cong E_2'$ and $\tau E_2' \cong E_1'$. Again, the indecomposables in $\mathcal{EXT}_A(E_1', E_2')$ are uniserial and form a standard stable tube $\mathcal{T}_3^A$ of rank $r_3^A = 2$ in $\Gamma(\operatorname{mod} A)$.

Because $n = |Q_0| = 5$ then, by (2.9), we get $\sum_{i=1}^{3}(r_i^A - 1) = 3 = n - 2$ and, hence, all the remaining tubes in $\Gamma(\operatorname{mod} A)$ are homogeneous. It follows that $\Gamma(\operatorname{mod} A) = \mathcal{P}(A) \cup \mathcal{R}(A) \cup \mathcal{Q}(A)$ consists of a postprojective component $\mathcal{P}(A)$, a preinjective component $\mathcal{Q}(A)$, and the regular part $\mathcal{R}(A)$; consisting of the three tubes $\mathcal{T}_1^A$, $\mathcal{T}_2^A$, $\mathcal{T}_3^A$ of rank 2, and a family $\{\mathcal{T}_\lambda^A\}_{\lambda \in \Lambda} \setminus \{\mathcal{T}_1^A, \mathcal{T}_2^A, \mathcal{T}_3^A\}$ of homogeneous tubes.

XI.3. The category of regular modules over a concealed algebra of Euclidean type

We start this section by deriving some consequences for concealed algebras. As seen in Chapter VIII, the module category over a concealed algebra is very close to the module category over the hereditary algebra it originates from.

Throughout this section we assume that $Q = (Q_0, Q_1)$ is an acyclic quiver whose underlying graph is Euclidean, $n = |Q_0|$,

$$A = KQ$$

is the path K-algebra of Q, and B is a **concealed K-algebra of Euclidean type** Q, that is, B is the endomorphism algebra

$$B = \operatorname{End} T_A \qquad (3.1)$$

of a postprojective tilting A-module T in $\operatorname{mod} A$, see Chapter VIII.

We recall from (III.3.13) that, for an algebra R of finite global dimension, the Euler $\mathbb{Z}$-bilinear form $\langle -, - \rangle_R : K_0(R) \times K_0(R) \longrightarrow \mathbb{Z}$ of R coincides with the **Euler characteristic** $\chi_R : K_0(R) \times K_0(R) \longrightarrow \mathbb{Z}$ of R defined by the formula

$$\chi_R(\mathbf{dim}\, M, \mathbf{dim}\, N) = \sum_{s=0}^{\infty}(-1)^s \dim_K \operatorname{Ext}_R^s(M, N), \qquad (3.2)$$

for any pair M, N of modules in $\operatorname{mod} R$.

Because, by (VIII.3.2), any concealed algebra B of Euclidean type has gl.dim $B \leq 2$, then its Euler characteristic reduces to the first three terms, that is,

$$\chi_B(\mathbf{dim}\, M, \mathbf{dim}\, N) = \dim_K \operatorname{Hom}_B(M, N) - \dim_K \operatorname{Ext}_B^1(M, N)$$
$$+ \dim_K \operatorname{Ext}_B^2(M, N),$$

for any M and N in $\operatorname{mod} B$.

The following theorem collects the main facts on concealed algebras of Euclidean type we use throughout this chapter.

3.3. Theorem. *Let Q be an acyclic quiver whose underlying graph is Euclidean, let $A = KQ$ be the hereditary path K-algebra of Q, and let B be a concealed algebra (3.1) of the Euclidean type Q.*

(a) gl.dim $B \leq 2$. *Moreover*, pd $Z \leq 1$ *and* id $Z \leq 1$, *for all but finitely many non-isomorphic indecomposable B-modules Z that are postprojective or preinjective.*

(b) *The Euler $\mathbb{Z}$-bilinear form $\langle -, - \rangle_B : K_0(B) \times K_0(B) \longrightarrow \mathbb{Z}$ of B is $\mathbb{Z}$-congruent to the Euler form $\langle -, - \rangle_A : K_0(A) \times K_0(A) \longrightarrow \mathbb{Z}$ of A, and the Euler quadratic form $q_B : K_0(B) \longrightarrow \mathbb{Z}$ of B is $\mathbb{Z}$-congruent to the Euler form $q_A : K_0(A) \longrightarrow \mathbb{Z}$ of A, that is, there exists a group isomorphism $f : K_0(A) \xrightarrow{\cong} K_0(B) \cong \mathbb{Z}^n$ making the following two diagrams commutative*

$$
\begin{array}{ccc}
K_0(B) \times K_0(B) & \xrightarrow{\langle -,- \rangle_B} & \mathbb{Z} \\
{\scriptstyle f \times f} \uparrow {\scriptstyle \cong} & \nearrow {\scriptstyle \langle -,- \rangle_A} & \\
K_0(A) \times K_0(A) & &
\end{array}
\qquad \text{and} \qquad
\begin{array}{ccc}
K_0(B) & \xrightarrow{q_B} & \mathbb{Z} \\
{\scriptstyle f} \uparrow {\scriptstyle \cong} & \nearrow {\scriptstyle q_A} & \\
K_0(A) & &
\end{array}
$$

(c) *The Euler quadratic form $q_B : K_0(B) \longrightarrow \mathbb{Z}$ is positive semidefinite of corank one.*

(d) *The Auslander–Reiten quiver $\Gamma(\text{mod } B)$ of B has the disjoint union form*

$$\Gamma(\text{mod } B) = \mathcal{P}(B) \cup \mathcal{R}(B) \cup \mathcal{Q}(B)$$

of Figure (2.1), with A and B interchanged, where $\mathcal{P}(B)$ is a postprojective component, $\mathcal{Q}(B)$ is a preinjective component $\mathcal{Q}(B)$, and $\mathcal{R}(B)$ is a family of regular components. The non-zero homomorphisms can only go from the left to the right, that is, we have

$$\text{Hom}_B(\mathcal{Q}(B), \mathcal{P}(B)) = 0, \ \text{Hom}_B(\mathcal{R}(B), \mathcal{P}(B)) = 0, \ \text{Hom}_B(\mathcal{Q}(B), \mathcal{R}(B)) = 0.$$

(e) *The regular part $\mathcal{R}(B)$ of $\Gamma(\text{mod } B)$ is not empty. Moreover*, pd $Z \leq 1$ *and* id $Z \leq 1$, *for any regular module Z in $\mathcal{R}(B)$.*

(f) *Let* add $\mathcal{R}(B)$ *be the full subcategory of* mod B *whose objects are all the regular B-modules. The functor* $\text{Hom}_A(T, -) : \text{mod } A \longrightarrow \text{mod } B$ *restricts to the equivalences* $\text{Hom}_A(T, -) : \text{add } \mathcal{R}(A) \longrightarrow \text{add } \mathcal{R}(B)$ *of categories such that the following diagram is commutative*

$$
\begin{array}{ccc}
\text{add } \mathcal{R}(A) & \underset{\tau_A^{-1}}{\overset{\tau_A}{\rightleftarrows}} & \text{add } \mathcal{R}(A) \\
{\scriptstyle \text{Hom}_A(T,-)} \downarrow {\scriptstyle \cong} & & {\scriptstyle \cong} \downarrow {\scriptstyle \text{Hom}_A(T,-)} \\
\text{add } \mathcal{R}(B) & \underset{\tau_B^{-1}}{\overset{\tau_B}{\rightleftarrows}} & \text{add } \mathcal{R}(B),
\end{array}
$$

where the horizontal functors are mutually inverse equivalences of categories induced by the Auslander–Reiten translates τ_A in mod A *and τ_B in* mod B, *respectively.*

Proof. For the convenience of the reader, we prove the statements (b) and (c). For the proof of the remaining statements the reader is referred to (VIII.3.2), (VIII.4.5), (VIII.2.9), (VI.4.7) and (VII.4.2).

Assume that B is a concealed algebra (3.1) of the Euclidean type Q, $n = |Q_0|$, and $A = KQ$. Because gl.dim $A = 1$ and gl.dim $B \leq 2$, by (VIII.3.2), then the Euler $\mathbb{Z}$-bilinear form $\langle -, - \rangle_B : K_0(B) \times K_0(B) \longrightarrow \mathbb{Z}$ of B coincides with the Euler characteristic $\chi_B : K_0(B) \times K_0(B) \longrightarrow \mathbb{Z}$ (3.2) of B, and the Euler $\mathbb{Z}$-bilinear form $\langle -, - \rangle_A : K_0(A) \times K_0(A) \longrightarrow \mathbb{Z}$ of A coincides with the Euler characteristic $\chi_A : K_0(A) \times K_0(A) \longrightarrow \mathbb{Z}$ of A, by (III.3.13).

By (VI.4.5), there exists a group isomorphism $f : K_0(A) \xrightarrow{\simeq} K_0(B) = \mathbb{Z}^n$ such that
$$\chi_A(\dim M, \dim N) = \chi_B(f(\dim M), f(\dim N)),$$
for any pair M, N of modules in mod A. It then follows that the equality

$$\langle \dim M, \dim N \rangle_A = \langle f(\dim M), f(\dim N) \rangle_B$$

holds, for any pair M, N of modules in mod A, and the Cartan matrices $\mathbf{C}_A$ and $\mathbf{C}_B$ are $\mathbb{Z}$-congruent, see (VI.4.6). Consequently, the equality

$$\langle \mathbf{x}, \mathbf{y} \rangle_A = \langle f(\mathbf{x}), f(\mathbf{y}) \rangle_B$$

holds, for all vectors $\mathbf{x}, \mathbf{y} \in K_0(A) \cong \mathbb{Z}^n$, or equivalently, the two diagrams in (b) are commutative. Hence easily follows that the Euler form q_B of B is positive semidefinite of corank one, because so is the Euler form q_A of A, by (VII.4.2). $\qquad\square$

Then we have the following immediate consequences of (2.4), (2.8), and (2.9).

3.4. Theorem. *Let $B = \operatorname{End} T_A$ be a concealed algebra of a Euclidean type Q, where Q is an acyclic Euclidean quiver, $A = KQ$ the path algebra of Q, and T_A a multiplicity-free postprojective tilting A-module. Let add $\mathcal{R}(B)$ be the full subcategory of mod B whose objects are all the regular B-modules in $\mathcal{R}(B)$.*

(a) *The subcategory add $\mathcal{R}(B)$ of mod B is serial, abelian and closed under extensions.*

(b) *The components of $\mathcal{R}(B)$ form a family $\boldsymbol{T}^B = \{T^B_\lambda\}_{\lambda \in \Lambda}$ of pairwise orthogonal standard stable tubes. Every tube T^A_λ in mod B is the image of a tube in mod A under the functor $\operatorname{Hom}_A(T, -) :$ mod $A \longrightarrow$ mod B. Moreover, if $r^B_\lambda \geq 1$ denotes the rank of T^B_λ and $n - |Q_0|$, then*

$$\sum_{\lambda \in \Lambda} (r^B_\lambda - 1) \leq n - 2.$$

(c) *All but finitely many of the tubes $\mathcal{T}_\lambda^B$ in $\mathcal{R}(A)$ are homogeneous, and there are at most $n - 2$ non-homogeneous tubes $\mathcal{T}_\lambda^B$.*

(d) *For each $\lambda \in \Lambda$, the tube $\mathcal{T}_\lambda^B$ consists of the indecomposable objects which are uniserial in an abelian subcategory of $\operatorname{mod} B$ of the form*

$$\mathcal{E}_A = \mathcal{EXT}_A(E_1, \ldots, E_r),$$

where $E_1, \ldots, E_{r_\lambda}$ are pairwise orthogonal bricks in $\operatorname{mod} B$ such that there is an isomorphism $\tau E_{i+1} \cong E_i$, for all $i \in \{1, \ldots, r_\lambda\}$, and $E_1 = E_{r_\lambda + 1}$. □

We prove later, in (XII.3.5), that for any concealed algebra B of Euclidean type the equality $\sum_{\lambda \in \Lambda} (r_\lambda^B - 1) = n - 2$ holds, and that the index set Λ is the projective line $\mathbb{P}_1(K)$ over K.

3.5. Lemma. *Let B be a concealed algebra of Euclidean type, $\mathcal{T}$ be a stable tube of rank $r \geq 1$ in $\mathcal{R}(B)$, and M be an indecomposable module from $\mathcal{T}$.*

(a) *If $k \geq 0$ is an integer such that $k \cdot r < r\ell(M) \leq (k+1) \cdot r$, then $\dim_K \operatorname{End} M = k + 1$.*

(b) *If $k \geq 0$ is an integer such that $k \cdot r \leq r\ell(M) < (k+1) \cdot r$, then $\dim_K \operatorname{Ext}_A^1(M, M) = k$.*

Proof. Because M lies in $\mathcal{R}(B)$, we have $\operatorname{pd} M \leq 1$. Hence

$$\operatorname{Ext}_B^1(M, M) \cong D\operatorname{Hom}_B(M, \tau M).$$

The required equalities follow then from the standardness of the tube $\mathcal{T}$. □

The following two facts are frequently used.

3.6. Corollary. *Let B be a concealed algebra of Euclidean type, $q_B : K_0(B) \longrightarrow \mathbb{Z}$ the Euler quadratic form of B, $\mathcal{T}$ a stable tube of rank $r \geq 1$ in $\mathcal{R}(B)$, and M be an indecomposable module from $\mathcal{T}$.*

(a) *$q_B(\mathbf{dim}\, M) \in \{0, 1\}$.*

(b) *$q_B(\mathbf{dim}\, M) = 0$ if and only if the regular length $r\ell(M)$ of M is a multiple of the rank r of $\mathcal{T}$.*

(c) *If $\mathcal{T}$ is a homogeneous tube, then $q_B(\mathbf{dim}\, M) = 0$.*

Proof. Because B is a concealed algebra and M is an indecomposable module from $\mathcal{T}$, then (VIII.4.5) yields $\operatorname{pd} M \leq 1$. It follows that

$$q_B(\mathbf{dim}\, M) = \dim_K \operatorname{End} M - \dim_K \operatorname{Ext}_B^1(M, M).$$

Hence we get (a) and (b), by applying (3.5). The statement (c) follows from (b). □

3.7. Proposition. *Assume that Q is an acyclic Euclidean quiver, $n = |Q_0|$, and $A = KQ$ is the path algebra of Q. Let B be a concealed algebra of the Euclidean type Q.*

(a) *The Euler quadratic form $q_B : K_0(B) \longrightarrow \mathbb{Z}$ is positive semidefinite of corank one, and there exists a unique positive vector $\mathbf{h}_B \in K_0(B) \cong \mathbb{Z}^n$ such that the radical $\operatorname{rad} q_B = \{x \in K_0(B), q_B(x) = 0\}$ of q_B is an infinite cyclic subgroup of $K_0(B)$ of the form*

$$\operatorname{rad} q_B = \mathbb{Z} \cdot \mathbf{h}_B.$$

Moreover, all coordinates of the vector $\mathbf{h}_B$ are positive.

(b) *If $f : K_0(A) \longrightarrow K_0(B)$ is a group isomorphism making the diagrams of (3.3)(b) commutative, then the following diagram*

$$
\begin{array}{ccc}
K_0(A) & \xrightarrow{\ \Phi_A\ } & K_0(A) \\
{\scriptstyle f}\downarrow{\scriptstyle \cong} & & {\scriptstyle f}\downarrow{\scriptstyle \cong} \\
K_0(B) & \xrightarrow{\ \Phi_B\ } & K_0(B)
\end{array}
$$

is commutative, where Φ_A and Φ_B is the Coxeter transformation of the algebra A and B, respectively.

Proof. (a) Assume that Q is an acyclic Euclidean quiver, $n = |Q_0|$, and $A = KQ$ is the path algebra of Q. Let $B = \operatorname{End} T_A$ be a concealed algebra of the Euclidean type Q, where T is a postprojective tilting module in $\operatorname{mod} A$.

Then, by (VI.4.7) and (3.3), the Euler quadratic form $q_B : K_0(B) \longrightarrow \mathbb{Z}$ of B is $\mathbb{Z}$-congruent to the Euler form $q_A : K_0(A) \longrightarrow \mathbb{Z}$ of the hereditary algebra A. By (VII.4.2), the form q_A is positive semidefinite of corank one and $\operatorname{rad} q_A = \mathbb{Z} \cdot \mathbf{h}_Q$, where $\mathbf{h}_Q$ is a corresponding positive vector shown in (1.1). By the commutativity of the right hand diagram in (3.3)(b), with a group isomorphism $f : K_0(A) \longrightarrow K_0(B)$, the Euler form q_B is also positive semidefinite of corank one and $\operatorname{rad} q_B = \mathbb{Z} \cdot \mathbf{h}$, where $\mathbf{h} = f(\mathbf{h}_Q) \in K_0(B) \cong \mathbb{Z}^n$.

On the other hand, we know from (VIII.2.9) and (3.3)(c) that the regular part $\mathcal{R}(B)$ of $\Gamma(\operatorname{mod} B)$ is not empty, and, according to (3.4), $\mathcal{R}(B)$ contains at least one stable tube $\mathcal{T}$, say of rank $r \geq 1$. By (1.3) and (3.5), there exists an indecomposable module $M[r]$ in $\mathcal{T}$ of regular length $r \geq 1$; and then

$$q_B(\operatorname{\mathbf{dim}} M[r]) = \dim_K \operatorname{End}_B(M[r]) - \dim_K \operatorname{Ext}_B^1(M[r], M[r]) = 1 - 1 = 0.$$

Hence $\operatorname{\mathbf{dim}} M[r] \in \operatorname{rad} q_B = \mathbb{Z} \cdot \mathbf{h}$, that is, there exists $m \in \mathbb{Z}$ such that $\operatorname{\mathbf{dim}} M[r] = m \cdot \mathbf{h}$. Because $\operatorname{\mathbf{dim}} M[r] \neq 0$ and all coordinates of $\operatorname{\mathbf{dim}} M[r]$ are non-negative, then there exists a positive vector $\mathbf{h}_B$ such that $\operatorname{rad} q_B = \mathbb{Z} \cdot \mathbf{h}_B$. It is clear that such a positive vector $\mathbf{h}_B$ is unique.

Now, we show that all coordinates of the vector $\mathbf{h}_B$ are positive by proving that the regular B-module $M[r]$ is sincere. By (3.3)(f), there exists an indecomposable regular A-module $E[r]$ of regular length $r \geq 1$ in a stable tube $\mathcal{T}'$ of rank r in $\Gamma(\text{mod } A)$ such that

$$M[r] = \text{Hom}_A(T, E[r]).$$

Let $T_1, \ldots, T_n$ be the indecomposable pairwise non-isomorphic direct summands of T, and

$$P_1 = \text{Hom}_A(T, T_1), \quad \ldots, \quad P_n = \text{Hom}_A(T, T_n)$$

be the associated indecomposable projective B-modules. To prove our claim, it is sufficient to show that $\text{Hom}_B(P_j, M[r]) \cong \text{Hom}_A(T_j, E[r])$ is non-zero, for any $j \in \{1, \ldots, n\}$.

We know that $\mathbf{dim}\, E[r] = s \cdot \mathbf{h}_Q$, for a positive integer s, and all coordinates of the vector $\mathbf{h}_Q$ are positive, see (1.1). This shows that the indecomposable A-module $E[r]$ is sincere. Fix an index $j \in \{1, \ldots, n\}$. Because T_j is a postprojective A-module, then there exist an indecomposable projective A-module P and an integer $t_j \geq 0$ such that $T_j \cong \tau^{-t_j} P$, see (VIII.2.2).

Observe also that $\tau^{t_j} E[r]$ is an indecomposable regular A-module of regular length $r \geq 1$ in the stable tube $\mathcal{T}'$ of rank r containing the module $E[r]$. Then, by applying (3.5) again, we get $q_A(\mathbf{dim}\, \tau^{t_j} E[r]) = 0$, that is, $\mathbf{dim}\, \tau^{t_j} E[r] = p \cdot \mathbf{h}_Q$, for a positive integer $p \geq 1$. It then follows that the module $\tau^{t_j} E[r]$ is sincere. Then, by applying (IV.2.15), we get

$$\text{Hom}_A(T_j, E[r]) \cong \text{Hom}_A(\tau^{-t_j} P, E[r]) \cong \text{Hom}_A(P, \tau^{t_j} E[r]) \neq 0.$$

This finishes the proof of (a).

(b) Let $f : K_0(A) \longrightarrow K_0(B)$ be a group isomorphism making the diagrams of (3.3)(b) commutative. It follows that there exists a matrix $U \in \mathbb{M}_n(\mathbb{Z})$, where $n = |Q_0|$, such that $\det U \in \{-1, 1\}$, $f(\mathbf{x}) = U \cdot \mathbf{x}$, for any $\mathbf{x} \in K_0(A) = \mathbb{Z}^n$, and the equality holds

$$\mathbf{C}_B = U \cdot \mathbf{C}_A \cdot U^t.$$

Hence we get the following matrix equalities

$$\mathbf{\Phi}_A = -\mathbf{C}_A \cdot \mathbf{C}_A^{-1} = -U^{-1} \cdot \mathbf{C}_B \cdot \mathbf{C}_B^{-1} \cdot U = U^{-1} \cdot \mathbf{\Phi}_B \cdot U.$$

This implies the equality $\mathbf{\Phi}_B \cdot U = U \cdot \mathbf{\Phi}_A$ of matrices and, equivalently, the commutativity of the diagram in (b). This finishes the proof. $\qquad\square$

3.8. Definition.   Assume that B is a concealed algebra of Euclidean type and let $q_B : K_0(B) \longrightarrow \mathbb{Z}$ be the Euler quadratic form of the algebra B. The unique positive vector $\mathbf{h}_B \in K_0(B)$ such that

$$\operatorname{rad} q_B = \mathbb{Z} \cdot \mathbf{h}_B$$

is called a **positive generator of the group** $\operatorname{rad} q_B$.

The following corollary is an immediate consequence of the proof of (3.7).

3.9. Corollary. *Let B be a concealed algebra of Euclidean type. If $\mathcal{T}$ is a stable tube in $\Gamma(\operatorname{mod} B)$ of rank $r \geq 1$ and $M[r]$ is an indecomposable regular module in $\mathcal{T}$ of regular length r then $M[r]$ is sincere and $\dim M[r] \in \operatorname{rad} q_B = \mathbb{Z} \cdot \mathbf{h}_B$, where $\mathbf{h}_B$ is the positive generator of $\operatorname{rad} q_B$.* $\square$

We show in Chapter XII that, for any concealed algebra B of Euclidean type and for each B-module X of regular length r lying in a stable tube $\mathcal{T}$ of $\mathcal{R}(B)$ of rank r, we have in fact $\dim X = \mathbf{h}_B$.

For the final result of this section we need the following description of the dimension vectors of indecomposable regular modules over concealed algebras of Euclidean type.

3.10. Lemma. *Assume that B is a concealed algebra of Euclidean type. Let $\mathcal{T}$ be a stable tube in the Auslander–Reiten quiver $\Gamma(\operatorname{mod} B)$ of B of rank $r \geq 2$ with the τ_B-cycle $(E_1, E_2, \ldots, E_r)$ of modules lying on the mouth of $\mathcal{T}$. We set $\mathbf{h} = \dim E_1 + \dim E_2 + \ldots + \dim E_r$.*

If X is an indecomposable regular module in $\mathcal{T}$, E_i is the regular socle of X and $r\ell(X) = mr + s$, where $m \geq 0$ and $0 \leq s \leq r_1$, then

(a) $\dim X = m \cdot \mathbf{h} + \sum\limits_{p=0}^{s-1} \dim E_{i+p}$, *where we set $E_{i+r} = E_i$ and $E_{-1} = 0$.*

(b) *for a positive integer $m' \geq 1$, $r\ell(X) = m' \cdot r$ if and only if $\dim X = m' \cdot \mathbf{h}$.*

Proof. (a) Because $(E_1, E_2, \ldots, E_r)$ is a τ_B-cycle then there are isomorphisms $\tau E_1 \cong E_r$, $\tau E_2 \cong E_1$, $\ldots$, $\tau E_r \cong E_{r-1}$. Let X be an indecomposable regular B-module in the tube $\mathcal{T}$ with the regular socle E_i. In the notation of (X.2.2), there is an isomorphism $X \cong E_i[mr + s]$.

If $r\ell(X) = 1$, then $X \cong E_i = E_i[1]$, $m = 0$, $s = 1$, and the required equality holds. Assume that $r\ell(X) \geq 2$. Then, by (X.2.2), there exists a short exact sequence

$$0 \longrightarrow E_i[1] \longrightarrow E_i[mr+s] \longrightarrow E_{i+1}[mr+s-1] \longrightarrow 0.$$

Hence we get $\dim X = \dim E_i[mr+s] = \dim E_{i+1}[mr+s-1] + \dim E_i$. Then the required formula follows by induction on the regular length of the modules in the tube $\mathcal{T}$, because $r\ell(E_{i+1}[mr+s-1]) < r\ell(X)$ and the module

E_{i+1} is the regular socle of the module $E_{i+1}[mr+s-1]$. This finishes the proof of (a).

Because (b) is an immediate consequence of (a), the proof is complete. $\square$

3.11. Corollary. *Let B be a concealed algebra of Euclidean type, $\mathcal{T}$ a stable tube $\Gamma(\mathrm{mod}\,B)$ of rank $r \geq 2$, and X, Y two indecomposable regular modules in $\mathcal{T}$, such that $r\ell(X) = m \cdot r = r\ell(Y)$, for some $m \geq 1$. Then $\dim X = \dim Y$.*

Proof. Apply (3.10). $\square$

We show in Chapter XII that if B is a concealed algebra of Euclidean type such that the rank of the group $K_0(B)$ is at least 3, then $\Gamma(\mathrm{mod}\,B)$ contains at least one non-homogeneous tube, and hence, by (3.11), it contains two non-isomorphic indecomposable modules X and Y such that $\dim X = \dim Y$. Moreover, we show in the following section that this is also the case for the Kronecker algebra $A = \begin{bmatrix} K & 0 \\ K^2 & K \end{bmatrix}$. On the other hand, we have the following result.

3.12. Proposition. *Let B be a concealed algebra of Euclidean type, let $q_B : K_0(B) \longrightarrow \mathbb{Z}$ be the Euler quadratic form of the algebra B, and X, Y a pair of indecomposable B-modules such that $\dim X = \dim Y$. If X is regular and $q_B(\dim X) = 1$, then $X \cong Y$.*

Proof. Assume that X and Y are indecomposable regular B-modules such that $\dim X = \dim Y$ and $q_B(\dim X) = 1$. By (3.6), the module X belongs to a standard stable tube $\mathcal{T}$ in $\Gamma(\mathrm{mod}\,B)$ of rank $r \geq 2$ and the regular length $r\ell(X)$ of X is not a multiple of r. Moreover, by (VIII.3.2), (VIII.4.5), and (XI.3.3) we have gl.dim $B \leq 2$, and pd $Z \leq 1$ and id $Z \leq 1$, for any indecomposable module in $\mathcal{T}$. By applying this to $Z = X$, we get pd $X \leq 1$ and id $X \leq 1$, and hence $\mathrm{Ext}_B^2(X,Y) = 0$ and $\mathrm{Ext}_B^2(Y,X) = 0$. Then, by applying (III.3.13) and (3.2), we get

$$1 = q_B(\dim X) = \langle \dim X, \dim X \rangle_B = \langle \dim X, \dim Y \rangle_B$$
$$= \dim_K \mathrm{Hom}_B(X,Y) - \dim_K \mathrm{Ext}_B^1(X,Y),$$
$$1 = q_B(\dim X) = \langle \dim X, \dim X \rangle_B = \langle \dim Y, \dim X \rangle_B$$
$$= \dim_K \mathrm{Hom}_B(Y,X) - \dim_K \mathrm{Ext}_B^1(Y,X),$$

where $\langle -, - \rangle_B : K_0(B) \times K_0(B) \longrightarrow \mathbb{Z}$ is the Euler $\mathbb{Z}$-bilinear form of B. It follows that $\mathrm{Hom}_B(X,Y) \neq 0$, $\mathrm{Hom}_B(Y,X) \neq 0$, and therefore the module Y also belongs to the tube $\mathcal{T}$, because $\mathrm{Hom}_B(\mathcal{R}(B), \mathcal{P}(B)) = 0$, $\mathrm{Hom}_B(\mathcal{Q}(B), \mathcal{R}(B)) = 0$, and the regular part $\mathcal{R}(B)$ of $\Gamma(\mathrm{mod}\,B)$ is a disjoint union of pairwise orthogonal standard stable tubes, see (3.3) and (3.4).

Let $(E_1, E_2, \ldots, E_r)$ be a τ_B-cycle of B-modules lying on the mouth of the tube $\mathcal{T}$, that is, $\tau E_1 \cong E_r, \tau E_2 \cong E_1, \ldots, \tau E_r \cong E_{r-1}$. We set

$$\mathbf{h} = \dim E_1 + \dim E_2 + \ldots + \dim E_r.$$

Let $r\ell(X) = mr + s$ and $r\ell(Y) = m'r + s'$, where $m, m' \geq 0$ and $0 \leq s, s' \leq r_1$. Assume that E_i is the regular socle of X and E_j is the regular socle of Y. By (X.2.2), there are isomorphisms $X \cong E_i[mr + s]$ and $Y \cong E_j[m'r + s']$. By applying (3.4), we get

$$\dim X = m \cdot \mathbf{h} + \mathbf{a} \quad \text{and} \quad \dim Y = m' \cdot \mathbf{h} + \mathbf{a}',$$

where

$$\mathbf{a} = \sum_{p=0}^{s-1} \dim E_{i+p} \quad \text{and} \quad \mathbf{a}' = \sum_{q=0}^{s'-1} \dim E_{j+q}$$

and we set $E_{k+r} = E_k$ and $E_{-1} = 0$. Because, by our assumption, $\dim X = \dim Y$, $\mathbf{a} < \mathbf{h}$, and $\mathbf{a}' < \mathbf{h}$, then $m = m'$ and, consequently, $\mathbf{a} = \mathbf{a}'$. Moreover, by (3.6)(b), the equalities $1 = q_B(\dim X) = q_B(\dim Y)$ imply that the regular ranks $r\ell(X)$ and $r\ell(Y)$ of X and Y are not multiples of r; hence $s \geq 1$ and $s' \geq 1$. Observe also that

$$\dim E_i[s] = \mathbf{a} = \mathbf{a}' = \dim E_j[s'].$$

Further, $M = E_i[s]$ and $N = E_j[s']$ are modules in the tube $\mathcal{T}$ of regular length smaller than r; hence $q_B(\dim M) = 1$. Hence we conclude, as in the first part of the proof, that $\mathrm{Hom}_B(M, N) \neq 0$, $\mathrm{Hom}_B(M, N) \neq 0$. Because the tube $\mathcal{T}$ is standard, we get the inequalities $i \leq j \leq i + s - 1$ and $j \leq i \leq i + s' - 1$. It follows that $i = j$ and $s = s'$, because $\mathbf{a} = \mathbf{a}'$. Consequently, there are isomorphisms $X \cong E_i[mr+s] = E_j[m'r+s'] \cong Y$, and the proof is complete. $\qquad\qquad$ $\square$

XI.4. The category of modules over the Kronecker algebra

In this section, we classify all indecomposable modules over the **Kronecker algebra**

$$A = \begin{bmatrix} K & 0 \\ K^2 & K \end{bmatrix}$$

that is, the path algebra of the so-called **Kronecker quiver**

$$1 \circ \underset{\beta}{\overset{\alpha}{\longleftarrow\!\!\!=\!\!\!=}} \circ 2 \,.$$

The indecomposable projective A-modules are

$$P(1) = (K \overset{}{\longleftarrow\!\!\!=\!\!\!=} 0) \quad \text{and} \quad P(2) = (K^2 \underset{\left[\begin{smallmatrix}0\\1\end{smallmatrix}\right]}{\overset{\left[\begin{smallmatrix}1\\0\end{smallmatrix}\right]}{\longleftarrow\!\!\!\longleftarrow}} K)$$

and the indecomposable injective A-modules are

$$I(1) = (K \xLeftarrow[\text{[0 1]}]{\text{[1 0]}} K^2) \qquad \text{and} \qquad I(2) = (0 \Longleftarrow K),$$

viewed as K-linear representations of the Kronecker quiver. Computing the Cartan matrix $\mathbf{C}_A$, its inverse $\mathbf{C}_A^{-1}$, the Coxeter matrix $\mathbf{\Phi}_A = -\mathbf{C}_A^t \mathbf{C}_A^{-1}$ and its inverse $\mathbf{\Phi}_A^{-1}$ yields

$$\mathbf{C}_A = \begin{bmatrix} 1 & 2 \\ 0 & 1 \end{bmatrix}, \quad \mathbf{C}_A^{-1} = \begin{bmatrix} 1 & -2 \\ 0 & 1 \end{bmatrix}, \quad \mathbf{\Phi}_A = \begin{bmatrix} -1 & 2 \\ -2 & 3 \end{bmatrix}, \quad \mathbf{\Phi}_A^{-1} = \begin{bmatrix} 3 & -2 \\ 2 & -1 \end{bmatrix}.$$

The Euler quadratic form $q_A : \mathbb{Z}^2 \longrightarrow \mathbb{Z}$ of the Kronecker algebra A is defined on a vector $\mathbf{x} = \begin{bmatrix} x_1 \\ x_2 \end{bmatrix} \in K_0(A) = \mathbb{Z}^2$ by the formula

$$q_A(\mathbf{x}) = \mathbf{x}^t (\mathbf{C}_A^{-1})^t \mathbf{x} = [x_1 \; x_2] \begin{bmatrix} 1 & 0 \\ -2 & 1 \end{bmatrix} \begin{bmatrix} x_1 \\ x_2 \end{bmatrix} = x_1^2 + x_2^2 - 2x_1 x_2 = (x_1 - x_2)^2.$$

Hence $\operatorname{rad} q_A = \mathbb{Z} \cdot \mathbf{h}$, where $\mathbf{h} = \begin{bmatrix} 1 \\ 1 \end{bmatrix}$. Moreover, for $\mathbf{x} = \begin{bmatrix} x_1 \\ x_2 \end{bmatrix} \in K_0(A)$, we have

$$\begin{aligned}
\mathbf{\Phi}_A(\mathbf{x}) &= \begin{bmatrix} -1 & 2 \\ -2 & 3 \end{bmatrix} \cdot \begin{bmatrix} x_1 \\ x_2 \end{bmatrix} = \begin{bmatrix} -x_1 + 2x_2 \\ -2x_1 + 3x_2 \end{bmatrix} \\
&= \begin{bmatrix} x_1 \\ x_2 \end{bmatrix} + (-2x_1 + 2x_2) \begin{bmatrix} 1 \\ 1 \end{bmatrix} = \mathbf{x} + (-2x_1 + 2x_2) \cdot \mathbf{h}.
\end{aligned}$$

Hence the defect $\partial_A : \mathbb{Z}^2 \longrightarrow \mathbb{Z}$ is given by the formula $\partial_A(\mathbf{x}) = 2(x_2 - x_1)$.

(4.1) Postprojective modules. By (VII.2.3), the postprojective component $\mathcal{P}(A)$ of A is of the form

$$P(2) \quad ---- \quad \tau^{-1}P(2) \quad ---- \quad \tau^{-2}P(2) \quad ----$$

$$P(1) \quad ---- \quad \tau^{-1}P(1) \quad ---- \quad \tau^{-2}P(1) \quad ---- \quad \cdots$$

It follows that, for each $m \geq 0$, the postprojective A-modules $\tau^{-m} P(1)$ and $\tau^{-m} P(2)$ are directing and, according to (IX.3.1), they are uniquely determined by their dimension vectors. Since $\mathbf{dim}\, P(1) = \begin{bmatrix} 1 \\ 0 \end{bmatrix}$, $\mathbf{dim}\, P(2) = \begin{bmatrix} 2 \\ 1 \end{bmatrix}$, we get, from (IV.2.9), for any $m \geq 0$,

$$\mathbf{dim}\, \tau^{-m} P(1) = \mathbf{\Phi}_A^{-m}(\mathbf{dim}\, P(1)) = \begin{bmatrix} 3 & -2 \\ 2 & -1 \end{bmatrix}^m \cdot \begin{bmatrix} 1 \\ 0 \end{bmatrix} = \begin{bmatrix} 2m+1 \\ 2m \end{bmatrix},$$

$$\mathbf{dim}\, \tau^{-m} P(2) = \mathbf{\Phi}_A^{-m}(\mathbf{dim}\, P(2)) = \begin{bmatrix} 3 & -2 \\ 2 & -1 \end{bmatrix}^m \cdot \begin{bmatrix} 2 \\ 1 \end{bmatrix} = \begin{bmatrix} 2m+2 \\ 2m+1 \end{bmatrix}.$$

Now, it is easy to check that, for each $t \geq 0$, the A-module

$$P_t = (K^{t+1} \underset{\varphi_{\beta_t}}{\overset{\varphi_{\alpha_t}}{\longleftarrow}} K^t),$$

with φ_{α_t} and φ_{β_t} given by the matrices

$$\varphi_{\alpha_t} = \begin{bmatrix} 1 & 0 & 0 & \cdots & 0 \\ 0 & 1 & 0 & \cdots & 0 \\ 0 & 0 & 1 & \cdots & 0 \\ \vdots & \vdots & \vdots & \ddots & \vdots \\ 0 & 0 & 0 & \cdots & 1 \\ 0 & 0 & 0 & \cdots & 0 \end{bmatrix}, \qquad \varphi_{\beta_t} = \begin{bmatrix} 0 & 0 & 0 & \cdots & 0 \\ 1 & 0 & 0 & \cdots & 0 \\ 0 & 1 & 0 & \cdots & 0 \\ \vdots & \vdots & \vdots & \ddots & \vdots \\ 0 & 0 & 0 & \cdots & 0 \\ 0 & 0 & 0 & \cdots & 1 \end{bmatrix},$$

is a brick, and therefore indecomposable. Because, for each $m \geq 0$, we have $\dim \tau^{-m} P(1) = \dim P_{2m}$ and $\dim \tau^{-m} P(2) = \dim P_{2m+1}$, then there are isomorphisms

$$\tau^{-m} P(1) \cong P_{2m} \quad \text{and} \quad \tau^{-m} P(2) \cong P_{2m+1}$$

of A-modules.

(4.2) Preinjective modules. By (VIII.2.3), the preinjective component $\mathcal{Q}(A)$ of A is of the form

$$\cdots \quad - - - - \quad \tau^2 I(2) \quad - - - - \quad \tau I(2) \quad - - - - \quad I(2)$$
$$\searrow \quad \nearrow\nearrow \quad \searrow \quad \nearrow\nearrow \quad \searrow \quad \nearrow\nearrow$$
$$- - - - \quad \tau^2 I(1) \quad - - - - \quad \tau I(1) \quad - - - - \quad I(1)$$

It follows that, for each $m \geq 0$, the preinjective modules $\tau^m I(1)$ and $\tau^m I(2)$ are directing and, according to (IX.3.1), they are uniquely determined by their dimension vectors. Because $\dim I(1) = \begin{bmatrix} 1 \\ 2 \end{bmatrix}$, $\dim I(2) = \begin{bmatrix} 0 \\ 1 \end{bmatrix}$, we deduce from (IV.2.9) that, for each $m \geq 0$,

$$\dim \tau^m I(1) = \Phi_A^m(\dim I(1)) = \begin{bmatrix} -1 & 2 \\ -2 & 3 \end{bmatrix}^m \cdot \begin{bmatrix} 1 \\ 2 \end{bmatrix} = \begin{bmatrix} 2m+1 \\ 2m+2 \end{bmatrix},$$

$$\dim \tau^m I(2) = \Phi_A^m(\dim I(2)) = \begin{bmatrix} -1 & 2 \\ -2 & 3 \end{bmatrix}^m \cdot \begin{bmatrix} 0 \\ 1 \end{bmatrix} = \begin{bmatrix} 2m \\ 2m+1 \end{bmatrix}.$$

It is easy to check that, for each $t \geq 0$, the A-module

$$I_t = (K^t \underset{\varphi_{\beta_t}}{\overset{\varphi_{\alpha_t}}{\longleftarrow}} K^{t+1})$$

with φ_{α_t} and φ_{β_t} given by the matrices

$$\varphi_{\alpha_t} = \begin{bmatrix} 1 & 0 & 0 & \cdots & 0 & 0 \\ 0 & 1 & 0 & \cdots & 0 & 0 \\ 0 & 0 & 1 & \cdots & 0 & 0 \\ \vdots & \vdots & \vdots & \ddots & \vdots & \vdots \\ 0 & 0 & 0 & \cdots & 1 & 0 \end{bmatrix}, \qquad \varphi_{\beta_t} = \begin{bmatrix} 0 & 1 & 0 & \cdots & 0 & 0 \\ 0 & 0 & 1 & \cdots & 0 & 0 \\ 0 & 0 & 0 & \cdots & 0 & 0 \\ \vdots & \vdots & \vdots & \ddots & \vdots & \vdots \\ 0 & 0 & 0 & \cdots & 0 & 1 \end{bmatrix},$$

is a brick, and therefore indecomposable. Because, for each $m \geq 0$, we have $\dim \tau^m I(1) = \dim I_{2m+1}$ and $\dim \tau^m I(2) = \dim I_{2m}$, then there are isomorphisms of A-modules

$$\tau^m I(1) \cong I_{2m+1} \quad \text{and} \quad \tau^m I(2) \cong I_{2m}.$$

(4.3) Regular modules. By (III.3.5), the Grothendieck group $K_0(A)$ of A is free abelian of rank $n = 2$. It then follows from (2.8) and (2.9) that the regular part $\mathcal{R}(A)$ of $\Gamma(\operatorname{mod} A)$ consists of a family $\mathcal{T}^A = \{\mathcal{T}^A_\lambda\}_{\lambda \in \Lambda}$ of pairwise orthogonal standard homogeneous tubes. We show that Λ is the projective line $\mathbb{P}_1(K)$ and, for each $\lambda \in \mathbb{P}_1(K)$, we give an explicit form of the A-modules in the tube $\mathcal{T}^A_\lambda$.

Let M be an indecomposable regular A-module. Then M lies in a homogeneous tube and therefore there is an isomorphism $M \cong \tau M$. Note also that if $\dim M = \mathbf{x} = \begin{bmatrix} x_1 \\ x_2 \end{bmatrix}$, then $x_1 = x_2$. Indeed, because (2.3) yields $0 = \partial_A(\dim M) = 2(x_2 - x_1)$, then we get $x_1 = x_2$.

Now we describe all the indecomposable regular A-modules lying in a given homogeneous tube. We start by defining a $\mathbb{P}_1(K)$-family of simple regular A-modules.

For each vector $\mu = \begin{bmatrix} \mu_1 \\ \mu_2 \end{bmatrix} \in K^2 \setminus \left\{ \begin{bmatrix} 0 \\ 0 \end{bmatrix} \right\}$, let

$$E_\mu = (K \xleftarrow[\varphi_{\mu_2}]{\varphi_{\mu_1}} K),$$

where φ_{μ_1} (or φ_{μ_2}) denotes the multiplication by the scalar μ_1 (or μ_2, respectively). Clearly, E_μ is indecomposable. Moreover, for any pair of vectors $\mu = \begin{bmatrix} \mu_1 \\ \mu_2 \end{bmatrix}$ and $\mu' = \begin{bmatrix} \mu'_1 \\ \mu'_2 \end{bmatrix}$ in $K^2 \setminus \left\{ \begin{bmatrix} 0 \\ 0 \end{bmatrix} \right\}$, we have $E_\mu \cong E_{\mu'}$ if and only if there exists $\nu \in K \setminus \{0\}$ such that $\mu'_1 = \nu\mu_1$, $\mu'_2 = \nu\mu_2$ or, equivalently, if and only if μ and μ' define the same point $(\mu_1 : \mu_2)$ on the projective line $\mathbb{P}_1(K)$. Here we identify $\mathbb{P}_1(K)$ with $K \cup \{\infty\}$ via

(4.4) $K = \{(1 : \lambda) \in \mathbb{P}_1(K) \mid \lambda \in K\}$ and $\infty = (0 : 1)$.

For $\lambda = (\mu_1 : \mu_2)$ in $\mathbb{P}_1(K)$, we set $E_\lambda = E_\mu$. Then, for each $\lambda \in K$, we have

$$E_\lambda = (K \xleftarrow[\lambda]{1} K) \quad \text{and} \quad E_\infty = (K \xleftarrow[1]{0} K).$$

The reader might notice that this construction was performed in the proof of (VII.2.3) for a family of indecomposable representations of a quiver Q of the Euclidean type $\widetilde{\mathbb{A}}_m$, where $m \geq 1$.

Note that, for each $\lambda \in \mathbb{P}_1(K)$, the module E_λ is simple regular. Indeed, because obviously $\partial_A(\dim E_\lambda) = 2(1 - 1) = 0$ then, by (2.3), the module E_λ is regular. Moreover, if Y is an indecomposable regular submodule of E_λ and $\dim Y = \begin{bmatrix} y_1 \\ y_2 \end{bmatrix}$, then $y_1 = y_2$ and, because $0 < y_1 \leq 1$, we get $y_1 = y_2 = 1$ and, consequently, $Y = E_\lambda$. This shows that the module E_λ is simple regular.

For each $\lambda \in \mathbb{P}_1(K)$, we denote by $\mathcal{T}^A_\lambda$ the homogeneous tube of $\mathcal{R}(A)$ containing the simple regular module E_λ. Because, for $\lambda \neq \mu$ in $\mathbb{P}_1(K)$, $\operatorname{Hom}_A(E_\lambda, E_\mu) = 0$ then, by (2.8), the tubes $\mathcal{T}^A_\lambda$ and $\mathcal{T}^A_\mu$ are orthogonal.

By the results of Section X.2, for any $\lambda \in \mathbb{P}_1(K)$, the simple regular modules E_λ lie on the mouth of the homogeneous tube $\mathcal{T}_\lambda^A$, and all indecomposable modules in $\mathcal{T}_\lambda^A$ lie on the ray

$$E_\lambda = E_\lambda[1] \longrightarrow E_\lambda[2] \longrightarrow E_\lambda[3] \longrightarrow \cdots \longrightarrow E_\lambda[d] \longrightarrow E_\lambda[d+1] \longrightarrow \cdots$$

such that $E_\lambda[d+1]/E_\lambda[d] \cong E_\lambda$, for any $d \geq 1$. Using the short exact sequence

$$0 \longrightarrow E_\lambda[d] \longrightarrow E_\lambda[d+1] \longrightarrow E_\lambda \longrightarrow 0$$

and the uniseriality of $E_\lambda[d+1]$ in $\mathcal{R}(A)$, we see easily that, if $\lambda = (1 : \lambda) \in K$, then

$$E_\lambda[d] = (K^d \xleftarrow{\substack{\begin{bmatrix} 1 & 0 & \cdots & 0 \\ 0 & 1 & & 0 \\ \vdots & & \ddots & \vdots \\ 0 & 0 & \cdots & 1 \end{bmatrix} \\ \begin{bmatrix} \lambda & 1 & 0 & \cdots & 0 \\ 0 & \lambda & 1 & \cdots & 0 \\ \vdots & & & \ddots & \vdots \\ 0 & 0 & 0 & & 1 \\ 0 & 0 & 0 & \cdots & \lambda \end{bmatrix}}} K^d)$$

and, for $\lambda = \infty$, we get

$$E_\infty[d] = (K^d \xleftarrow{\substack{\begin{bmatrix} 0 & 1 & 0 & \cdots & 0 \\ 0 & 0 & 1 & \cdots & 0 \\ \vdots & & & \ddots & \vdots \\ 0 & 0 & 0 & \cdots & 1 \\ 0 & 0 & 0 & \cdots & 0 \end{bmatrix} \\ \begin{bmatrix} 1 & 0 & 0 & \cdots & 0 \\ 0 & 1 & 0 & \cdots & 0 \\ 0 & 0 & 1 & \cdots & 0 \\ \vdots & & & \ddots & \vdots \\ 0 & 0 & 0 & \cdots & 1 \end{bmatrix}}} K^d).$$

4.5. Proposition. *Let A be the Kronecker K-algebra, and let E_λ and $E_\lambda[d]$ be the A-modules defined above.*

(a) *For any indecomposable regular A-module M, there exist $\lambda \in \mathbb{P}_1(K)$ and $d \geq 1$ such that $M \cong E_\lambda[d]$.*

(b) *Every simple regular A-module is of the form E_λ, up to isomorphism, where $\lambda \in \mathbb{P}_1(K)$.*

Proof. Assume that M is an indecomposable regular A-module and $\dim M = \begin{bmatrix} x_1 \\ x_2 \end{bmatrix}$. Then $x_1 = x_2$ and therefore M, viewed as a representation of the Kronecker quiver, has the form $M = (K^d \xleftarrow[\varphi_\beta]{\varphi_\alpha} K^d)$, up to isomorphism, where $d = x_1 = x_2$.

Because the modules E_λ, with $\lambda \in \mathbb{P}_1(K)$, are simple regular and, by (2.8), the regular part $\mathcal{R}(A)$ of $\Gamma(\mathrm{mod}\, A)$ consists of pairwise orthogonal stable tubes, it suffices to show that $\mathrm{Hom}_A(E_\lambda, M) \neq 0$ for some $\lambda \in \mathbb{P}_1(K)$.

Suppose first that φ_α is not an isomorphism, and let $x \in K^d$ be a non-zero vector in $\operatorname{Ker} \varphi_\alpha$. Then there is a non-zero morphism $f : E_\infty \longrightarrow M$ given by the commutative square

$$
\begin{array}{ccc}
K & \xLeftarrow[\;1\;]{\;0\;} & K \\
{\scriptstyle f_1}\downarrow & & \downarrow{\scriptstyle f_2} \\
K^d & \xLeftarrow[\varphi_\beta]{\varphi_\alpha} & K^d
\end{array}
$$

with f_1 and f_2 defined by the formulae $f_1(1) = \varphi_\beta(x)$ and $f_2(1) = x$, respectively.

Assume now that φ_α is an isomorphism, and let $\mathbf{z} \in K^d$ be an eigenvector of $\varphi_\beta \varphi_\alpha^{-1} : K^d \longrightarrow K^d$, say with eigenvalue λ. Then, for $\lambda = (1 : \lambda)$, we get a non-zero morphism $f : E_\lambda \longrightarrow M$ given by the commutative square

$$
\begin{array}{ccc}
K & \xLeftarrow[\;\lambda\;]{\;1\;} & K \\
{\scriptstyle f_1}\downarrow & & \downarrow{\scriptstyle f_2} \\
K^d & \xLeftarrow[\varphi_\beta]{\varphi_\alpha} & K^d
\end{array}
$$

with f_1 and f_2 defined by the formulae $f_1(1) = \mathbf{z}$ and $f_2(1) = \varphi_\alpha^{-1}(\mathbf{z})$. $\quad\square$

We summarise these considerations in the following theorem.

4.6. Theorem. *Let* $A = \left[\begin{smallmatrix} K & 0 \\ K^2 & K \end{smallmatrix} \right]$ *be the Kronecker K-algebra, $\mathcal{P}(A)$ be the postprojective component* (4.1) *and $\mathcal{Q}(A)$ be the preinjective component* (4.2).

(a) *The A-modules P_t, I_t (for $t \geq 0$), and $E_\lambda[d]$ (for $\lambda \in \mathbb{P}_1(K)$ and $d \geq 1$), form a complete set of representatives of the isomorphism classes of indecomposable A-modules.*

(b) *Every simple regular A-module is of the form E_λ, up to isomorphism, where $\lambda \in \mathbb{P}_1(K)$.*

(c) *The Auslander–Reiten quiver $\Gamma(\operatorname{mod} A)$ of A consists of the postprojective component $\mathcal{P}(A)$, the preinjective component $\mathcal{Q}(A)$, and the $\mathbb{P}_1(K)$-family $\boldsymbol{\mathcal{T}}^A = \{\mathcal{T}_\lambda^A\}_{\lambda \in \mathbb{P}_1(K)}$ of pairwise orthogonal standard homogeneous tubes defined above.*

In view of (4.6), the Auslander–Reiten quiver $\Gamma(\operatorname{mod} A)$ of $\operatorname{mod} A$ may be visualised as follows.

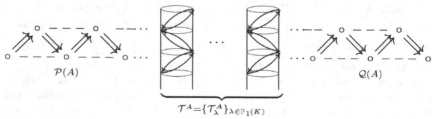

$$\mathcal{T}^A = \{\mathcal{T}_\lambda^A\}_{\lambda \in \mathbb{P}_1(K)}$$

The following proposition expresses what is known as the separation property of the family $\boldsymbol{\mathcal{T}}^A = \{\mathcal{T}_\lambda^A\}_{\lambda \in \mathbb{P}_1(K)}$ of tubes defined above.

4.7. Proposition. *Let* $A = \begin{bmatrix} K & 0 \\ K^2 & K \end{bmatrix}$ *be the Kronecker K-algebra, P be a postprojective A-module, I be a preinjective A-module, and $f : P \longrightarrow I$ be any homomorphism. Then, for any $\lambda \in \mathbb{P}_1(K)$, there exists $R_\lambda \in \operatorname{add} \mathcal{T}_\lambda^A$ such that f factors through R_λ.*

Proof. We first show that any non-zero postprojective A-module P maps non-trivially into a module from $\mathcal{T}_\lambda^A$. The module P has an indecomposable direct summand of the form $\tau^{-t} P(i)$, with $i \in \{1, 2\}$ and $t \geq 0$. On the other hand, the homogeneity of $\mathcal{T}_\lambda^A$ implies that $\tau E_\lambda \cong E_\lambda$. Hence, by applying (IV.2.15), we get the isomorphisms

$$\operatorname{Hom}_A(\tau^{-t} P(i), E_\lambda) \cong \operatorname{Hom}_A(\tau^{-t} P(i), \tau^{-t} E_\lambda) \cong \operatorname{Hom}_A(P(i), E_\lambda) \neq 0.$$

We next prove that P may in fact be embedded into a module from $\operatorname{add} \mathcal{T}_\lambda^A$. Let $u : P \longrightarrow E_\lambda$ be a non-zero homomorphism. We proceed by induction on the dimension of $P' = \operatorname{Ker} u$. If $P' = 0$, there is nothing to prove. Assume that $P' \neq 0$. By induction, there is a monomorphism $u' : P' \longrightarrow R_\lambda'$, with $R_\lambda' \in \operatorname{add} \mathcal{T}_\lambda^A$. Consider the induced commutative diagram with exact rows

$$
\begin{array}{ccccccccc}
0 & \longrightarrow & P' & \longrightarrow & P & \longrightarrow & P/P' & \longrightarrow & 0 \\
& & \Big\downarrow{\scriptstyle u'} & & \Big\downarrow{\scriptstyle v'} & & \Big\| & & \\
0 & \longrightarrow & R_\lambda' & \longrightarrow & M & \longrightarrow & P/P' & \longrightarrow & 0.
\end{array}
$$

Applying the snake lemma yields that v' is a monomorphism. If u is an epimorphism, then $P/P' \cong E_\lambda$ and consequently $M \in \operatorname{add} \mathcal{T}_\lambda^A$, because $\operatorname{add} \mathcal{T}_\lambda^A$ is closed under extensions, thus v' is the required embedding.

If u is not an epimorphism, then P/P' is a proper submodule of E_λ. Because E_λ is simple regular, we have $P/P' \in \operatorname{add} \mathcal{P}(A)$, hence

$$\operatorname{Ext}_A^1(P/P', R_\lambda') \cong D\operatorname{Hom}_A(R_\lambda', \tau(P/P')) = 0.$$

Therefore the lower exact sequence yields $M \cong R_\lambda' \oplus P/P'$, and the embedding $P/P' \longrightarrow E_\lambda$ induces an embedding $P \longrightarrow M \longrightarrow R_\lambda' \oplus E_\lambda$, with $R_\lambda' \oplus E_\lambda \in \operatorname{add} \mathcal{T}_\lambda^A$. This establishes our claim.

Let now $f : P \longrightarrow I$ be a non-zero homomorphism with P postprojective and I preinjective. We may clearly assume that I is indecomposable. Assume first that I is injective. We have shown that there is a monomorphism $j : P \longrightarrow R_\lambda$, with $R_\lambda \in \operatorname{add} \mathcal{T}_\lambda^A$. The injectivity of I yields a non-zero homomorphism $g : R_\lambda \longrightarrow I$ such that $f = gj$, so we are done in this case. If, on the other hand, I is not injective, there exist $t \geq 1$ and $i \in \{1, 2\}$ such that $I \cong \tau^t I(i)$. But then the homomorphism $\tau^{-t} f : \tau^{-t} P \longrightarrow \tau^{-t} I \cong I(i)$ factors through a module $R_\lambda \in \operatorname{add} \mathcal{T}_\lambda^A$, and so f factors through $\tau^t R_\lambda \cong R_\lambda$. $\qquad\square$

XI.5. A characterisation of concealed algebras of Euclidean type

The main aim of this section is to establish the following useful characterisation of concealed algebras of Euclidean type in terms of properties of their module categories. The section ends with an important consequence of the criterion.

5.1. Theorem. *An algebra B is a concealed algebra of Euclidean type if and only if the quiver $\Gamma(\mathrm{mod}\, B)$ contains two different components $\mathcal{P}$ and $\mathcal{Q}$ satisfying the following conditions:*

    (a) *$\mathcal{P}$ is a postprojective component containing all indecomposable projective B-modules,*

    (b) *$\mathcal{Q}$ is a preinjective component containing all indecomposable injective B-modules, and*

    (c) *one of the components $\mathcal{P}$ or $\mathcal{Q}$ contains a section Σ of Euclidean type.*

Proof. To prove the necessity, assume that B is a concealed algebra of type Q, where Q is an acyclic quiver whose underlying graph is Euclidean. Then, by definition (VIII.4.6), $B = \mathrm{End}\, T_A$ is the endomorphism algebra of a postprojective tilting A-module T over the path algebra $A = KQ$ of the quiver Q.

Applying (VIII.4.5)(c), we easily conclude that the connecting component $\mathcal{C}_T$ of $\Gamma(\mathrm{mod}\, B)$ determined by T is a preinjective component $\mathcal{Q}(B)$ containing all indecomposable injective B-modules but no projective B-module.

Moreover, $\mathcal{Q}(B) = \mathcal{C}_T$ contains the section $\Sigma = Q^{\mathrm{op}}$ of Euclidean type given by the indecomposable B-modules $\mathrm{Hom}_A(T, I)$, where I runs through the pairwise non-isomorphic indecomposable injective A-modules. Further, by (VIII.4.5)(e), the images under the functor $\mathrm{Hom}_A(T, -)$ of the postprojective indecomposable A-modules in the torsion class $\mathcal{T}(T)$ form a postprojective component $\mathcal{P}(B)$ of $\Gamma(\mathrm{mod}\, B)$ containing all indecomposable projective B-modules but no injective B-module. Clearly then $\mathcal{P}(B) \neq \mathcal{Q}(B)$.

Finally, we note that $\mathcal{P}(B)$ contains a section of Euclidean type Q^{op}. Indeed, by (VIII.4.5)(a), the torsion class $\mathcal{T}(T)$ contains all but finitely many non-isomorphic indecomposable A-modules, and any indecomposable A-module not in $\mathcal{T}(T)$ is postprojective. Then the postprojective component $\mathcal{P}(A)$ of $\Gamma(\mathrm{mod}\, B)$ contains a full translation subquiver $\mathcal{P}'(A)$, isomorphic to $(-\mathbb{N})Q^{\mathrm{op}}$, closed under successors in $\mathcal{P}(A)$, and consisting entirely of modules from $\mathcal{T}(T)$. Then the image $\mathcal{P}'(B)$ of $\mathcal{P}'(A)$ under the functor $\mathrm{Hom}_A(T, -)$ is a full translation subquiver of the postprojective component

$\mathcal{P}(B)$ which is closed under successors. Obviously, $\mathcal{P}'(B)$ is isomorphic to $(-\mathbb{N})Q^{\mathrm{op}}$, and hence $\mathcal{P}(B)$ contains a section of type Q^{op}. This finishes the proof of the necessity part.

To prove the sufficiency, assume that B is an algebra such that the quiver $\Gamma(\mathrm{mod}\,B)$ contains two different components $\mathcal{P}$ and $\mathcal{Q}$ satisfying the conditions (a), (b), and (c). The condition (c) forces two cases to consider.

Case 1° Assume that the preinjective component $\mathcal{Q}$ contains a section Σ of Euclidean type. Observe, that the section Σ is faithful, that is, the direct sum T_B of all modules lying on Σ is a faithful B-module. Indeed, the preinjective component $\mathcal{Q}$ contains, by (b), all indecomposable injective B-modules, which are clearly successors of Σ in $\mathcal{Q}$.

To show that T_B is faithful, it is enough to prove that the module $D(B)$ is generated by T_B (see Section VI.1 and (VI.2.2)). For, consider a projective cover $f : P \to D(B)$ of $D(B)$ in $\mathrm{mod}\,B$. Because $\mathcal{P} \neq \mathcal{Q}$, the preinjective component $\mathcal{Q}$ has no projective B-module, is acyclic, and admits only a finite number of τ_B-orbits. Then, by applying (IV.5.1)(b), we conclude that f factorises through a direct sum M of indecomposable modules lying on the section Σ. Therefore, the module T_B generates the module $D(B)$.

Observe also that $\mathrm{Hom}_B(U, \tau_B V) = 0$, for all modules U and V from Σ, because Σ is a section of the preinjective component $\mathcal{Q}$, that is closed under successors in $\mathrm{mod}\,B$, see (VIII.2.5)(b).

Applying now the criterion (VIII.5.6) we conclude that T_B is a tilting B-module such that $A = \mathrm{End}\,T_B$ is a hereditary algebra, $\mathcal{Q}$ is the connecting component $\mathcal{C}_{T^*}$ determined by the tilting A-module $T_A^* = D(_A T)$. Moreover, A is the path algebra KQ of the Euclidean quiver $Q = \Sigma^{\mathrm{op}}$.

Further, because $\mathcal{Q} = \mathcal{C}_{T^*}$ has no projective B-module, applying (VIII.4.1)(a), we conclude that T^* has no preinjective direct summand. We claim that in fact T^* is a postprojective A-module, and consequently $B = \mathrm{End}\,T_A^*$ is a concealed algebra of Euclidean type. Because the connecting component $\mathcal{C}_{T^*} = \mathcal{Q}$ is preinjective, it follows from (VIII.4.2) that T^* is not a regular A-module, and consequently admits at least one postprojective direct summand.

Assume that the module T^* admits also a regular direct summand. We know that $B = \mathrm{End}\,T_A^*$ is a connected algebra, because A is the path algebra of the connected quiver Q (see (VI.3.5)). Moreover, $\mathrm{Hom}_A(\mathcal{R}(A), \mathcal{P}(A)) = 0$. Therefore, there are two indecomposable direct summands T_1^* and T_2^* of T^* such that $T_1^* \in \mathcal{P}(A)$, $T_2^* \in \mathcal{R}(A)$, and $\mathrm{Hom}_A(T_1^*, T_2^*) \neq 0$. It follows from (2.8) and (2.9) that the regular part $\mathcal{R}(A)$ consists of pairwise orthogonal standard stable tubes, and only finitely many of them are non-homogeneous. Denote by m a common multiple of the ranks of stable tubes in $\mathcal{R}(A)$. Then, for any indecomposable module $X_A \in \mathcal{R}(A)$, we have, $\tau_A^m X \cong X$. Applying now (IV.2.15), we obtain

$$\operatorname{Hom}_A(\tau_A^{-pm}T_1^*, T_2^*) \cong \operatorname{Hom}_A(T_1^*, \tau_A^{pm}T_2^*) = \operatorname{Hom}_A(T_1^*, T_2^*) \neq 0,$$

for any $p \in \mathbb{N}$.

On the other hand, the modules $\tau_A^{-pm}T_1^*$ belong to the postprojective component $\mathcal{P}(A)$. Then, because the quiver $\mathcal{P}(A)$ is acyclic, there exists a positive integer s such that no postprojective indecomposable direct summand of T^* is a successor of a module of the form $\tau_A^{-pm}T_1^*$ with $p \geq s$.

Denote also by T_{reg}^* the direct sum of all regular direct summands of T^*. Then, applying the Auslander–Reiten formula (IV.2.13) and (IV.2.15), we obtain

$$D\operatorname{Ext}_A^1(T^*, \tau_A^{-pm}T_1^*) \cong \operatorname{Hom}_A(\tau_A^{-pm}T_1^*, \tau_A T^*) \cong \operatorname{Hom}_A(\tau_A^{-pm}T_1^*, \tau_A T_{\text{reg}}^*)$$

$$\cong \operatorname{Hom}_A(T_1^*, \tau_A^{pm+1}T_{\text{reg}}^*) \cong \operatorname{Hom}_A(T_1^*, \tau_A T_{\text{reg}}^*)$$

$$\cong D\operatorname{Ext1}_A(T_{\text{reg}}^*, T_1^*) = 0,$$

for any $p > s$. Hence, the postprojective modules $\tau_A^{-pm}T_1^*$, for $p > s$, belong to the torsion class $\mathcal{T}(T)$, and they are predecessors of T_2^* in $\mathcal{T}(T)$. Then, by (VI.3.8), the B-modules $\operatorname{Hom}_A(T, \tau_A^{-pm}T_1^*)$, $p > s$, are pairwise non-isomorphic and predecessors of the indecomposable projective B-module $\operatorname{Hom}_A(T, T_2^*)$ in $\operatorname{mod} B$. But, by the assumption (a), all indecomposable projective B-modules belong to the postprojective component $\mathcal{P}$, and hence have only finitely many predecessors in $\operatorname{mod} B$, see (VIII.2.5)(a). Therefore, the module T_A^* is indeed postprojective, and A is a concealed algebra of Euclidean type.

Case 2° Assume that the postprojective component $\mathcal{P}$ contains a section Σ of Euclidean type. Then $\operatorname{Hom}_B(U, \tau_B V) = 0$, for all modules U and V from Σ. Because $\mathcal{P} \neq \mathcal{Q}$ and (b) holds, the component $\mathcal{P}$ has no injective B-modules. Then taking the injective envelope $f : B \to I(B)$ of B in $\operatorname{mod} B$ and applying (IV.5.1)(a), we conclude that B is cogenerated by the direct sum T_B of all modules lying on Σ. Hence, applying (VI.2.2), we obtain that the section Σ is faithful.

In view of the criterion (VIII.5.6), it follows that T_B is a tilting B-module such that $A = \operatorname{End} T_B$ is a hereditary algebra, and $\mathcal{P}$ is the connecting component $\mathcal{C}_{T^*}$ determined by the tilting A-module $T_A^* = D({}_A T)$. Moreover, A is the path algebra of the Euclidean quiver $Q = \Sigma^{\text{op}}$.

Applying dual arguments to the above ones, we conclude that T^* is a preinjective A-module. Applying now (VIII.4.7)(a), we deduce that the torsion-free part $\mathcal{F}(T^*)$ contains all but finitely many non-isomorphic indecomposable A-modules and any indecomposable A-module not in $\mathcal{F}(T^*)$ is preinjective. Moreover, by (VIII.4.7)(e), the images under the functor $\operatorname{Ext}_A^1(T^*, -)$ of the preinjective torsion-free A-modules form a preinjective component $\mathcal{Q}(B)$ containing all indecomposable injective B-modules but no projective B-module, and consequently $Q = \mathcal{Q}(B)$.

Further, the preinjective component $\mathcal{Q}(A)$ of $\Gamma(\mathrm{mod}\,A)$ contains a full translation subquiver $\mathcal{Q}'(A)$, isomorphic to $\mathbb{N}Q^{\mathrm{op}} = \mathbb{N}\Sigma$, closed under predecessors in $\mathcal{Q}(A)$, and consisting entirely of modules from $\mathcal{F}(T^*)$. Then the image $\mathcal{Q}'(B)$ of $\mathcal{Q}'(A)$ under the functor $\mathrm{Ext}_A^1(T^*, -)$ is a full translation subquiver of the preinjective component $Q = \mathcal{Q}(B)$ which is closed under predecessors. Clearly, $\mathcal{Q}'(B)$ is isomorphic to $\mathbb{N}Q^{\mathrm{op}}$, and hence the component $\mathcal{Q}(B)$ contains a section of Euclidean type $Q^{\mathrm{op}} = \Sigma$. Therefore, it follows (as in the proof above) that A is a concealed algebra of Euclidean type. This finishes the proof.   □

The following useful fact is an immediate consequence of (VIII.4.5), (VIII.4.7), and the above proof.

5.2. Corollary. *Let Q be an acyclic Euclidean quiver. Then an algebra B is a concealed algebra of type Q if and only if $B = \mathrm{End}\,T_A$ for a preinjective tilting A-module over the path algebra $A = KQ$.*   □

We finish this section by the following two simple corollaries.

5.3. Corollary. *Let B be a concealed algebra of a Euclidean type Q. Then the algebra B^{op} opposite to B is a concealed algebra of type Q^{op}.*

Proof. Let Q be an acyclic Euclidean quiver, $A = KQ$ the path algebra of Q, T a postprojective tilting A-module, and $B = \mathrm{End}\,T_A$. Then there is an isomorphism $A^{\mathrm{op}} = K(Q^{\mathrm{op}})$ of algebras, $T^* = D(T)$ is a preinjective tilting A^{op}-module, and there is an isomorphism $B^{\mathrm{op}} \cong \mathrm{End}\,T_{A^{\mathrm{op}}}^*$ of algebras. It follows from (5.2) that B^{op} is a concealed algebra of type Q^{op}.   □

We also note that a tilted algebra B of a Euclidean type Q is representation-finite if and only if there is an isomorphism $B \cong \mathrm{End}\,T_A$, where T is a tilting A-module having both a postprojective and a preinjective direct summand, see (VIII.4.3) and (VIII.4.4).

In (XII.5.12) and (XII.5.13), we present examples of representation-infinite tilted algebras of Euclidean type that are not concealed.

XI.6. Exercises

1. Assume that $q : \mathbb{Z}^n \longrightarrow \mathbb{Z}$ is a positive definite quadratic form on the free abelian group $\mathbb{Z}^n$ and let $||-|| : \mathbb{R}^n \longrightarrow \mathbb{R}$ be the Euclidean norm on $\mathbb{R}^n$.
Prove that

(a) the set $\mathcal{R}_q = \{v \in \mathbb{Z}^n; q(v) = 1\}$ of all roots of q is finite and
(b) $||v|| \leq \frac{1}{\sqrt{\mu}}$, for any vector $v \in \mathcal{R}_q$, where $\mu = \inf\{q(u);\, u \in \mathbb{R}^n, ||u|| = 1\}$ is the minimum of the values $q(u)$ of the quadratic

function $q : \mathbb{R}^n \longrightarrow \mathbb{R}$ restricted to the unit sphere $\mathcal{S}^{n-1} = \{u \in \mathbb{R}^n, ||u|| = 1\}$ of $\mathbb{R}^n$.

2. Let Q be an acyclic quiver whose underlying graph is Euclidean, $A = KQ$ the path algebra of Q, $n = |Q_0|$, $\Phi_A : \mathbb{Z}^n \longrightarrow \mathbb{Z}^n$ the Coxeter transformation of A, $n = |Q_0|$, $\partial_A : \mathbb{Z}^n \longrightarrow \mathbb{Z}$ the defect of A, and $d_Q \geq 1$ the defect number of Q. Show that, for any $m \geq 1$ and $\mathbf{x} \in \mathbb{Z}^n$, the following equality holds

$$\Phi_A^{m d_Q}(\mathbf{x}) = \mathbf{x} + m \cdot \partial_A(\mathbf{x}) \cdot \mathbf{h}_Q$$

3. Let A denote the Kronecker algebra. Show that, for any $\lambda \in \mathbb{P}_1(K)$, and for an arbitrary preinjective A-module M there exists an epimorphism

$$h : X_1 \oplus \ldots \oplus X_m \longrightarrow M,$$

where $X_1, \ldots, X_m$ are indecomposable modules in the tube $\mathcal{T}_\lambda^A$.

4. Let $A = K[t_1, t_2]/(t_1^2, t_2^2)$, see Exercise (X.5.8).
   (a) Prove that $\dim_K \Lambda = 4$, $K_0(\Lambda) \cong \mathbb{Z}$, and $\mathrm{gl.dim}\,\Lambda = \infty$.
   (b) Compute the Cartan matrix $\mathbf{C}_A$ of the algebra A and note that $\mathbf{C}_A$ is not invertible (over $\mathbb{Z}$).

5. Let Q be an acyclic quiver $Q = (Q_0, Q_1)$, whose underlying graph $\overline{Q}$ is one of the Dynkin diagrams $\mathbb{A}_n$, $\mathbb{D}_n$, $\mathbb{E}_6$, $\mathbb{E}_7$, and $\mathbb{E}_8$, or one of the Euclidean diagrams $\widetilde{\mathbb{D}}_n$, $\widetilde{\mathbb{E}}_6$, $\widetilde{\mathbb{E}}_7$, and $\widetilde{\mathbb{E}}_8$. Let $m = |Q_0|$, and define the incidence matrix

$$C_Q = [c_{ij}]_{i,j \in Q_0} \in \mathbb{M}_m(\mathbb{Z})$$

of the quiver Q by setting

$$c_{ij} = \begin{cases} 1, & \text{if } i = j, \\ 1, & \text{if there is a path from the vertex } i \text{ to the vertex } j \text{ in } Q, \\ 0, & \text{otherwise.} \end{cases}$$

Let $A = KQ$ be the path K-algebra of Q.
   (a) Show that the Cartan matrix $\mathbf{C}_A \in \mathbb{M}_m(\mathbb{Z})$ of the algebra A is the transpose of the matrix C_Q.
   (b) Show that the inverse of C_Q is the matrix $C'_Q = [c'_{ij}]_{i,j \in Q_0} \in \mathbb{M}_m(\mathbb{Z})$ defined by the formula

$$c'_{ij} = \begin{cases} 1, & \text{if } i = j, \\ -1, & \text{if there is an arrow } i \to j \text{ in } Q, \\ 0, & \text{otherwise.} \end{cases}$$

6. Let A be the path K-algebra of the following Euclidean quiver

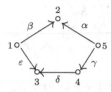

of type $\widetilde{\mathbb{A}}_4$.

(a) Compute the Coxeter matrix $\mathbf{\Phi}_A \in \mathbb{M}_5(\mathbb{Z})$ of A and the defect $\partial_A : \mathbb{Z}^5 \longrightarrow \mathbb{Z}$ of A.

(b) Compute the Auslander–Reiten quiver $\Gamma(\bmod A)$ of A. In particular, show that the regular part $\mathcal{R}(A)$ of $\Gamma(\bmod A)$ consists of a $\mathbb{P}_1(K)$-family $\boldsymbol{\mathcal{T}}^A = \{\mathcal{T}_\lambda^A\}_{\lambda \in \mathbb{P}_1(K)}$ of tubes $\mathcal{T}_\lambda^A$, and two of them are non-homogeneous of rank 2 and 3.

7. Let A be the path K-algebra of the following Euclidean quiver

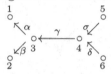

(a) Compute the Coxeter matrix $\mathbf{\Phi}_A \in \mathbb{M}_6(\mathbb{Z})$ of A and the defect $\partial_A : \mathbb{Z}^6 \longrightarrow \mathbb{Z}$ of A.

(b) Show that the regular part $\mathcal{R}(A)$ of $\Gamma(\bmod A)$ contains two tubes of rank 2, one tube of rank 3, and the remaining tubes of $\mathcal{R}(A)$ are of rank 1.

8. Let B be the path K-algebra of the following quiver

bound by the commutativity relation $\alpha\gamma = \beta\delta$.

(a) Show that B is a concealed algebra $\operatorname{End} T_A$, where $A = KQ$ is the hereditary algebra of Exercise 5 and T is a postprojective module in $\bmod A$.

(b) Compute the Euler quadratic form $q_B : \mathbb{Z}^6 \longrightarrow \mathbb{Z}$ of B.

(c) Find the positive generator $\mathbf{h}_B$ of $\operatorname{rad} q_B$.

(d) Describe the non-homogeneous stable tubes of $\Gamma(\bmod B)$.

9. Let B be the path K-algebra of the following quiver

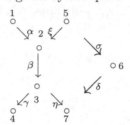

bound by the two commutativity relations $\omega\xi = \lambda\mu$ and $\alpha\gamma = \beta\delta$.

    (a) Show that B is a concealed algebra of Euclidean type $\widetilde{\mathbb{D}}_6$.

    (b) Compute the Euler quadratic form $q_B : \mathbb{Z}^7 \longrightarrow \mathbb{Z}$ of B.

    (c) Find the positive generator $\mathbf{h}_B$ of $\operatorname{rad} q_B$.

    (d) Describe the non-homogeneous stable tubes of $\Gamma(\operatorname{mod} B)$.

10. Let B be the algebra given by the quiver

bound by the commutativity relation $\xi\beta\eta = \sigma\delta$.

    (a) Show that B is a concealed algebra of Euclidean type $\widetilde{\mathbb{E}}_6$.

    (b) Compute the Euler quadratic form $q_B : \mathbb{Z}^7 \longrightarrow \mathbb{Z}$ of B.

    (c) Show that the vector

$$\begin{smallmatrix} & & 1 & & 1 & 1 \\ & & & 2 & & \\ & & & 2 & & \\ & & 1 & & 1 & \end{smallmatrix}$$

    is the positive generator $\mathbf{h}_B$ of $\operatorname{rad} q_B$.

    (d) Show that $\Gamma(\operatorname{mod} B)$ admits a stable tube of rank 2, whose modules on the mouth have the dimension vectors

$$\begin{smallmatrix} 0 & & 0 & 1 \\ & 1 & & \\ & 1 & & \\ 1 & & 1 & \end{smallmatrix} \quad \text{and} \quad \begin{smallmatrix} 1 & & 1 & 0 \\ & 1 & & \\ & 1 & & \\ 0 & & 0 & \end{smallmatrix}.$$

    (e) Show that $\Gamma(\operatorname{mod} B)$ admits a stable tube of rank 3, whose modules on the mouth have the dimension vectors

$$\begin{smallmatrix} 1 & & 1 & 1 \\ & 1 & & \\ & 1 & & \\ 1 & & 1 & \end{smallmatrix}, \quad \begin{smallmatrix} 0 & & 0 & 0 \\ & 0 & & \\ & 1 & & \\ 0 & & 0 & \end{smallmatrix}, \quad \begin{smallmatrix} 0 & & 0 & 0 \\ & 1 & & \\ & 0 & & \\ 0 & & 0 & \end{smallmatrix}.$$

    (f) Show that $\Gamma(\operatorname{mod} B)$ admits a stable tube of rank 3, whose modules on the mouth have the dimension vectors

$$\begin{smallmatrix} 0 & & 1 & 0 \\ & 1 & & \\ & 1 & & \\ 1 & & 0 & \end{smallmatrix}, \quad \begin{smallmatrix} 1 & & 0 & 0 \\ & 1 & & \\ & 1 & & \\ 0 & & 1 & \end{smallmatrix}, \quad \begin{smallmatrix} 0 & & 0 & 1 \\ & 0 & & \\ & 0 & & \\ 0 & & 0 & \end{smallmatrix}.$$

Chapter XII

Regular modules and tubes over concealed algebras of Euclidean type

The main aim of this chapter is to describe the structure of the family of stable tubes in the Auslander–Reiten quiver

$$\Gamma(\operatorname{mod} B) = \mathcal{P}(B) \cup \mathcal{R}(B) \cup \mathcal{Q}(B)$$

of any concealed algebra B of Euclidean type (over an algebraically closed field K). A prominent rôle is played by the class of canonical K-algebras introduced and studied by Ringel in [215].

In Section 1 we introduce the canonical algebras of Euclidean type and we prove that they are concealed algebras of the form $\operatorname{End}_{K\Delta}(T)$, where T is a postprojective tilting module over the path algebra $K\Delta$ of a canonically oriented quiver Δ of Euclidean type.

In Section 2 we provide a detailed description of the indecomposable regular modules over any canonical algebra C of Euclidean type, and the structure of the stable tubes in the regular part $\mathcal{R}(C)$ of $\Gamma(\operatorname{mod} C)$.

In Section 3, by applying the tilting theory, we prove general results on the structure of the module category $\operatorname{mod} B$ over an arbitrary concealed algebra B of Euclidean type. In particular, we show that

- the regular part $\mathcal{R}(B)$ of the Auslander–Reiten quiver $\Gamma(\operatorname{mod} B)$ is a disjoint union of the $\mathbb{P}_1(K)$-family

$$\boldsymbol{\mathcal{T}}^B = \{\mathcal{T}_\lambda^B\}_{\lambda \in \mathbb{P}_1(K)}$$

  of pairwise orthogonal standard stable tubes $\mathcal{T}_\lambda^B$, where $\mathbb{P}_1(K)$ is the projective line over K,
- the family $\boldsymbol{\mathcal{T}}^B$ separates the postprojective component $\mathcal{P}(B)$ of $\Gamma(\operatorname{mod} B)$ from the preinjective component $\mathcal{Q}(B)$,
- the module category $\operatorname{mod} B$ is controlled by the Euler quadratic form $q_B : K_0(B) \longrightarrow \mathbb{Z}$ of the algebra B.

Throughout, we assume that K is an algebraically closed field, and by an algebra we mean a finite dimensional K-algebra.

XII.1. Canonical algebras of Euclidean type

In this section we describe the category of regular modules over a special type of concealed K-algebras of Euclidean type, called the canonical algebras of Euclidean type.

For each pair of integers $q \geq p \geq 1$, we define $C(p,q)$ to be the path algebra

$$C(p,q) = K\Delta(p,q) \qquad (1.1)$$

of the following acyclic Euclidean quiver

$$\Delta(p,q) = \Delta(\widetilde{\mathbb{A}}_{p,q}) : \qquad$$

Observe that $C(1,1)$ is isomorphic to the Kronecker algebra

$$\begin{bmatrix} K & 0 \\ K^2 & K \end{bmatrix}.$$

For any triple (p,q,r) of integers such that $r \geq q \geq p \geq 1$, we set

$$C(p,q,r) = K\Delta(p,q,r)/I(p,q,r), \qquad (1.2)$$

that is, $C(p,q,r)$ is the bound quiver algebra $K\Delta(p,q,r)/I(p,q,r)$, where $\Delta(p,q,r)$ is the quiver

$$\Delta(p,q,r) : \qquad$$

and $I(p,q,r)$ is the two-sided ideal of the path K-algebra $K\Delta(p,q,r)$ generated by the element $\alpha_p \ldots \alpha_1 + \beta_q \ldots \beta_1 + \gamma_r \ldots \gamma_1$.

In this chapter we are only interested in the triples (p,q,r) of integers such that $r \geq q \geq p \geq 2$ and $\frac{1}{p} + \frac{1}{q} + \frac{1}{r} > 1$. It is easy to check that (p,q,r) is such a triple if and only if it is one of the following triples

$$(2,2,m-2), \text{ with } m \geq 4, \quad (2,3,3), \quad (2,3,4) \quad \text{or} \quad (2,3,5).$$

1.3. Definition. The following finite dimensional algebras

- $C(p,q)$, with $q \geq p \geq 1$,
- $C(2,2,m-2)$, with $m \geq 4$, and
- $C(2,3,3)$, $C(2,3,4)$, $C(2,3,5)$

are called the **canonical algebras of the Euclidean type** $\widetilde{\mathbb{A}}_{p,q}$, $\widetilde{\mathbb{D}}_m$, $\widetilde{\mathbb{E}}_6$, $\widetilde{\mathbb{E}}_7$, $\widetilde{\mathbb{E}}_8$, respectively.

Clearly, the canonical algebra $C(p,q)$ is a hereditary path algebra of the quiver $\Delta(p,q) = \Delta(\widetilde{\mathbb{A}}_{p,q})$ whose underlying graph is the Euclidean graph $\widetilde{\mathbb{A}}_{p+q-1}$. It follows from the following four propositions that the remaining canonical algebras are concealed of Euclidean type defined in Chapter VIII. In contrast to the algebra $C(p,q)$, each of the algebras $C(p,q,r)$ is not hereditary, because of the following lemma.

1.4. Lemma. *Let* $C = C(p,q,r)$ *be the algebra* (1.2), *where* $r \geq q \geq p \geq 2$.

(a) *The projective dimension* $\mathrm{pd}\, S(\omega)$ *of the unique simple injective* C-*module* $S(\omega)$ *equals* 2.

(b) $\mathrm{gl.dim}\, C = 2$.

Proof. (a) It is easy to see that the projective cover of the simple injective C-module $S(\omega)$ has the form $\varepsilon : P(\omega) \longrightarrow S(\omega)$ and $\mathrm{Ker}\,\varepsilon$ is isomorphic to the C-module

viewed as a linear representation of the bound quiver

$$(\Delta(p,q,r), I(p,q,r)).$$

It follows that the minimal projective presentation of the C-module $\mathrm{Ker}\,\varepsilon$ is of the form

$$0 \longrightarrow P(0) \longrightarrow P(a_{p-1}) \oplus P(b_{q-1}) \oplus P(c_{r-1}) \longrightarrow \mathrm{Ker}\,\varepsilon \longrightarrow 0.$$

This shows that $\mathrm{pd}\, S(\omega) = 2$.

(b) Similarly as above we show that $\mathrm{pd}\, S(j) = 1$, for each vertex j of the quiver $\Delta(p,q,r)$ such that $j \neq 0$ and $j \neq \omega$, because the kernel of the projective cover $P(j) \longrightarrow S(j)$ of the simple C-module $S(j)$ is projective. It follows that $\mathrm{gl.dim}\, C = 2$ (see (A.4.8) of Volume 1), because the remaining simple C-module $S(0)$ is projective. $\qquad\square$

1.5. Proposition. *For each $m \geq 4$, the canonical algebra (1.3) $C = C(2, 2, m-2)$ is concealed of the Euclidean type*

$$\Delta(\widetilde{\mathbb{D}}_m): \qquad \begin{array}{c} 1 \\ \nwarrow \\ \searrow \\ 3 \leftarrow \cdots \leftarrow m-1 \\ \swarrow \\ 2 \end{array} \qquad \begin{array}{c} m \\ \swarrow \\ \nwarrow \\ m+1 \end{array}$$

and gl.dim $C = 2$.

Proof. Let $\Delta = \Delta(\widetilde{\mathbb{D}}_m)$. The standard calculation technique shows that the left hand part of the component $\mathcal{P}$ of $\Gamma(\mathrm{mod}\, C)$ containing the simple projective module $S(0) = P(0)$ looks as follows

where

$$P(0) = \underset{00\ldots 00}{1\overset{0}{0}0}, \quad \text{and} \quad P(\omega) = \underset{11\ldots 11}{2\overset{1}{1}1},$$

and the indecomposable modules are represented by their dimension vectors. Observe that $\mathcal{P}$ is a postprojective component containing all the indecomposable projective modules and a section of the form Δ^{op} given by the modules

$$P(\omega) \qquad (1.6)$$

$$P(a_1) \longrightarrow \tau^{-1}P(0) \longrightarrow \tau^{-1}P(c_1) \longrightarrow \ldots\ldots \longrightarrow \tau^{-1}P(c_{m-4}) \longrightarrow \tau^{-1}P(c_{m-3})$$

$$P(b_1)$$

Dually, $\Gamma(\operatorname{mod} C)$ has a preinjective component $\mathcal{Q}$ containing all the indecomposable injective modules and a section of the form Δ given by the modules

$$I(a_1)$$

$$\tau I(c_1) \longrightarrow \tau I(c_2) \longrightarrow \ldots\ldots \longrightarrow \tau I(c_{m-3}) \longrightarrow \tau I(\omega) \longrightarrow I(b_1)$$

$$I(0)$$

Hence, by applying (XI.5.1), we conclude that C is a concealed algebra of the Euclidean type $\Delta(\widetilde{\mathbb{D}}_m)$. Finally, by (1.4) and (VIII.3.2), gl.dim $C = 2$.   □

The reader may have observed that we have shown in the course of the proof the following useful fact.

1.7. Corollary. *Let $C = C(2, 2, m-2)$ be the canonical algebra of type $\Delta(\widetilde{\mathbb{D}}_m)$.*

(a) *The family (1.6) is a section in the postprojective component $\mathcal{P}(C)$ of the quiver $\Gamma(\operatorname{mod} C)$ containing all the indecomposable projective C-modules.*

(b) *The module*

$$T = P(a_1) \oplus P(b_1) \oplus \tau^{-1}P(0) \oplus \tau^{-1}P(c_1) \oplus \ldots \oplus \tau^{-1}P(c_{m-3}) \oplus P(\omega)$$

is a postprojective tilting C-module and $A = \operatorname{End} T_C$ is a hereditary algebra isomorphic to the path algebra of the quiver $\Delta(\widetilde{\mathbb{D}}_m)$.   □

1.8. Proposition. *The canonical algebra $C = C(2, 3, 3)$ is concealed of the Euclidean type*

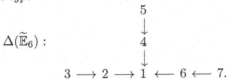

$$\Delta(\widetilde{\mathbb{E}}_6): \qquad \begin{array}{c} 5 \\ \downarrow \\ 4 \\ \downarrow \\ 3 \longrightarrow 2 \longrightarrow 1 \longleftarrow 6 \longleftarrow 7. \end{array}$$

and gl.dim $C = 2$.

Proof. Let $\Delta = \Delta(\widetilde{\mathbb{E}}_6)$. The standard calculation technique shows that the left hand part of the component $\mathcal{P}$ of $\Gamma(\mathrm{mod}\,C)$ containing the simple projective module $S(0) = P(0)$ looks as follows

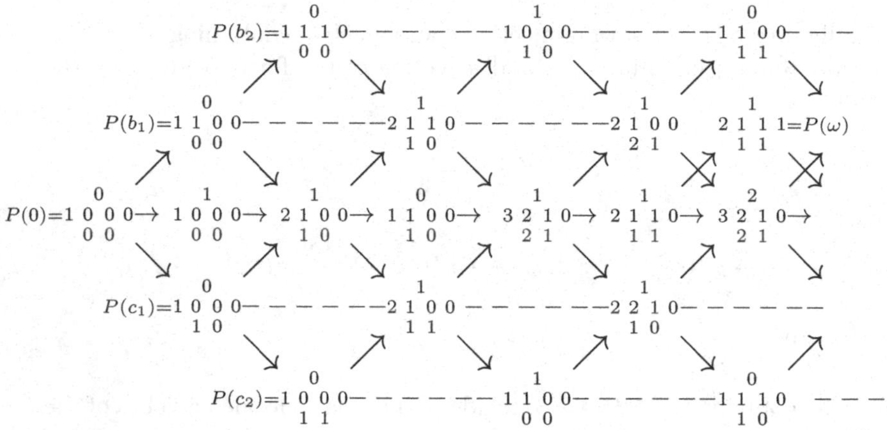

$$
\begin{array}{c}
\overset{0}{P(b_2)=1\ 1\ 1\ 0}\ \underset{0\ 0}{}\ {-}\ {-}\ {-}\ {-}\ {-}\ \overset{1}{1\ 0\ 0\ 0}\underset{1\ 0}{}\ {-}\ {-}\ {-}\ {-}\ {-}\overset{0}{1\ 1\ 0\ 0}\underset{1\ 1}{}\ {-}\ {-}\ {-}
\end{array}
$$

where

$$
P(a_1) = \overset{1}{1\ 0}\ \underset{0\ 0}{0\ 0},
$$

and the indecomposable modules are represented by their dimension vectors. Hence $\mathcal{P}$ is a postprojective component containing all the indecomposable projective modules and a section of the form Δ^{op} given by the modules

$$
\tau^{-2}P(b_2)
$$

$$
\nearrow
$$

$$
\tau^{-2}P(b_1)
$$

$$
\nearrow
$$

$$
\tau^{-2}P(0) \longrightarrow \tau^{-2}P(a_1) \longrightarrow P(\omega) \qquad\qquad (1.9)
$$

$$
\searrow
$$

$$
\tau^{-2}P(c_1)
$$

$$
\searrow
$$

$$
\tau^{-2}P(c_2)
$$

Dually, $\Gamma(\mathrm{mod}\,C)$ has a preinjective component $\mathcal{Q}$ containing all the indecomposable injective modules and a section of the form Δ given by the modules

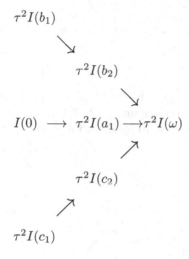

$$\tau^2 I(b_1)$$

$$\tau^2 I(b_2)$$

$$I(0) \longrightarrow \tau^2 I(a_1) \longrightarrow \tau^2 I(\omega)$$

$$\tau^2 I(c_2)$$

$$\tau^2 I(c_1)$$

Hence, by applying (XI.5.1), we conclude that C is a concealed algebra of the Euclidean type $\Delta(\widetilde{\mathbb{E}}_6)$. Finally, by (1.4) and (VIII.3.2), gl.dim $C = 2$. $\square$

We have shown in the course of the proof the following useful fact.

1.10. Corollary. *Let* $C = C(2,3,3)$ *be the canonical algebra of type* $\Delta(\widetilde{\mathbb{E}}_6)$.

(a) *The family (1.9) is a section in the postprojective component* $\mathcal{P}(C)$ *of the quiver* $\Gamma(\mathrm{mod}\,C)$ *containing all the indecomposable projective C-modules.*

(b) *The module*

$$T = \tau^{-2} P(a_1) \oplus \tau^{-2} P(b_1) \oplus \tau^{-2} P(b_2) \oplus \tau^{-2} P(0) \oplus \tau^{-2} P(c_1)$$
$$\oplus \tau^{-2} P(c_2) \oplus P(\omega)$$

is a postprojective tilting C-module and $A = \mathrm{End}\,T_C$ *is a hereditary algebra isomorphic to the path algebra of the quiver* $\Delta(\widetilde{\mathbb{E}}_6)$. $\square$

1.11. Proposition. *The canonical algebra* $C = C(2, 3, 4)$ *is concealed of the Euclidean type*

$$\Delta(\widetilde{\mathbb{E}}_7) : \qquad \begin{array}{c} 5 \\ \downarrow \\ 4 \longrightarrow 3 \longrightarrow 2 \longrightarrow 1 \longleftarrow 6 \longleftarrow 7 \longleftarrow 8. \end{array}$$

and gl.dim $C = 2$.

Proof. Let $\Delta = \Delta(\widetilde{\mathbb{E}}_7)$. The standard calculation technique shows that the left hand part of the component $\mathcal{P}$ of $\Gamma(\mathrm{mod}\, C)$ containing the simple projective module $S(0) = P(0)$ looks as follows

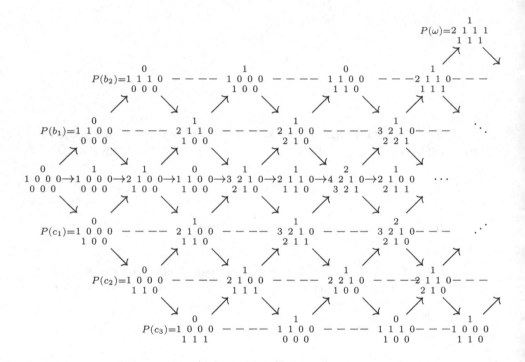

where

$$P(0) = 1 \begin{smallmatrix} 0 \\ 0 \\ 0 \end{smallmatrix} \begin{smallmatrix} 0 \\ 0 \end{smallmatrix} 0, \qquad P(a_1) = 1 \begin{smallmatrix} 1 \\ 0 \\ 0 \end{smallmatrix} \begin{smallmatrix} 0 \\ 0 \end{smallmatrix} 0,$$

and the indecomposable modules are represented by their dimension vectors. Hence $\mathcal{P}$ is a postprojective component containing all the indecomposable projective modules and a section of the form Δ^{op} given by the modules

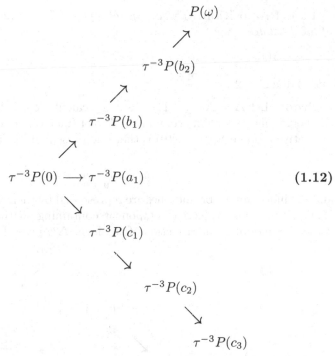

$$P(\omega)$$

$$\tau^{-3}P(b_2)$$

$$\tau^{-3}P(b_1)$$

$$\tau^{-3}P(0) \longrightarrow \tau^{-3}P(a_1) \qquad\qquad (1.12)$$

$$\tau^{-3}P(c_1)$$

$$\tau^{-3}P(c_2)$$

$$\tau^{-3}P(c_3)$$

Dually, $\Gamma(\operatorname{mod} C)$ has a preinjective component $\mathcal{Q}$ containing all the indecomposable injective modules and a section of the form Δ given by the modules

$$\tau^3 I(\omega), \ \tau^3 I(a_1), \ \tau^3 I(b_2), \ \tau^3 I(b_1), \ I(0), \ \tau^3 I(c_3), \ \tau^3 I(c_2), \ \text{and } \tau^3 I(c_1).$$

Hence, by applying (XI.5.1), we conclude that C is a concealed algebra of the Euclidean type $\Delta = \Delta(\widetilde{\mathbb{E}}_7)$. Finally, by (1.4) and (VIII.3.2), gl.dim $C = 2$. $\square$

We have shown in the course of the proof the following useful fact.

1.13. Corollary. *Let $C = C(2,3,4)$ be the canonical algebra of type $\Delta(\widetilde{\mathbb{E}}_7)$.*

(a) *The family (1.12) is a section in the postprojective component $\mathcal{P}(C)$ of the quiver $\Gamma(\operatorname{mod} C)$ containing all the indecomposable projective C-modules.*

(b) *The module*

$$T = \tau^{-3}P(a_1) \oplus \tau^{-3}P(b_1) \oplus \tau^{-3}P(b_2) \oplus \tau^{-3}P(0) \oplus \tau^{-3}P(c_1)$$
$$\oplus \tau^{-3}P(c_2) \oplus \tau^{-3}P(c_3) \oplus P(\omega)$$

is a postprojective tilting C-module and $A = \operatorname{End} T_C$ is a hereditary algebra isomorphic to the path algebra of the quiver $\Delta(\widetilde{\mathbb{E}}_7)$. $\square$

1.14. Proposition. *The canonical algebra* $C = C(2, 3, 5)$ *is concealed of the Euclidean type*

$$\Delta(\widetilde{\mathbb{E}}_8) : \qquad \begin{array}{c} 4 \\ \downarrow \\ 3 \longrightarrow 2 \longrightarrow 1 \longleftarrow 5 \longleftarrow 6 \longleftarrow 7 \longleftarrow 8 \longleftarrow 9 \end{array}$$

and $\mathrm{gl.dim}\, C = 2$.

Proof. Let $\Delta = \Delta(\widetilde{\mathbb{E}}_8)$. The standard calculation technique shows that the beginning part of the component $\mathcal{P}$ of $\Gamma(\mathrm{mod}\, C)$ containing the simple projective module $S(0) = P(0)$ is that one presented in Figure 1.15a below, where

$$P(a_1) = 1 \;\; \begin{smallmatrix} 1 \\ 0 \; 0 \; 0 \\ 0 \; 0 \; 0 \; 0 \end{smallmatrix}$$

and the indecomposable modules are represented by their dimension vectors. Hence $\mathcal{P}$ is a postprojective component containing all the indecomposable projective modules and a section of the form Δ^{op} given by the modules

$$\tau^{-5} P(0) \qquad\qquad\qquad (1.15)$$

$$\swarrow \qquad\qquad \searrow$$

$$\tau^{-5} P(c_1) \qquad \downarrow \qquad \tau^{-5} P(b_1)$$

$$\swarrow \qquad\qquad\qquad\qquad \searrow$$

$$\tau^{-5} P(c_2) \qquad\qquad \tau^{-5} P(a_1) \qquad\qquad \tau^{-5} P(b_2).$$

$$\swarrow$$

$$\tau^{-5} P(c_3)$$

$$\swarrow$$

$$\tau^{-5} P(c_4)$$

$$\swarrow$$

$$P(\omega)$$

Dually, $\Gamma(\mathrm{mod}\, C)$ has a preinjective component $\mathcal{Q}$ containing all the indecomposable injective modules and a section of the form Δ given by the modules

$$\tau^5 I(\omega),\ \tau^5 I(a_1),\ \tau^5 I(b_2),\ \tau^5 I(b_1),\ \tau^5 I(c_4),\ \tau^3 I(c_5),\ \tau^5 I(c_2),\ \tau^5 I(c_1),\ I(0).$$

Hence, by applying again (XI.5.1), we conclude that C is a concealed algebra of the Euclidean type $\Delta(\widetilde{\mathbb{E}}_8)$. Finally, by (1.4) and (VIII.3.2), $\mathrm{gl.dim}\, C = 2$. $\qquad\square$

1.15a. Figure

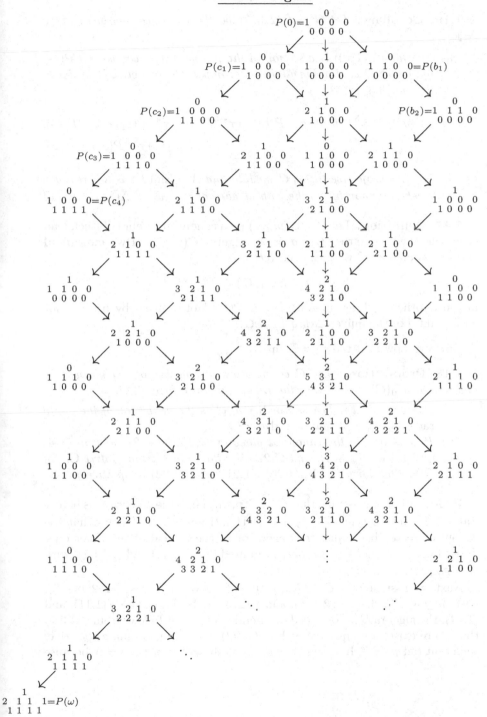

We have shown in the course of the proof the following useful fact.

1.16. Corollary. *Let $C = C(2,3,5)$ be the canonical algebra of type $\Delta(\widetilde{\mathbb{E}}_8)$.*

    (a) *The family (1.12) is a section in the postprojective component $\mathcal{P}(C)$ of $\Gamma(\mathrm{mod}\, C)$ containing all the indecomposable projective C-modules.*

    (b) *The module*

$$T = \tau^{-5}P(a_1) \oplus \tau^{-5}P(b_1) \oplus \tau^{-5}P(b_2) \oplus \tau^{-5}P(0) \oplus \tau^{-5}P(c_1) \oplus \tau^{-5}P(c_2)$$
$$\oplus\, \tau^{-5}P(c_3) \oplus \tau^{-5}P(c_4) \oplus P(\omega)$$

is a postprojective tilting C-module and $A = \mathrm{End}\, T_C$ is a hereditary algebra isomorphic to the path algebra of the quiver $\Delta(\widetilde{\mathbb{E}}_8)$. $\qquad\square$

1.17. Definition. Let $C = C(p,q,r)$ be a canonical algebra of Euclidean type, where we agree that $C(p,q)$ is the algebra $C(p,q,1)$. By a **canonical radical vector** of C we mean the vector

$$\mathbf{h}_C = (\mathbf{h}_j) \in K_0(C) \cong \mathbb{Z}^{p+q+r-1}$$

in the Grothendieck group $K_0(C) \cong \mathbb{Z}^{p+q+r-1}$ of C given by $\mathbf{h}_j = 1$, for each point j of the quiver $\Delta(p,q,r)$ of C.

Now we show that $\mathrm{rad}\, q_C = \mathbb{Z} \cdot \mathbf{h}_C$.

1.18. Proposition. *Let C be a canonical algebra of Euclidean type and let $q_C : K_0(C) \longrightarrow \mathbb{Z}$ be the Euler quadratic form of C.*

    (a) *The canonical radical vector $\mathbf{h}_C$ of C is the generator of the group $\mathrm{rad}\, q_C \subseteq K_0(C)$, that is, $\mathrm{rad}\, q_C = \mathbb{Z} \cdot \mathbf{h}_C$.*

    (b) *If C is one of the canonical algebras $C(2,2,m-2)$, with $m \geq 4$, $C(2,3,3)$, $C(2,3,4)$, and $C(2,3,5)$, then the Cartan matrix $\mathbf{C}_C$ of C has the form (1.19), (1.20), (1.21), and (1.22), respectively.*

Proof. First, we assume that $C = C(p,q)$. Then the algebra C is hereditary of Euclidean type $\widetilde{\mathbb{A}}_{p+q-1}$ and (VII.4.2) applies. It follows that the canonical vector $\mathbf{h}_C$ is just the generator of the radical of the Euler quadratic form $q_C : K_0(C) \longrightarrow \mathbb{Z}$ of C presented in (VII.4.2) and (XI.1.1), with $\overline{Q} = \widetilde{\mathbb{A}}_{p+q-1}$.

Next, we assume that $C = C(p,q,r)$ is of one of the types $C(2,2,m-2)$, with $m \geq 4$, $C(2,3,3)$, $C(2,3,4)$, and $C(2,3,5)$. By (1.5), (1.8), (1.11), and (1.14), the algebra C is concealed of Euclidean type. It follows from (XI.3.7) that there exists a unique vector $\mathbf{h} \in K_0(C)$, with all coordinates positive, such that $\mathrm{rad}\, q_C = \mathbb{Z}\cdot\mathbf{h}$. Then, it remains to show that $q_C(\mathbf{h}_C) = 0$, because this yields $\mathbf{h}_C = m \cdot \mathbf{h}$, for some $m \in \mathbb{Z}$, and hence we easily conclude that $m = 1$ and, consequently, $\mathbf{h} = \mathbf{h}_C$ and $\operatorname{rad} q_C = \mathbb{Z} \cdot \mathbf{h}_C$.

To finish the proof, we recall from (III.3.11) and Section XI.1 that the Euler quadratic form

$$q_C : K_0(C) \longrightarrow \mathbb{Z}$$

is given by the formula

$$q_C(x) = x^t (\mathbf{C}_C^{-1})^t x,$$

where $\mathbf{C}_C$ is the Cartan matrix of the algebra C. For each of the types $C(2,2,m-2)$, with $m \geq 4$, $C(2,3,3)$, $C(2,3,4)$, $C(2,3,5)$, we now compute the Cartan matrix $\mathbf{C}_C$ and the Euler quadratic form q_C, and then we show that $\operatorname{rad} q_C = \mathbb{Z} \cdot \mathbf{h}_C$. Now we do it by a case by case inspection.

Case 1° Assume that C is the algebra

$$C(2,2,m-2) = K\Delta(2,2,m-2)/I(2,2,m-2),$$

where $m \geq 4$. We compute the matrices $\mathbf{C}_C$ and $(\mathbf{C}_C^{-1})^t$ with respect to the following ordering

$$0, a_1, b_1, c_1, \ldots, c_{m-3}, \omega$$

of vertices of the quiver $\Delta(2,2,m-2)$. Because, by (III.3.8), the j-th column of the matrix $\mathbf{C}_C$ is the dimension vector $\dim P(j)$ of the indecomposable projective right C-module $P(j) = e_j C$ corresponding to the j-th vertex of $\Delta(2,2,m-2)$, with respect to the ordering fixed above, then the square $(m+1) \times (m+1)$ matrices $\mathbf{C}_C$ and $(\mathbf{C}_C^{-1})^t$ have the forms

$$\mathbf{C}_C = \begin{bmatrix} 1 & 1 & 1 & 1 & 1 & \cdots & 1 & 1 & 1 & 2 \\ 0 & 1 & 0 & 0 & 0 & \cdots & 0 & 0 & 0 & 1 \\ 0 & 0 & 1 & 0 & 0 & \cdots & 0 & 0 & 0 & 1 \\ 0 & 0 & 0 & 1 & 1 & \cdots & 1 & 1 & 1 & 1 \\ 0 & 0 & 0 & 0 & 1 & \cdots & 1 & 1 & 1 & 1 \\ \vdots & \vdots & \vdots & \vdots & \vdots & \ddots & \vdots & \vdots & \vdots & \vdots \\ 0 & 0 & 0 & 0 & 0 & \cdots & 1 & 1 & 1 & 1 \\ 0 & 0 & 0 & 0 & 0 & \cdots & 0 & 1 & 1 & 1 \\ 0 & 0 & 0 & 0 & 0 & \cdots & 0 & 0 & 1 & 1 \\ 0 & 0 & 0 & 0 & 0 & \cdots & 0 & 0 & 0 & 1 \end{bmatrix}, \ (\mathbf{C}_C^{-1})^t = \begin{bmatrix} 1 & 0 & 0 & 0 & 0 & \cdots & 0 & 0 & 0 & 0 \\ -1 & 1 & 0 & 0 & 0 & \cdots & 0 & 0 & 0 & 0 \\ -1 & 0 & 1 & 0 & 0 & \cdots & 0 & 0 & 0 & 0 \\ -1 & 0 & 0 & 1 & 0 & \cdots & 0 & 0 & 0 & 0 \\ 0 & 0 & 0 & -1 & 1 & \ddots & 0 & 0 & 0 & 0 \\ 0 & 0 & 0 & 0 & -1 & \ddots & 0 & 0 & 0 & 0 \\ \vdots & \vdots & \vdots & \vdots & \vdots & \ddots & \ddots & \vdots & \vdots & \vdots \\ 0 & 0 & 0 & 0 & 0 & \ddots & -1 & 1 & 0 & 0 \\ 0 & 0 & 0 & 0 & 0 & \ddots & 0 & -1 & 1 & 0 \\ 0 & 0 & 0 & 0 & 0 & \cdots & 0 & 0 & -1 & 1 & 0 \\ 1 & -1 & -1 & 0 & 0 & \cdots & 0 & 0 & 0 & -1 & 1 \end{bmatrix} \quad (1.19)$$

We note that the coefficient $c_{1,m+1}$ of the matrix $\mathbf{C}_C$ equals 2, because it is the dimension of the K-vector space

$$e_\omega C e_0 \cong (K\alpha_p \ldots \alpha_1 \oplus K\beta_q \ldots \beta_1 \oplus K\gamma_r \ldots \gamma_1)/K(\alpha_p \ldots \alpha_1 + \beta_q \ldots \beta_1 + \gamma_r \ldots \gamma_1).$$

Then the Euler quadratic form $q_C : K_0(C) \longrightarrow \mathbb{Z}$ of C is given by the formula

$$q_C(x) = x^t (\mathbf{C}_C^{-1})^t x = x_0^2 + x_{a_1}^2 + x_{b_1}^2 + x_{c_1}^2 + \ldots + x_{c_{m-3}}^2 + x_\omega^2$$
$$- x_0(x_{a_1} + x_{b_1} + x_{c_1} - x_\omega) - x_{a_1} x_\omega - x_{b_1} x_\omega$$
$$- x_{c_1} x_{c_2} - x_{c_2} x_{c_3} - \ldots - x_{c_{m-4}} x_{c_{m-3}} - x_{c_{m-3}} x_\omega$$

$$= (x_0 - \frac{1}{2} x_{a_1} - \frac{1}{2} x_{b_1} - \frac{1}{2} x_{c_1} + \frac{1}{2} x_\omega)^2$$
$$+ \frac{3}{4} (x_{a_1} - \frac{1}{3} x_{b_1} - \frac{1}{3} x_\omega - \frac{1}{3} x_{c_1})^2 + \frac{2}{3} (x_{b_1} - \frac{1}{2} x_\omega - \frac{1}{2} x_{c_1})^2$$
$$+ \frac{1}{2} (x_{c_1} - x_{c_2})^2 + \frac{1}{2} (x_{c_2} - x_{c_3})^2 + \ldots + \frac{1}{2} (x_{c_{m-4}} - x_{c_{m-3}})^2$$
$$+ \frac{1}{2} (x_{c_{m-3}} - x_\omega)^2.$$

It follows that $q_C(\mathbf{h}_C) = 0$ and, given $\mathbf{x} \in \operatorname{rad} q_C$, there is an integer m such that $\mathbf{x} = m \cdot \mathbf{h}_C$. This shows that $\operatorname{rad} q_C = \mathbb{Z} \cdot \mathbf{h}_C$, and we are done.

Case 2° Assume that C is the algebra

$$C(2,3,3) = K\Delta(2,3,3)/I(2,3,3).$$

We compute the matrices $\mathbf{C}_C, (\mathbf{C}_C^{-1})^t$ with respect to the following ordering

$$0, a_1, b_1, b_2, c_1, c_2, \omega$$

of vertices of the quiver $\Delta(2,3,3)$. Because the j-th column of $\mathbf{C}_C$ is the dimension vector $\mathbf{dim}\, P(j)$ of the right C-module $P(j) = e_j C$ corresponding to the j-th vertex of $\Delta(2,3,3)$, then the square 7×7 matrices $\mathbf{C}_C$ and $(\mathbf{C}_C^{-1})^t$ have the forms

$$\mathbf{C}_C = \begin{bmatrix} 1&1&1&1&1&1&2 \\ 0&1&0&0&0&0&1 \\ 0&0&1&1&0&0&1 \\ 0&0&0&1&0&0&1 \\ 0&0&0&0&1&1&1 \\ 0&0&0&0&0&1&1 \\ 0&0&0&0&0&0&1 \end{bmatrix}, \quad (\mathbf{C}_C^{-1})^t = \begin{bmatrix} 1&0&0&0&0&0&0 \\ -1&1&0&0&0&0&0 \\ -1&0&1&0&0&0&0 \\ 0&0&-1&1&0&0&0 \\ -1&0&0&0&1&0&0 \\ 0&0&0&0&-1&1&0 \\ 1&-1&0&-1&0&-1&1 \end{bmatrix} \in \mathbb{M}_7(\mathbb{Z}). \qquad (1.20)$$

Hence, the Euler quadratic form $q_C : K_0(C) \longrightarrow \mathbb{Z}$ of C is given by the formula

$$q_C(x) = x^t (\mathbf{C}_C^{-1})^t x = x_0^2 + x_{a_1}^2 + x_{b_1}^2 + x_{b_2}^2 + x_{c_1}^2 + x_{c_2}^2 + x_\omega^2$$
$$- x_0(x_{a_1} + x_{b_1} + x_{c_1} - x_\omega) - x_{a_1} x_\omega - x_{b_1} x_{b_2} - x_{b_2} x_\omega - x_{c_1} x_{c_2} - x_{c_2} x_\omega$$

$$= (x_0 - \frac{1}{2} x_{a_1} - \frac{1}{2} x_{b_1} - \frac{1}{2} x_{c_1} + \frac{1}{2} x_\omega)^2 + \frac{3}{4} (x_{a_1} - \frac{1}{3} x_\omega - \frac{1}{3} x_{c_1} - \frac{1}{3} x_{b_1})^2$$
$$+ \frac{2}{3} (x_{b_1} - \frac{3}{4} x_{b_2} - \frac{1}{2} x_{c_1} + \frac{1}{4} x_\omega)^2 + \frac{5}{8} (x_{b_2} - \frac{3}{5} x_\omega - \frac{2}{5} x_{c_1})^2$$
$$+ \frac{2}{5} (x_{c_1} - \frac{5}{4} x_{c_2} + \frac{1}{4} x_\omega)^2 + \frac{3}{8} (x_{c_2} - x_\omega)^2.$$

It follows that $q_C(\mathbf{h}_C) = 0$ and, given $\mathbf{x} \in \operatorname{rad} q_C$, there is an integer m such that $\mathbf{x} = m \cdot \mathbf{h}_C$. This shows that $\operatorname{rad} q_C = \mathbb{Z} \cdot \mathbf{h}_C$, and we are done.

Case 3° Assume that C is the algebra

$$C(2,3,4) = \Delta(2,3,4)/I(2,3,4).$$

We compute the matrices $\mathbf{C}_C$ and $(\mathbf{C}_C^{-1})^t$ with respect to the following ordering

$$0, a_1, b_1, b_2, c_1, c_2, c_3, \omega$$

of vertices of the quiver $\Delta(2,3,4)$. Because the j-th column of $\mathbf{C}_C$ is the dimension vector $\mathbf{dim}\, P(j)$ of the right C-module $P(j) = e_j C$ corresponding to the j-th vertex of $\Delta(2,3,4)$, then the square $8{\times}8$ matrices $\mathbf{C}_C$ and $(\mathbf{C}_C^{-1})^t$ have the forms

$$\mathbf{C}_C = \begin{bmatrix} 1&1&1&1&1&1&1&2 \\ 0&1&0&0&0&0&0&1 \\ 0&0&1&1&0&0&0&1 \\ 0&0&0&1&0&0&0&1 \\ 0&0&0&0&1&1&1&1 \\ 0&0&0&0&0&1&1&1 \\ 0&0&0&0&0&0&1&1 \\ 0&0&0&0&0&0&0&1 \end{bmatrix}, \ (\mathbf{C}_C^{-1})^t = \begin{bmatrix} 1&0&0&0&0&0&0&0 \\ -1&1&0&0&0&0&0&0 \\ -1&0&1&0&0&0&0&0 \\ 0&0&-1&1&0&0&0&0 \\ -1&0&0&0&1&0&0&0 \\ 0&0&0&0&-1&1&0&0 \\ 0&0&0&0&0&-1&1&0 \\ 1&-1&0&-1&0&0&-1&1 \end{bmatrix} \in \mathbb{M}_8(\mathbb{Z}).\ \textbf{(1.21)}$$

Hence, the Euler quadratic form $q_C : K_0(C) \longrightarrow \mathbb{Z}$ of C is given by the formula

$$q_C(x) = x^t (\mathbf{C}_C^{-1})^t x = x_0^2 + x_{a_1}^2 + x_{b_1}^2 + x_{b_2}^2 + x_{c_1}^2 + x_{c_2}^2 + x_{c_3}^2 + x_\omega^2$$
$$- x_0(x_{a_1} + x_{b_1} + x_{c_1} - x_\omega) - x_{a_1} x_\omega - x_{b_1} x_{b_2} - x_{b_2} x_\omega$$
$$- x_{c_1} x_{c_2} - x_{c_2} x_{c_3} - x_{c_3} x_\omega$$

$$= \left(x_0 - \frac{1}{2}x_{a_1} - \frac{1}{2}x_{b_1} - \frac{1}{2}x_{c_1} + \frac{1}{2}x_\omega\right)^2 + \frac{3}{4}\left(x_{a_1} - \frac{1}{3}x_{c_1} - \frac{1}{3}x_{b_1} - \frac{1}{3}x_\omega\right)^2$$
$$+ \frac{2}{3}\left(x_{b_1} - \frac{3}{4}x_{b_2} - \frac{1}{2}x_{c_1} + \frac{1}{4}x_\omega\right)^2 + \frac{5}{8}\left(x_{b_2} - \frac{3}{5}x_\omega - \frac{2}{5}x_{c_1}\right)^2$$
$$+ \frac{2}{5}\left(x_{c_1} - \frac{5}{4}x_{c_2} + \frac{1}{4}x_\omega\right)^2 + \frac{3}{8}\left(x_{c_2} - \frac{4}{3}x_{c_3} + \frac{1}{3}x_\omega\right)^2 + \frac{1}{3}(x_{c_3} - x_\omega)^2.$$

It follows that $q_C(\mathbf{h}_C) = 0$ and, given $\mathbf{x} \in \operatorname{rad} q_C$, there is an integer m such that $\mathbf{x} = m \cdot \mathbf{h}_C$. This shows that $\operatorname{rad} q_C = \mathbb{Z} \cdot \mathbf{h}_C$, and we are done.

Case 4° Assume that C is the algebra

$$C(2,3,5) = \Delta(2,3,5)/I(2,3,5).$$

We compute the matrices $\mathbf{C}_C$ and $(\mathbf{C}_C^{-1})^t$ with respect to the following ordering

$$0, a_1, b_1, b_2, c_1, c_2, c_3, c_4, \omega$$

of vertices of the quiver $\Delta(2,3,5)$. Because the j-th column of $\mathbf{C}_C$ is the dimension vector $\dim P(j)$ of the right C-module $P(j) = e_j C$ corresponding to the j-th vertex of $\Delta(2,3,5)$, then the square 9×9 matrices $\mathbf{C}_C$ and $(\mathbf{C}_C^{-1})^t$ have the forms

$$
\mathbf{C}_C = \begin{bmatrix}
1 & 1 & 1 & 1 & 1 & 1 & 1 & 1 & 2 \\
0 & 1 & 0 & 0 & 0 & 0 & 0 & 0 & 1 \\
0 & 0 & 1 & 1 & 0 & 0 & 0 & 0 & 1 \\
0 & 0 & 0 & 1 & 0 & 0 & 0 & 0 & 1 \\
0 & 0 & 0 & 0 & 1 & 1 & 1 & 1 & 1 \\
0 & 0 & 0 & 0 & 0 & 1 & 1 & 1 & 1 \\
0 & 0 & 0 & 0 & 0 & 0 & 1 & 1 & 1 \\
0 & 0 & 0 & 0 & 0 & 0 & 0 & 1 & 1 \\
0 & 0 & 0 & 0 & 0 & 0 & 0 & 0 & 1
\end{bmatrix}, \quad
(\mathbf{C}_C^{-1})^t = \begin{bmatrix}
1 & 0 & 0 & 0 & 0 & 0 & 0 & 0 & 0 \\
-1 & 1 & 0 & 0 & 0 & 0 & 0 & 0 & 0 \\
-1 & 0 & 1 & 0 & 0 & 0 & 0 & 0 & 0 \\
0 & 0 & -1 & 1 & 0 & 0 & 0 & 0 & 0 \\
-1 & 0 & 0 & 0 & 1 & 0 & 0 & 0 & 0 \\
0 & 0 & 0 & 0 & -1 & 1 & 0 & 0 & 0 \\
0 & 0 & 0 & 0 & 0 & -1 & 1 & 0 & 0 \\
0 & 0 & 0 & 0 & 0 & 0 & -1 & 1 & 0 \\
1 & -1 & 0 & -1 & 0 & 0 & 0 & -1 & 1
\end{bmatrix}. \tag{1.22}
$$

Hence, the Euler quadratic form $q_C : K_0(C) \longrightarrow \mathbb{Z}$ of C is given by the formula

$$
\begin{aligned}
q_C(x) &= x^t (\mathbf{C}_C^{-1})^t x = x_0^2 + x_{a_1}^2 + x_{b_1}^2 + x_{b_2}^2 + x_{c_1}^2 + x_{c_2}^2 + x_{c_3}^2 + x_{c_4}^2 + x_\omega^2 \\
&\quad - x_0 (x_{a_1} + x_{b_1} + x_{c_1} - x_\omega) - x_{a_1} x_\omega \\
&\quad - x_{b_1} x_{b_2} - x_{b_2} x_\omega - x_{c_1} x_{c_2} - x_{c_2} x_{c_3} - x_{c_3} x_{c_4} - x_{c_4} x_\omega \\
&= (x_0 - \tfrac{1}{2} x_{a_1} - \tfrac{1}{2} x_{b_1} - \tfrac{1}{2} x_{c_1} + \tfrac{1}{2} x_\omega)^2 + \tfrac{3}{4}(x_{a_1} - \tfrac{1}{3} x_{c_1} - \tfrac{1}{3} x_{b_1} - \tfrac{1}{3} x_\omega)^2 \\
&\quad + \tfrac{2}{3}(x_{b_1} - \tfrac{3}{4} x_{b_2} - \tfrac{1}{2} x_{c_1} + \tfrac{1}{4} x_\omega)^2 + \tfrac{5}{8}(x_{b_2} - \tfrac{3}{5} x_\omega - \tfrac{2}{5} x_{c_1})^2 \\
&\quad + \tfrac{2}{5}(x_{c_1} - \tfrac{5}{4} x_{c_2} + \tfrac{1}{4} x_\omega)^2 + \tfrac{3}{8}(x_{c_2} - \tfrac{4}{3} x_{c_4} + \tfrac{1}{3} x_\omega)^2 \\
&\quad + \tfrac{1}{3}(x_{c_4} - \tfrac{3}{2} x_{c_3} + \tfrac{1}{2} x_\omega)^2 + \tfrac{1}{4}(x_{c_3} - x_\omega)^2.
\end{aligned}
$$

It follows that $q_C(\mathbf{h}_C) = 0$ and, given $\mathbf{x} \in \operatorname{rad} q_C$, there is an integer m such that $\mathbf{x} = m \cdot \mathbf{h}_C$. This shows that $\operatorname{rad} q_C = \mathbb{Z} \cdot \mathbf{h}_C$, and we are done. This completes the proof of the proposition. $\qquad\square$

XII.2. Regular modules and tubes over canonical algebras of Euclidean type

Our next aim is to give a precise description of connected components of the Auslander–Reiten quiver $\Gamma(\operatorname{mod} C)$ of any canonical algebra C of Euclidean type.

It follows from (VIII.4.5), (XI.3.4), and the preceding propositions that, if C is a canonical algebra of Euclidean type, then $\Gamma(\operatorname{mod} C)$ consists of the following three types of connected components:

- the postprojective component $\mathcal{P}(C)$ containing all the indecomposable projective modules,
- the preinjective component $\mathcal{Q}(C)$ containing all the indecomposable injective modules, and
- the family $\mathcal{R}(C)$ of pairwise orthogonal standard stable tubes.

Moreover, we know from (XI.3.4) that the full subcategory add $\mathcal{R}(C)$ of mod C is abelian and serial. In the proofs of (1.5), (1.8), (1.11) and (1.14) we also have indicated the structure of the components $\mathcal{P}(C)$ and $\mathcal{Q}(C)$.

To describe a detailed structure of the components in regular part $\mathcal{R}(C)$ of C, we establish a relationship between the regular C-modules and the regular modules over the Kronecker algebra $C(1,1)$.

Let C be a canonical algebra of Euclidean type. Denote by A the algebra eCe defined by the idempotent $e = e_0 + e_\omega \in C$, where e_0 and e_ω are the primitive idempotents corresponding to the vertices 0 and ω. It is easy to see that there is a K-algebra isomorphism

$$A = eCe \cong \begin{bmatrix} K & 0 \\ K^2 & K \end{bmatrix} \tag{2.1}$$

of A with the Kronecker algebra given by the Kronecker quiver

$$0 \xleftarrow{} \omega.$$

We may then consider the idempotent embedding functors (see (I.6.6))

$$\operatorname{mod} A \xleftrightarrow[\operatorname{res}_e]{T_e, L_e} \operatorname{mod} C$$

given by the formulae

$$\operatorname{res}_e(X) = Xe, \quad T_e(Y) = Y \otimes_A eC, \quad L_e(Y) = \operatorname{Hom}_A(Ce, Y)$$

for $X \in \operatorname{mod} C$ and $Y \in \operatorname{mod} A$. By (I.6.8), res_e is the restriction functor, T_e and L_e are full and faithful embeddings, L_e is right adjoint to res_e, T_e is left adjoint to res_e, and $\operatorname{res}_e T_e \cong 1_{\operatorname{mod} A} \cong \operatorname{res}_e L_e$.

We also need the following two lemmata.

2.2. Lemma. *Let C be a canonical algebra of Euclidean type and let $A = eCe$, where $e = e_0 + e_\omega$. Let X be an indecomposable C-module.*

(a) *If $X \in \mathcal{P}(C)$, then $\operatorname{res}_e(X) \in \operatorname{add} \mathcal{P}(A)$.*

(b) *If $X \in \mathcal{Q}(C)$, then $\operatorname{res}_e(X) \in \operatorname{add} \mathcal{Q}(A)$.*

Proof. (a) Assume that $X \in \mathcal{P}(C)$. To prove that $\operatorname{res}_e X \in \operatorname{add} \mathcal{P}(A)$, it is enough to show that every indecomposable direct summand of $\operatorname{res}_e(X)$ has only finitely many indecomposable predecessors in mod A. Let

$$M_r \longrightarrow M_{r-1} \longrightarrow \cdots \longrightarrow M_1$$

be a path in $\operatorname{mod} A$ with $\operatorname{Hom}_A(M_1, \operatorname{res}_e(X)) \neq 0$. Then we have a path

$$T_e(M_r) \longrightarrow T_e(M_{r-1}) \longrightarrow \cdots \longrightarrow T_e(M_1)$$

in $\operatorname{mod} C$ with $\operatorname{Hom}_C(T_e(M_1), X) \cong \operatorname{Hom}_A(M_1, \operatorname{res}_e(X)) \neq 0$. Because X has only finitely many predecessors in $\operatorname{mod} C$ and any indecomposable A-module M is of the form $M \cong \operatorname{res}_e(T_e(M))$, the required claim follows. Hence the A-module $\operatorname{res}_e(X)$ lies in $\operatorname{add} \mathcal{P}(A)$.

(b) If $X \in \mathcal{Q}(C)$ then, applying the functor L_e, we show similarly that every indecomposable direct summand of $\operatorname{res}_e X$ has only finitely many successors in $\operatorname{mod} A$, and consequently the A-module $\operatorname{res}_e(X)$ lies in $\operatorname{add} \mathcal{Q}(A)$. $\quad\square$

We prove later that $X \in \mathcal{R}(C)$ also implies $\operatorname{res}_e(X) \in \operatorname{add} \mathcal{R}(A)$.

2.3. Lemma. *Let C be a canonical algebra of Euclidean type, let $\mathbf{h}_C \in K_0(C)$ be the canonical radical vector of C (1.17), and let $A = eAe$, where $e = e_0 + e_\omega$. If E is a simple regular C-module satisfying the following three conditions:*

- *$\dim E = m \cdot \mathbf{h}_C$ for some $m \geq 1$,*
- *$\operatorname{res}_e(E) \in \operatorname{add} \mathcal{R}(A)$, and*
- *$\operatorname{Hom}_C(E, M) = 0$, for any indecomposable C-module M, with $\operatorname{res}_e M = 0$,*

then $\dim E = \mathbf{h}_C$.

Proof. Because the functor $L_e : \operatorname{mod} A \longrightarrow \operatorname{mod} C$ is right adjoint to the restriction functor $\operatorname{res}_e : \operatorname{mod} C \longrightarrow \operatorname{mod} A$, then there is a functorial isomorphism of K-vector spaces

$$\operatorname{Hom}_C(E, L_e(\operatorname{res}_e(E))) \xrightarrow[\simeq]{\psi} \operatorname{Hom}_A(\operatorname{res}_e(E), \operatorname{res}_e(E)).$$

Take the unique homomorphism $\eta : E \longrightarrow L_e(\operatorname{res}_e(E))$ in $\operatorname{mod} C$ such that $\psi(\eta)$ is the identity homomorphism of $\operatorname{res}_e(E)$. Clearly, $\eta \neq 0$.

We claim that $L_e(\operatorname{res}_e(E))$ belongs to $\operatorname{add} \mathcal{R}(C)$. To prove it, suppose to the contrary that $L_e(\operatorname{res}_e(E))$ admits a postprojective indecomposable direct summand P. Then, by (2.2), the module $\operatorname{res}_e(P)$ is a postprojective direct summand of $\operatorname{res}_e(E)$, and we get a contradiction with our assumption. Similarly, we prove that the module $L_e(\operatorname{res}_e(E))$ has no preinjective direct summands. Therefore, $L_e(\operatorname{res}_e(E))$ is a regular C-module, and consequently $\eta : E \longrightarrow L_e(\operatorname{res}_e(E))$ is a monomorphism, because E is a simple object of the abelian category $\operatorname{add} \mathcal{R}(C)$.

Consider now the induced exact sequence

$$0 \longrightarrow E \xrightarrow{\eta} L_e(\operatorname{res}_e(E)) \xrightarrow{\xi} M \longrightarrow 0,$$

where $M = \operatorname{Coker}\eta$. Because the functor res_e is exact and $\operatorname{res}_e(\eta)$ is the identity on $\operatorname{res}_e(E)$, we get $\operatorname{res}_e M = 0$. Hence, invoking our assumption, we have $\operatorname{Hom}_C(E, M) = 0$. Take now $\varphi \in \operatorname{End}_A(\operatorname{res}_e(E))$. Because $\xi L_e(\varphi)\eta = 0$, we get a commutative diagram with exact rows

$$
\begin{array}{ccccccccc}
0 & \longrightarrow & E & \xrightarrow{\;\eta\;} & L_e(\operatorname{res}_e(E)) & \xrightarrow{\;\xi\;} & M & \longrightarrow & 0 \\
 & & \varphi' \downarrow & & \downarrow L_e(\varphi) & & \downarrow \varphi'' & & \\
0 & \longrightarrow & E & \xrightarrow{\;\eta\;} & L_e(\operatorname{res}_e(E)) & \xrightarrow{\;\xi\;} & M & \longrightarrow & 0.
\end{array}
$$

Suppose that $\varphi' = 0$. Then $L_e(\varphi) = \varrho\xi$, for some $\varrho : M \longrightarrow L_e(\operatorname{res}_e(E))$. But

$$\operatorname{Hom}_C(M, L_e(\operatorname{res}_e(E))) \cong \operatorname{Hom}_A(\operatorname{res}_e(M), \operatorname{res}_e(E)) = 0,$$

and so $L_e(\varphi) = 0$. Then $\varphi = 0$, because the functor $L_e : \operatorname{mod} A \longrightarrow \operatorname{mod} C$ is a full and faithful embedding. Therefore, we get an injective algebra homomorphism

$$\operatorname{End}_A(\operatorname{res}_e(E)) \longrightarrow \operatorname{End}_C(E)$$

which assigns to $\varphi \in \operatorname{End}_A(\operatorname{res}_A(E))$ the endomorphism $\varphi' \in \operatorname{End}_C(E)$, defined above. Because E is a simple object of $\operatorname{add}\mathcal{R}(C)$, we know that E is a brick. Hence $\operatorname{res}_e(E)$ is a brick. Finally, observe that the hypothesis $\mathbf{dim}\, E = m \cdot \mathbf{h}_C$, where $m \geq 1$, implies $\mathbf{dim}\,\operatorname{res}_e(E) = [m, m]^t$. By (XI.4.5), $\operatorname{res}_e(E)$ is isomorphic to the Kronecker module $E_\lambda[m]$ for some $\lambda \in \mathbb{P}_1(K)$. Because the module $\operatorname{res}_e(E)$ is a brick, then $m = 1$ and, consequently, $\mathbf{dim}\, E = \mathbf{h}_C$. $\qquad\square$

2.4. An identification

(a) Throughout, we identify any $C(p, q)$-module M with the K-linear representation

$$
\begin{array}{c}
M_{a_1} \xleftarrow{\varphi_{\alpha_2}} M_{a_2} \xleftarrow{\quad} \cdots \xleftarrow{\varphi_{\alpha_{p-1}}} M_{a_{p-1}} \\
\end{array}
$$

with vertices M_0, M_ω and arrows φ_{α_1}, φ_{α_p}, φ_{β_1}, φ_{β_q}, and

$$
M_{b_1} \xleftarrow{\varphi_{\beta_2}} M_{b_2} \xleftarrow{\quad} \cdots \xleftarrow{\varphi_{\beta_{q-1}}} M_{b_{q-1}}
$$

of the quiver $\Delta(p, q)$, see (1.1).

(b) Similarly, we identify any $C(p, q, r)$-module M with the K-linear representation

$$
M_{a_1} \xleftarrow{\varphi_{\alpha_2}} M_{a_2} \xleftarrow{\quad} \cdots \xleftarrow{\varphi_{\alpha_{p-1}}} M_{a_{p-1}}
$$

$$
M_0 \xleftarrow{\varphi_{\beta_1}} M_{b_1} \xleftarrow{\varphi_{\beta_2}} M_{b_2} \xleftarrow{\quad} \cdots \xleftarrow{\varphi_{\beta_{q-1}}} M_{b_{q-1}} \xleftarrow{\varphi_{\beta_q}} M_\omega
$$

$$
M_{c_1} \xleftarrow{\varphi_{\gamma_2}} M_{c_2} \xleftarrow{\quad} \cdots \xleftarrow{\varphi_{\gamma_{r-1}}} M_{c_{r-1}}
$$

with arrows φ_{α_1}, φ_{α_p}, φ_{γ_1}, φ_{γ_r}.

of the quiver $\Delta(p, q, r)$ satisfying the relation (see (1.1))

$$\varphi_{\alpha_1} \cdots \varphi_{\alpha_p} + \varphi_{\beta_1} \cdots \varphi_{\beta_q} + \varphi_{\gamma_1} \cdots \varphi_{\gamma_r} = 0.$$

(c) Denote by C one of the algebras $C(p, q)$ or $C(p, q, r)$. Let $e = e_0 + e_\omega \in C$, and let $A = eCe \cong \begin{bmatrix} K & 0 \\ K^2 & K \end{bmatrix}$ be the Kronecker algebra, see (2.1). Then, under the preceding identifications, the restriction functor

$$\mathrm{res}_e : \mathrm{mod}\, C \longrightarrow \mathrm{mod}\, A$$

assigns to each C-module M, viewed as a K-linear representation, the A-module

$$\mathrm{res}_e(M) = \left(M_0 \underset{\varphi_{\beta_1} \varphi_{\beta_2} \cdots \varphi_{\beta_p}}{\overset{\varphi_{\alpha_1} \varphi_{\alpha_2} \cdots \varphi_{\alpha_p}}{\rightleftarrows}} M_\omega \right),$$

viewed as a K-linear representation of the Kronecker quiver

$$1 \; \circ \; \rightleftarrows \; \circ \; 2.$$

We exhibit now a class of indecomposable modules over the canonical algebras of Euclidean type and show that they are periodic with respect to the action of the Auslander–Reiten translation $\tau_C = D\mathrm{Tr}$ of C, and, as a consequence, they belong to stable tubes. To formulate the results we need a family of canonical forms of modules presented in the following table.

2.5. Table

Indecomposable $C(p, q)$-modules of the dimension vector $\leq \mathbf{h}_C$

Let p and q be integers such that $q \geq p \geq 1$. Consider the following family of indecomposable $C(p, q)$-modules:

(1) $E_i^{(\infty)} = S(a_i)$, with $1 \leq i \leq p-1$, $E_p^{(\infty)}$:

(2) $E_j^{(0)} = S(b_j)$, with $1 \leq j \leq q - 1$, $E_q^{(0)}$:

(3) *for each* s, t *such that* $1 \le s \le p$ *and* $1 \le t \le q$, *the modules*

$$K \xleftarrow{\ 1\ } \cdots \xleftarrow{\ 1\ } K \xleftarrow{\varphi_{\alpha_s} = 0} K \xleftarrow{\ 1\ } \cdots \xleftarrow{\ 1\ } K$$

$$R^{\alpha_s} : \quad K \qquad\qquad\qquad\qquad\qquad\qquad K,$$

$$K \xleftarrow{\ 1\ } K \xleftarrow{\quad} \cdots \cdots \xleftarrow{\quad} K \xleftarrow{\ 1\ } K$$

$$K \xleftarrow{\ 1\ } K \xleftarrow{\quad} \cdots \cdots \xleftarrow{\quad} K \xleftarrow{\ 1\ } K$$

$$R^{\beta_t} : \quad K \qquad\qquad\qquad\qquad\qquad\qquad K,$$

$$K \xleftarrow{\ 1\ } \cdots \xleftarrow{\ 1\ } K \xleftarrow{\varphi_{\beta_t} = 0} K \xleftarrow{\ 1\ } \cdots \xleftarrow{\ 1\ } K$$

(4) *for each* $\lambda \in K \setminus \{0\}$, *the modules*

$$K \xleftarrow{\ 1\ } K \xleftarrow{\ 1\ } \cdots \xleftarrow{\ 1\ } K$$

$$E^{(\lambda)} : \quad K \qquad\qquad\qquad\qquad K.$$

$$K \xleftarrow{\ 1\ } K \xleftarrow{\quad} \cdots \xleftarrow{\ 1\ } K$$

We now state the following useful fact.

2.6. Lemma. *Let* p *and* q *be integers such that* $q \ge p \ge 1$, *let* $C = C(p, q)$ *and let* $\mathbf{h}_C \in K_0(C)$ *be the canonical radical vector of* C. *The* C-*modules* R^{α_s}, *with* $1 \le s \le p$, R^{β_t}, *with* $1 \le t \le q$, *and the modules* $E^{(\lambda)}$, *with* $\lambda \in K \setminus \{0\}$, *presented in Table (2.5), form a complete set of pairwise non-isomorphic indecomposable* C-*modules of the dimension vector* $\mathbf{h}_C$.

Proof. (a) Because the vector $\mathbf{h}_C$ has the identity coordinate over each vertex of the quiver $\Delta(p, q)$ of C then any module M (viewed as a K-linear representation of $\Delta(p, q)$), with $\dim M = \mathbf{h}_C$, has $M_j \cong K$ for all points j of $\Delta(p, q)$. Hence the lemma easily follows. □

Assume that $C = C(p, q)$ and let $\mathbb{P}_1(K) = K \cup \{\infty\}$ be the projective line over K. Consider the following $\mathbb{P}_1(K)$-family

$$\boldsymbol{\mathcal{T}}^C = \{\mathcal{T}_\lambda^C\}_{\lambda \in \mathbb{P}_1(K)} \qquad\qquad (2.7)$$

of connected components of the Auslander–Reiten quiver $\Gamma(\operatorname{mod} C)$ of C, where

- $\mathcal{T}_\infty^C$ is the component containing $E^{(\infty)} = E_p^{(\infty)}$,
- $\mathcal{T}_0^C$ is the component containing $E^{(0)} = E_q^{(0)}$, and
- for each $\lambda \in K \setminus \{0\}$, $\mathcal{T}_\lambda^C$ is the component containing the module $E^{(\lambda)}$.

The following proposition shows that the $\mathbb{P}_1(K)$-family $\boldsymbol{\mathcal{T}}^C$ consists of stable tubes lying in $\mathcal{R}(C)$.

2.8. Proposition. *Let p and q be integers such that $q \geq p \geq 1$, let $C = C(p,q)$ and let $\boldsymbol{\mathcal{T}}^C = \{\mathcal{T}_\lambda^C\}_{\lambda \in \mathbb{P}_1(K)}$ be the $\mathbb{P}_1(K)$-family (2.7).*

(a) *The component $\mathcal{T}_\infty^C$ of $\Gamma(\operatorname{mod} C)$ is a standard stable tube in $\mathcal{R}(C)$ of rank $r_\infty^C = p$ and, for each $i \in \{1, \dots, p\}$, there is an isomorphism $\tau_C E_{i+1}^{(\infty)} \cong E_i^{(\infty)}$, where we set $E_{p+1}^{(\infty)} = E_1^{(\infty)}$.*

(b) *The component $\mathcal{T}_0^C$ of $\Gamma(\operatorname{mod} C)$ is a standard stable tube in $\mathcal{R}(C)$ of rank $r_0^C = q$ and, for each $j \in \{1, \dots, q\}$, there is an isomorphism $\tau_C E_{j+1}^{(0)} \cong E_j^{(0)}$, where we set $E_{q+1}^{(0)} = E_1^{(0)}$.*

(c) *For each $\lambda \in K \setminus \{0\}$, the component $\mathcal{T}_\lambda^C$ of $\Gamma(\operatorname{mod} C)$ is a standard stable tube in $\mathcal{R}(C)$ of rank $r_\lambda^C = 1$ and there is an isomorphism $\tau_C E^{(\lambda)} \cong E^{(\lambda)}$.*

(d) *Each tube in $\mathcal{R}(C) \setminus (\mathcal{T}_\infty^C \cup \mathcal{T}_0^C)$ is of rank 1.*

(e) *The C-modules $E_1^{(\infty)}, \dots, E_p^{(\infty)}, E_1^{(0)}, \dots, E_q^{(0)}$ and $E^{(\lambda)}$, with $\lambda \in K \setminus \{0\}$, are simple regular.*

Proof. By a direct calculation of the Auslander–Reiten translations of the modules

$$E_1^{(\infty)}, \dots, E_p^{(\infty)}, E_1^{(0)}, \dots, E_q^{(0)}, \text{ and } E_j^{(\lambda)}, \text{ with } \lambda \in K \setminus \{0\},$$

we easily get the isomorphisms $\tau_C E_{i+1}^{(\infty)} \cong E_i^{(\infty)}$, $\tau_C E_{j+1}^{(0)} \cong E_j^{(0)}$ and $\tau_C E^{(\lambda)} \cong E^{(\lambda)}$ as follows.

First, we prove that $\tau_C E^{(\lambda)} \cong E^{(\lambda)}$, for any $\lambda \neq 0$. Note that, for each $\lambda \neq 0$, there exists an exact sequence in $\operatorname{mod} C$

$$0 \longrightarrow e_0 C \overset{u}{\longrightarrow} e_\omega C \overset{\pi}{\longrightarrow} E^{(\lambda)} \longrightarrow 0.$$

We know from (III.2.4) that the indecomposable projective C-module $e_\omega C$, viewed as a representation of the quiver $\Delta(p,q)$, is given by the diagram

$$e_\omega C: \quad
\begin{array}{c}
K \overset{1}{\longleftarrow} K \overset{1}{\longleftarrow} \cdots \overset{1}{\longleftarrow} K \\
{\scriptstyle \left[\begin{smallmatrix}1\\0\end{smallmatrix}\right]} \nearrow \qquad\qquad\qquad \nwarrow {\scriptstyle 1} \\
K^2 \qquad\qquad\qquad\qquad K. \\
{\scriptstyle \left[\begin{smallmatrix}0\\1\end{smallmatrix}\right]} \nwarrow \qquad\qquad\qquad \swarrow {\scriptstyle 1} \\
K \underset{1}{\longleftarrow} K \underset{1}{\longleftarrow} \cdots \underset{1}{\longleftarrow} K
\end{array}$$

The epimorphism $\pi = (\pi_j) : e_\omega C \longrightarrow E^{(\lambda)}$ is defined by setting $\pi_j = 1_K$, for $j \neq 0$, and $\pi_0 = [1 \ \ \lambda] : K^2 \longrightarrow K$. Further, $u : e_0 C \longrightarrow e_\omega C$ is given by the obvious isomorphism $e_0 C \cong \operatorname{Ker} \pi$. The homomorphism u induces the commutative diagram (with exact rows)

$$
\begin{array}{ccccccccc}
0 & \longrightarrow & \operatorname{Hom}_C(e_\omega C, C) & \xrightarrow{\operatorname{Hom}_C(u,C)} & \operatorname{Hom}_C(e_0 C, C) & \longrightarrow & \operatorname{Tr} E^{(\lambda)} & \longrightarrow & 0 \\
& & f_\omega \downarrow {\scriptstyle \cong} & & f_0 \downarrow {\scriptstyle \cong} & & & & \\
0 & \longrightarrow & C e_\omega & \xrightarrow{\ \ v\ \ } & C e_0 & & \longrightarrow & \operatorname{Coker} v \longrightarrow 0
\end{array}
$$

of left C-modules, where v is the diagonal embedding of the simple left C-module Ce_ω into the socle soc $Ce_0 \cong Ce_\omega \oplus Ce_\omega$ of Ce_0 and the vertical homomorphisms f_ω and f_0 are the bijective functorial maps given by (I.4.2). Hence, there is an isomorphism $f : \text{Tr } E^{(\lambda)} \xrightarrow{\simeq} \text{Coker } v$ induced by f_0. On the other hand, it is easy to see that Coker $v \cong DE^{(\lambda)}$. Hence we get the isomorphisms $\tau_C E^{(\lambda)} = D\text{Tr } E^{(\lambda)} \cong D\text{Coker } v \cong DDE^{(\lambda)} \cong E^{(\lambda)}$, and we are done.

Next, we prove that, for each $j \in \{1, \ldots, q\}$, there is an isomorphism $\tau_C E_{j+1}^{(0)} \cong E_j^{(0)}$. We consider three cases.

<u>Case 1°</u> $j = q - 1$. By the arguments given above, there exists an exact sequence in mod C

$$0 \longrightarrow e_{b_{q-1}}C \xrightarrow{u} e_\omega C \longrightarrow E_q^{(0)} \longrightarrow 0$$

and the homomorphism u induces the commutative diagram (with exact rows)

$$
\begin{array}{ccccccccc}
0 & \longrightarrow & \text{Hom}_C(e_\omega C, C) & \xrightarrow{\text{Hom}_C(u,C)} & \text{Hom}_C(e_{b_{q-1}}C, C) & \longrightarrow & \text{Tr } E_q^{(0)} & \longrightarrow & 0 \\
& & f_1 \downarrow \cong & & f_0 \downarrow \simeq & & f \downarrow \cong & & \\
0 & \longrightarrow & Ce_\omega & \xrightarrow{v} & Ce_{b_{q-1}} & \longrightarrow & \text{Coker } v & \longrightarrow & 0
\end{array}
$$

of left C-modules, where the vertical homomorphisms f_1 and f_0 are the bijective functorial maps given by (I.4.2), whereas $f : \text{Tr } E_q^{(0)} \xrightarrow{\simeq} \text{Coker } v$ is the isomorphism induced by f_0. On the other hand, it is easy to see that Coker $v \cong DS(b_{q-1}) = DE_{q-1}^{(0)}$. Hence we get the isomorphisms

$$\tau_C E_q^{(0)} = D\text{Tr } E_q^{(0)} \cong D\text{Coker } v \cong DDE_{q-1}^{(0)} \cong E_{q-1}^{(0)},$$

and we are done.

<u>Case 2°</u> $1 < j < q - 1$. By the arguments given above, there exists an exact sequence in mod C

$$0 \longrightarrow e_{b_{j-1}}C \xrightarrow{u} e_{b_j}C \longrightarrow S(b_j) \longrightarrow 0$$

and the homomorphism u induces the commutative diagram (with exact rows)

$$
\begin{array}{ccccccccc}
0 & \longrightarrow & \text{Hom}_C(e_{b_j}C, C) & \xrightarrow{\text{Hom}_C(u,C)} & \text{Hom}_C(e_{b_{j-1}}C, C) & \longrightarrow & \text{Tr } S(b_j) & \longrightarrow & 0 \\
& & f_1 \downarrow \cong & & f_0 \downarrow \simeq & & f \downarrow \cong & & \\
0 & \longrightarrow & Ce_{b_j} & \xrightarrow{v} & Ce_{b_{j-1}} & \longrightarrow & \text{Coker } v & \longrightarrow & 0
\end{array}
$$

of left C-modules, where the vertical homomorphisms f_1 and f_0 are bijective functorial maps, whereas $f : \mathrm{Tr}\, S(b_j) \xrightarrow{\;\simeq\;} \mathrm{Coker}\, v$ is the isomorphism induced by f_0. On the other hand, it is easy to see that $\mathrm{Coker}\, v \cong DS(b_{j-1}) = DE_{j-1}^{(0)}$. Hence we get a sequence of isomorphisms

$$\tau_C E_j^{(0)} = D\mathrm{Tr}\, E_j^{(0)} = D\mathrm{Tr}\, S(b_j) \cong D\mathrm{Coker}\, v \cong DDE_{j-1}^{(0)} \cong E_{j-1}^{(0)},$$

and we are done.

Case 3° $\;1 = j < q - 1$. By the arguments given above, there exists an exact sequence in $\mathrm{mod}\, C$

$$0 \longrightarrow e_0 C \xrightarrow{\;u\;} e_{b_1} C \longrightarrow S(b_1) \longrightarrow 0$$

and the homomorphism u induces the commutative diagram, with exact rows,

$$
\begin{array}{ccccccccc}
0 & \longrightarrow & \mathrm{Hom}_C(e_{b_1}C, C) & \xrightarrow{\mathrm{Hom}_C(u,C)} & \mathrm{Hom}_C(e_0C, C) & \longrightarrow & \mathrm{Tr}\, S(b_1) & \longrightarrow & 0 \\
& & \Big\downarrow{\scriptstyle f_1}\;{\scriptstyle\cong} & & \Big\downarrow{\scriptstyle f_0}\;{\scriptstyle\cong} & & \Big\downarrow{\scriptstyle f}\;{\scriptstyle\cong} & & \\
0 & \longrightarrow & Ce_{b_1} & \xrightarrow{\;v\;} & Ce_0 & \longrightarrow & \mathrm{Coker}\, v & \longrightarrow & 0
\end{array}
$$

of left C-modules, where the vertical homomorphisms f_1 and f_0 are bijective functorial maps, whereas $f : \mathrm{Tr}\, S(b_1) \xrightarrow{\;\simeq\;} \mathrm{Coker}\, v$ is the isomorphism induced by f_0. On the other hand, it is easy to see that $\mathrm{Coker}\, v \cong DE_q^{(0)}$. Hence we get a sequence of isomorphisms

$$\tau_C E_{q+1}^{(0)} = \tau_C E_1^{(0)} = D\mathrm{Tr}\, E_1^{(0)} = D\mathrm{Tr}\, S(b_1) \cong D\mathrm{Coker}\, v \cong DDE_q^{(0)} \cong E_q^{(0)},$$

and we are done.

The proof of the existence of an isomorphism $\tau_C E_{i+1}^{(\infty)} \cong E_i^{(\infty)}$, for each $i \in \{1, \dots, p\}$, is similar to the above one, and we leave it as an exercise.

This shows that the modules $E_1^{(\infty)}, \dots, E_p^{(\infty)}, E_1^{(0)}, \dots, E_q^{(0)}$ and $E_j^{(\lambda)}$, with $\lambda \in K \setminus \{0\}$, are τ_C-periodic. It follows that they are neither postprojective nor preinjective, hence regular and, consequently, the family $\boldsymbol{\mathcal{T}}^C = \{\mathcal{T}_\lambda^C\}_{\lambda \in \mathbb{P}_1(K)}$ consists of regular modules.

Because $C = C(p, q)$ is a hereditary algebra of the Euclidean type $\widetilde{\mathbb{A}}_{p+q-1}$ then, by (XI.2.8) and (XI.2.9), the components of $\mathcal{R}(C)$ form a family $\boldsymbol{\mathcal{T}} = \{\mathcal{T}_\lambda\}_{\lambda \in \Lambda}$, of pairwise orthogonal standard stable tubes, and if we let r_λ be the rank of $\mathcal{T}_\lambda$, then $\sum_{\lambda \in \Lambda} (r_\lambda - 1) \leq p + q - 2$, where $p + q = n$ is the rank of the group $K_0(C)$. It follows that, for each $\lambda \in \mathbb{P}_1(K)$, the component $\mathcal{T}_\lambda^C$ is a standard stable tube.

Because $q \geq p \geq 1$ and we have constructed above the stable tube $\mathcal{T}_\infty^C$ of rank $r_\infty^C = p$, the stable tube $\mathcal{T}_0^C$ of rank $r_0^C = q$ and, clearly,

$$r_\infty^C - 1 + r_0^C - 1 = (p-1) + (q-1) = p + q - 2,$$

then the remaining tubes of $\Gamma(\mathrm{mod}\, C)$ are of rank 1.

Because the regular C-modules $E_1^{(\infty)}, \ldots, E_{p-1}^{(\infty)}, E_1^{(0)}, \ldots, E_{q-1}^{(0)}$ are simple, then it remains to show that the remaining regular C-modules $E_p^{(\infty)}$, $E_q^{(0)}$ and $E^{(\lambda)}$, with $\lambda \in K \setminus \{0\}$, are simple regular. Let E be one of them, and let X be a proper submodule of E. It follows that $X_\omega = 0$ and a direct calculation shows that X is preprojective, hence it is not regular. This shows that E is simple regular and finishes the proof of the proposition. $\square$

2.9. Table

Indecomposable $C(p,q,r)$-modules of the dimension vector $\leq \mathbf{h}_C$

Let now (p,q,r) be any of the triples $(2,2,m-2)$, with $m \geq 4$, $(2,3,3)$, $(2,3,4)$ or $(2,3,5)$. Consider the following family of indecomposable $C(p,q,r)$-modules:

(1) $E_1^{(\infty)} = S(a_1)$ *and* $E_2^{(\infty)} : $

(2) $E_j^{(0)} = S(b_j)$, *with* $1 \leq j \leq q-1$, *and* $E_q^{(0)} : $

(3) $E_k^{(1)} = S(c_k)$, *with* $1 \leq k \leq r-1$, *and* $E_r^{(1)} : $

(4) *for* s, t, m *such that* $1 \leq s \leq 2$, $1 \leq t \leq q$ *and* $1 \leq m \leq r$, *the modules*

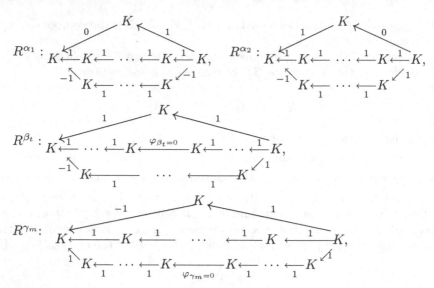

(5) *for* $\lambda \in K \setminus \{0,1\}$, *the modules*

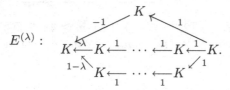

We also state the following useful fact.

2.10. Lemma. *Let* (p, q, r) *be any of the triples* $(2, 2, m-2)$, *with* $m \geq 4$, $(2, 3, 3)$, $(2, 3, 4)$, *or* $(2, 3, 5)$, *let* $C = C(p, q, r)$ *and let* $\mathbf{h}_C \in K_0(C)$ *be the canonical radical vector of* C. *The* C-*modules* R^{α_s}, *with* $1 \leq s \leq p$, R^{β_t}, *with* $1 \leq t \leq q$, R^{γ_m}, *with* $1 \leq m \leq r$, *and* $E^{(\lambda)}$, *with* $\lambda \in K \setminus \{0, 1\}$, *presented in* (2.9), *form a complete set of pairwise non-isomorphic indecomposable* C-*modules of the dimension vector* $\mathbf{h}_C$.

Proof. Apply the arguments used in the proof of (2.6). $\qquad\square$

Assume now that (p, q, r) is any of the triples $(2, 2, m - 2)$, with $m \geq 4$, $(2, 3, 3)$, $(2, 3, 4)$, or $(2, 3, 5)$, and $C = C(p, q, r)$. Consider the following $\mathbb{P}_1(K)$-family

$$\boldsymbol{\mathcal{T}}^C = \{\mathcal{T}^C_\lambda\}_{\lambda \in \mathbb{P}_1(K)} \qquad (2.11)$$

of connected components of the Auslander–Reiten quiver $\Gamma(\mathrm{mod}\, C)$ of C, where

- $\mathcal{T}^C_\infty$ is the component containing $E^{(\infty)} = E_p^{(\infty)} = E_2^{(\infty)}$,
- $\mathcal{T}^C_0$ is the component containing $E^{(0)} = E_q^{(0)}$,

- $\mathcal{T}_1^C$ is the component containing $E^{(1)} = E_r^{(1)}$, and
- for each $\lambda \in K \backslash \{0, 1\}$, $\mathcal{T}_\lambda^C$ is the component containing the module $E^{(\lambda)}$.

The following proposition shows that the $\mathbb{P}_1(K)$-family (2.11) consists of stable tubes lying in $\mathcal{R}(C)$.

2.12. Proposition. *Let (p, q, r) be any of the triples $(2, 2, m - 2)$, with $m \geq 4$, $(2, 3, 3)$, $(2, 3, 4)$, or $(2, 3, 5)$, let $C = C(p, q, r)$ and let $\mathcal{T}^C = \{\mathcal{T}_\lambda^C\}_{\lambda \in \mathbb{P}_1(K)}$ be the $\mathbb{P}_1(K)$-family (2.11).*

- (a) *The component $\mathcal{T}_\infty^C$ of $\Gamma(\operatorname{mod} C)$ is a standard stable tube in $\mathcal{R}(C)$ of rank $r_\infty^C = 2$ and there are isomorphisms $\tau_C E_2^{(\infty)} \cong E_1^{(\infty)}$ and $\tau_C E_1^{(\infty)} \cong E_2^{(\infty)}$.*
- (b) *The component $\mathcal{T}_0^C$ of $\Gamma(\operatorname{mod} C)$ is a standard stable tube in $\mathcal{R}(C)$ of rank $r_0^C = q$ and, for each $j \in \{1, \dots, q\}$, there is an isomorphism $\tau_C E_{j+1}^{(0)} \cong E_j^{(0)}$, where we set $E_{q+1}^{(0)} = E_1^{(0)}$.*
- (c) *The component $\mathcal{T}_1^C$ of $\Gamma(\operatorname{mod} C)$ is a standard stable tube in $\mathcal{R}(C)$ of rank $r_1^C = r$ and, for each $t \in \{1, \dots, r\}$, there is an isomorphism $\tau_C E_{t+1}^{(1)} \cong E_t^{(1)}$, where we set $E_{r+1}^{(1)} = E_1^{(1)}$.*
- (d) *For each $\lambda \in K \backslash \{0, 1\}$, the component $\mathcal{T}_\lambda^C$ of $\Gamma(\operatorname{mod} C)$ is a standard stable tube in $\mathcal{R}(C)$ of rank $r_\lambda^C = 1$ and there is an isomorphism $\tau_C E^{(\lambda)} \cong E^{(\lambda)}$.*
- (e) *The tubes $\mathcal{T}_\infty^C$, $\mathcal{T}_0^C$ and $\mathcal{T}_1^C$ of rank p, q and r, respectively, are the only non-homogeneous tubes in $\mathcal{R}(C)$.*
- (f) *The C-modules $E_1^{(\infty)}, E_2^{(\infty)}, E_1^{(0)}, \dots, E_q^{(0)}, E_1^{(1)}, \dots, E_r^{(1)},$ and $E^{(\lambda)}$, with $\lambda \in K \backslash \{0\}$, are simple regular.*

Proof. By a direct calculation of the Auslander–Reiten translation of the modules $E_1^{(\infty)}, E_2^{(\infty)}, E_1^{(0)}, \dots, E_q^{(0)}, E_1^{(1)}, \dots, E_r^{(1)}$, like in the proof of (2.8), we easily get isomorphisms

$$\tau_C E_2^{(\infty)} \cong E_1^{(\infty)}, \ \tau_C E_1^{(\infty)} \cong E_2^{(\infty)}, \ \tau_C E_{j+1}^{(0)} \cong E_j^{(0)}, \ \tau_C E_{t+1}^{(1)} \cong E_t^{(1)},$$

and $\tau_C E^{(\lambda)} \cong E^{(\lambda)}$, for each $\lambda \in K \backslash \{0, 1\}$, because the projective resolutions of these modules are very simple.

Because, by (1.5), (1.8), (1.11), and (1.14), C is a concealed algebra of Euclidean type, it follows from (XI.3.4) that the regular part $\mathcal{R}(C)$ of $\Gamma(\operatorname{mod} C)$ consists of a family $\mathcal{T} = \{\mathcal{T}_\lambda\}_{\lambda \in \Lambda}$, of pairwise orthogonal standard stable tubes and, if we let r_λ be the rank of $\mathcal{T}_\lambda$, then

$$\sum_{\lambda \in \Lambda} (r_\lambda - 1) \leq n - 2,$$

where n is the rank of the group $K_0(C)$. Further, add $\mathcal{R}(C)$ is a serial, abelian and full subcategory of $\operatorname{mod} C$ whose simple objects are the modules lying on the mouth of the tubes $\mathcal{T}_\lambda$, with $\lambda \in \Lambda$. Moreover, it follows from (VIII.4.5) that $\mathrm{pd}\, Z \leq 1$, for any module Z in $\mathcal{R}(C)$, and consequently $\mathrm{Ext}_C^2(X,Y) = 0$, for all modules X and Y in $\mathcal{R}(C)$.

A simple calculation shows that the modules
$$E_1^{(\infty)},\ E_2^{(\infty)},\ E_1^{(0)},\dots,E_q^{(0)},\ E_1^{(1)},\dots,E_r^{(1)} \text{ and } E_j^{(\lambda)}, \text{ with } \lambda \in K \setminus$$
$\{0,1\}$,
form a family of pairwise orthogonal bricks. Therefore, by applying (X.2.6), we show that

- the modules $E_1^{(\infty)}$ and $E_2^{(\infty)}$ form the mouth of a stable tube $\mathcal{T}_\infty^C$ of rank $r_\infty^C = 2$,
- the modules $E_1^{(0)}, E_2^{(0)}, \dots, E_q^{(0)}$ form the mouth of a stable tube $\mathcal{T}_0^C$ of rank $r_0^C = q$, and
- the modules $E_1^{(1)}, E_2^{(1)}, \dots, E_r^{(1)}$ form the mouth of a stable tube $\mathcal{T}_1^C$ of rank $r_1^C = r$.

Moreover, we have
$$r_\infty^C - 1 + r_1^C - 1 + r_0^C - 1 = (p-1) + (q-1) + (r-1) = n - 2.$$
Hence, by the inequality $\sum_{\lambda \in \Lambda}(r_\lambda - 1) \leq n - 2$, the remaining tubes of $\Gamma(\mathrm{mod}\, C)$ are homogeneous. Moreover, the C-modules
$$E_1^{(\infty)}, E_2^{(\infty)},\ E_1^{(0)},\dots,E_q^{(0)}, \text{ and } E_1^{(1)},\dots,E_r^{(1)}$$
are clearly simple regular. Finally, for each $\lambda \in K \setminus \{0,1\}$, the brick $E^{(\lambda)}$ forms the mouth of a stable tube $\mathcal{T}_\lambda^C$ of rank $r_\lambda^C = 1$, again by (X.2.6); hence $E^{(\lambda)}$ is simple regular. This finishes the proof. $\qquad\square$

We now give a description of all indecomposable regular modules over any canonical algebra of Euclidean type.

2.13. Theorem. *Let C be a canonical algebra of Euclidean type.*

(a) *The regular part $\mathcal{R}(C)$ of $\Gamma(\mathrm{mod}\, C)$ consists of the pairwise orthogonal stable tubes $\mathcal{T}_\lambda^C$ of the $\mathbb{P}_1(K)$-family*
$$\boldsymbol{\mathcal{T}}^C = \{\mathcal{T}_\lambda^C\}_{\lambda \in \mathbb{P}_1(K)}$$
described in (2.8) for $C = C(p,q)$, and in (2.12) for $C = C(p,q,r)$, where (p,q,r) is any of the triples $(2,2,m-2)$, with $m \geq 4$, $(2,3,3)$, $(2,3,4)$, or $(2,3,5)$. The tubes $\mathcal{T}_\infty^C$, $\mathcal{T}_0^C$ and $\mathcal{T}_1^C$ are of rank $r_\infty^C = p$, $r_0^C = q$ and $r_1^C = r$, respectively, and the remaining tubes $\mathcal{T}_\lambda^C$ are of rank $r_\lambda^C = 1$.

(b) *The C-modules R^{α_s}, with $1 \leq s \leq p$, lie in $\mathcal{T}_\infty^C$ and are of regular length p, the C-modules R^{β_t}, with $1 \leq t \leq q$, lie in $\mathcal{T}_0^C$ and have regular length q, and the C-modules R^{γ_m}, with $1 \leq m \leq r$, lie in $\mathcal{T}_1^C$ (if $C = C(p,q,r)$) and have regular length r.*

(c) *Every simple regular C-module lying in one of the tubes $\mathcal{T}_\infty^C$, $\mathcal{T}_0^C$*
and $\mathcal{T}_1^C$ is isomorphic to one of the modules

$$E_1^{(\infty)}, \ldots, E_p^{(\infty)}, \quad E_1^{(0)}, \ldots, E_q^{(0)}, \quad E_1^{(1)}, \ldots, E_r^{(1)}$$

defined in Tables (2.5) and (2.9), respectively. The remaining simple
regular C-modules are of the form $E^{(\lambda)}$, where $\lambda \in K \setminus \{0, 1\}$, up to
isomorphism.

Proof. The statement (a) follows from (c), (2.8) and (2.12).

(b) First we observe that the C-modules R^{α_s}, R^{β_t}, and R^{γ_m} are neither
postprojective nor preinjective; hence, they are regular C-modules. More-
over, we have

- $\operatorname{Hom}_C(E_s^{(\infty)}, R^{\alpha_s}) \neq 0$ and $\operatorname{Hom}_C(R^{\alpha_s}, E_{s-1}^{(\infty)}) \neq 0$, for each
  $s \in \{1, \ldots, p\}$,
- $\operatorname{Hom}_C(E_t^{(0)}, R^{\beta_t}) \neq 0$ and $\operatorname{Hom}_C(R^{\beta_t}, E_{t-1}^{(0)}) \neq 0$, for each
  $t \in \{1, \ldots, q\}$,
- $\operatorname{Hom}_C(E_m^{(1)}, R^{\gamma_m}) \neq 0$ and $\operatorname{Hom}_C(R^{\gamma_m}, E_{m-1}^{(1)}) \neq 0$, for each
  $m \in \{1, \ldots, r\}$ (if $C = C(p, q, r)$),

where $E_0^{(\infty)} = E_p^{(\infty)}$, $E_0^{(0)} = E_q^{(0)}$ and $E_0^{(1)} = E_r^{(1)}$. Hence, (b) follows.

To prove the statement (c), we consider the Kronecker algebra

$$A = eCe \cong \begin{bmatrix} K & 0 \\ K^2 & K \end{bmatrix}$$

and the restriction functor $\operatorname{res}_e : \operatorname{mod} C \longrightarrow \operatorname{mod} A$, where $e = e_0 + e_\omega$, see
(2.1).

We claim that if M is an indecomposable C-module with $\operatorname{res}_e(M) = 0$,
then M lies in one of the non-homogeneous tubes. Indeed, if $\operatorname{res}_e(M) = 0$,
then M is an indecomposable representation of the quiver obtained from
the quiver of C by removing the vertices 0 and ω, and our claim follows,
because

- the simple modules $S(a_i)$, $S(b_j)$ and $S(c_k)$ lie in non-homogeneous
  tubes of $\Gamma(\operatorname{mod} C)$, and
- we know the structure of the indecomposable modules over the path
  algebra

$$\mathbb{T}_n(K) = \begin{bmatrix} K & 0 & \cdots & 0 \\ K & K & \cdots & 0 \\ \vdots & \vdots & \ddots & \vdots \\ K & K & \cdots & K \end{bmatrix}$$

of the linear quiver $\overset{1}{\circ} \longleftarrow \overset{2}{\circ} \longleftarrow \overset{3}{\circ} \longleftarrow \cdots \longleftarrow \overset{n-1}{\circ} \longleftarrow \overset{n}{\circ}$ of Dynkin
type $\mathbb{A}_n$. We recall from Chapter V that $\mathbb{T}_n(K)$ is a Nakayama
algebra.

Take now a simple regular module E lying in a homogeneous tube of
$\Gamma(\operatorname{mod} C)$. Because the different tubes of $\Gamma(\operatorname{mod} C)$ are orthogonal, then

$\mathrm{Hom}_C(E, M) = 0$, for any indecomposable C-module with $\mathrm{res}_e(M) = 0$. Moreover, (XI.3.6), (XI.3.7), and (1.18) yield $\mathbf{dim}\, E \in \mathrm{rad}\, q_C = \mathbb{Z} \cdot \mathbf{h}_C$, where $\mathbf{h}_C$ is the canonical radical vector of C, see (1.17). Hence, $\mathbf{dim}\, E = m \cdot \mathbf{h}_C$, for some $m \geq 1$.

We prove now that the Kronecker A-module $\mathrm{res}_e(E)$ belongs to add $\mathcal{R}(A)$. It follows from our assumption on E, that E is orthogonal to any simple C-module except $S(0)$ and $S(\omega)$. This implies that all K-linear maps φ_{α_s}, φ_{β_t}, φ_{γ_m} in E, viewed as a K-linear representation, are bijective. Hence, the two maps in the representation $\mathrm{res}_e(E)$ of the Kronecker quiver $0 \mathrel{\substack{\longleftarrow \\[-0.6em] \longleftarrow}} \omega$ are also bijective, and, according to (XI.4.3), the module $\mathrm{res}_e(E)$ belongs to add $\mathcal{R}(A)$.

Consequently, (2.3) applies and we get $\mathbf{dim}\, E = \mathbf{h}_C$. It follows from (2.6) and (2.10) that E is isomorphic to one of the modules of Table 2.5 (or Table 2.9) if $C = C(p, q)$ (or $C = C(p, q, r)$, respectively). Hence, by (2.8), (2.12), and the statement (b) proved above, there is an isomorphism $E \cong E^{(\lambda)}$, for some $\lambda \in \mathbb{P}_1(K)$. We note that, for $C = C(1, q)$, there is an isomorphism $E \cong E^{(\infty)} = E_1^{(\infty)}$ and, for $C = C(1, 1)$, we have $E \cong E^{(0)} = E_1^{(0)}$.

Consequently, we have proved that the simple regular modules from the tubes $\mathcal{T}_\lambda^C$, with $\lambda \in \mathbb{P}_1(K)$, exhaust all the simple regular C-modules, and consequently the regular part $\mathcal{R}(C)$ of $\Gamma(\mathrm{mod}\, C)$ is the disjoint union of the stable tubes $\mathcal{T}_\lambda^C$, with $\lambda \in \mathbb{P}_1(K)$. This finishes the proof of (c) and completes the proof of the theorem. $\qquad \square$

As a byproduct of our considerations we get the following useful fact.

2.14. Corollary. *Assume that C is a canonical algebra of Euclidean type. Let $q_C : K_0(C) \longrightarrow \mathbb{Z}$ be the Euler quadratic form of C and let $\mathbf{h}_C$ be the canonical radical vector of C. If M is an indecomposable regular C-module lying in a tube of rank $\ell \geq 1$ then, for a positive integer $m \geq 1$, $r\ell(M) = m \cdot \ell$ if and only if $\mathbf{dim}\, M = m \cdot \mathbf{h}_C$.*

Proof. Let M be an indecomposable regular C-module of regular length $\ell \geq 1$ lying in a tube $\mathcal{T}$ of rank ℓ.

If $\ell \geq 2$ then, by (2.13), M is isomorphic to one of the modules:

- R^{α_s}, with $1 \leq s \leq p$,
- R^{β_t}, with $1 \leq t \leq q$, or
- R^{γ_m}, with $1 \leq m \leq r$ (if $C = C(p, q, r)$).

Furthermore, if $\ell = 1$, then M has the form $E^{(\lambda)}$, where $\lambda \in K \setminus \{0, 1\}$, up to isomorphism. In each of these cases we have $\mathbf{dim}\, M = \mathbf{h}_C$.

Because the vector $\mathbf{h}_C$ is the sum of the dimension vectors $\mathbf{dim}\, X$ of modules X lying on the mouth of the tube $\mathcal{T}$, then the corollary follows from (XI.3.10). $\qquad \square$

For a canonical algebra C of Euclidean type, we define a bijection $\xi_C : \mathbb{P}_1(K) \longrightarrow \mathbb{P}_1(K)$ as follows:

- $\xi_C(\lambda) = \lambda$, for any $\lambda \in \mathbb{P}_1(K)$, if $C = C(p, q)$, and
- $\xi_C(\infty) = \infty$ and $\xi_C(\lambda) = -\lambda$, for any $\lambda \in K$, if $C = C(p, q, r)$.

2.15. Proposition. *Let C be a canonical algebra of Euclidean type, $e = e_0 + e_\omega$, and let*

$$A = eCe \cong \begin{bmatrix} K & 0 \\ K^2 & K \end{bmatrix}$$

be the associated Kronecker algebra (see (2.1)), $\mathrm{res}_e : \mathrm{mod}\, C \longrightarrow \mathrm{mod}\, A$ the restriction functor, X a C-module and $\lambda \in \mathbb{P}_1(K)$. Then X belongs to $\mathrm{add}\,\mathcal{P}(C)$ (or to $\mathrm{add}\,\mathcal{T}_\lambda^C$, $\mathrm{add}\,\mathcal{Q}(C)$) if and only if $\mathrm{res}_e(X)$ belongs to $\mathrm{add}\,\mathcal{P}(A)$ (or to $\mathrm{add}\,\mathcal{T}_{\xi_C(\lambda)}^A$, $\mathrm{add}\,\mathcal{Q}(A)$, respectively). In particular, X lies in $\mathrm{add}\,\mathcal{R}(C)$ if and only if $\mathrm{res}_e(X)$ lies in $\mathrm{add}\,\mathcal{R}(A)$.

Proof. In view of (2.2), it is sufficient to show that $X \in \mathrm{add}\,\mathcal{T}_\lambda^C$ if and only if the module $\mathrm{res}_e(X)$ over the Kronecker algebra $A = eCe$ lies in $\mathrm{add}\,\mathcal{T}_{\xi_C(\lambda)}^A$. Let X be a simple regular C-module. Then it follows from our description of the simple regular C-modules and the description of simple regular A-modules given in (XI.4.5) that $X \in \mathcal{T}_\lambda^C$ if and only if $\mathrm{res}_e(X) \in \mathcal{T}_{\xi_C(\lambda)}^A$ or $\mathrm{res}_e(X) = 0$ (and X is a simple C-module). Hence $X \in \mathrm{add}\,\mathcal{T}_\lambda^C$ if and only if $\mathrm{res}_e(X) \in \mathrm{add}\,\mathcal{T}_{\xi_C(\lambda)}^A$.

In general, if $X \in \mathrm{add}\,\mathcal{R}(C)$ is a module of regular length $\ell \geq 2$, then $X \in \mathrm{add}\,\mathcal{T}_\lambda^C$ if and only if there exists an exact sequence

$$0 \longrightarrow E \longrightarrow X \longrightarrow Y \longrightarrow 0,$$

where E is a simple regular module from $\mathcal{T}_\lambda^C$ and Y is a module from $\mathrm{add}\,\mathcal{T}_\lambda^C$ of regular length $\ell - 1$. Then our claim follows by induction on the regular length of X due to the fact the res_e is an exact functor and the categories $\mathrm{add}\,\mathcal{T}_\lambda^C$ and $\mathrm{add}\,\mathcal{T}_{\xi_C(\lambda)}^A$ are closed under extensions and finite direct sums. $\square$

The preceding proposition is used in establishing the separation property of the tubular family $\boldsymbol{\mathcal{T}}^C = \{\mathcal{T}_\lambda^C\}_{\lambda \in \mathbb{P}_1(K)}$, by applying some arguments due to Ringel in [217]. We need the following technical lemma.

2.16. Lemma. *Let C be a canonical algebra of Euclidean type, $e = e_0 + e_\omega$, let*

$$A = eCe \cong \begin{bmatrix} K & 0 \\ K^2 & K \end{bmatrix}$$

be the associated Kronecker algebra see (2.1), $\mathrm{res}_e : \mathrm{mod}\, C \longrightarrow \mathrm{mod}\, A$ the restriction functor, and let $P \in \mathrm{add}\,\mathcal{P}(C)$. Then P can be embedded into a module $P' \in \mathrm{add}\,\mathcal{P}(C)$ such that $P'/P \in \mathrm{add}\,\mathcal{R}(C)$ and $\mathrm{Hom}_C(P', M) = 0$, for any C-module M with $\mathrm{res}_e(M) = 0$.

Proof. Let $X \subseteq P$ be minimal such that $\mathrm{res}_e(P/X) = 0$. Put $Z = P/X$. Then $Z \in \mathrm{add}\,\mathcal{R}(C)$ and we may embed Z into a module $R \in \mathrm{add}\,\mathcal{R}(C)$ such that $\mathrm{Hom}_C(R, M) = 0$ for any C-module M with $\mathrm{res}_e(M) = 0$. For, it is enough to choose a module R containing Z such that the regular top of R has only direct summands having non-zero restrictions to A. The canonical epimorphism $\varepsilon : P \longrightarrow P/X = Z$ induces an exact sequence

$$\mathrm{Ext}^1_C(R/Z, P) \longrightarrow \mathrm{Ext}^1_C(R/Z, Z) \longrightarrow \mathrm{Ext}^2_C(R/Z, X),$$

where $\mathrm{Ext}^2_C(R/Z, X) = 0$, because the module R/Z lies in $\mathrm{add}\,\mathcal{R}(C)$ and, by (XI.3.3) and (1.5)-(1.14), is of projective dimension at most 1. Hence, there exists a commutative diagram

$$
\begin{array}{ccccccccc}
0 & \longrightarrow & P & \overset{\mu}{\longrightarrow} & P'' & \overset{\pi}{\longrightarrow} & R/Z & \longrightarrow & 0 \\
& & {\scriptstyle \varepsilon}\downarrow & & {\scriptstyle \varepsilon'}\downarrow & & \| & & \\
0 & \longrightarrow & Z & \overset{\mu'}{\longrightarrow} & R & \overset{\pi'}{\longrightarrow} & R/Z & \longrightarrow & 0
\end{array}
$$

with exact rows, where $\mu' : Z \hookrightarrow R$ is the natural embedding.

Let M be a C-module with $\mathrm{res}_e(M) = 0$. We claim that $\mathrm{Hom}_C(P'', M) = 0$. Let $\alpha : P'' \to M$ be a morphism. Because the module X is minimal with respect to $\mathrm{res}_e(P/X) = 0$ and $Z = P/X$, we infer that the homomorphism $\alpha\mu$ factors through ε, say $\alpha\mu = \alpha'\varepsilon$, for some $\alpha' : Z \to M$. Further, because the above diagram is a pushout diagram, there is a homomorphism $\beta : R \to M$ such that $\beta\varepsilon' = \alpha$ and $\beta\mu' = \alpha'$. But $\mathrm{Hom}_C(R, M) = 0$ implies $\beta = 0$, and consequently $\alpha = 0$. Observe also that $P'' \in \mathrm{add}\,(\mathcal{P}(C) \vee \mathcal{R}(C))$, because $P \in \mathrm{add}\,\mathcal{P}(C)$ and $R/Z \in \mathrm{add}\,\mathcal{R}(C)$.

Let Y be a maximal submodule of P'' lying in the category $\mathrm{add}\,\mathcal{R}(C)$. Then the module

$$Y/(P \cap Y) \cong (P + Y)/P$$

embeds into the module $P''/P \cong R/Z$. Hence, the module $P \cap Y$ lies in $\mathrm{add}\,\mathcal{R}(C)$, because $P \cap Y$ is the kernel of the epimorphism $Y \longrightarrow Y/(P \cap Y)$ and $\mathrm{add}\,\mathcal{R}(C)$ is abelian. On the other hand, the module $P \cap Y$ lies in $\mathrm{add}\,\mathcal{P}(C)$, because it is a submodule of $P \in \mathrm{add}\,\mathcal{P}(C)$. Thus $P \cap Y = 0$.

Now we observe that the module $P' = P''/Y$ belongs to $\mathrm{add}\,\mathcal{P}(C)$, by our choice of Y. Moreover, M is a C-module with $\mathrm{res}_e(M) = 0$, and then the equality $\mathrm{Hom}_C(P'', M) = 0$ implies $\mathrm{Hom}_C(P', M) = 0$. Further, the module P embeds into P' and there is an isomorphism $P'/P \cong P''/(P+Y)$. Because $P''/(P + Y)$ is the cokernel of the composite monomorphism

$$Y = Y/(P \cap Y) \cong (P + Y)/P \hookrightarrow P''/P,$$

and the modules Y and $P''/P = R/Z$ lie in $\mathrm{add}\,\mathcal{R}(C)$, then the module P'/P also lies $\mathrm{add}\,\mathcal{R}(C)$. This completes the proof. $\qquad\square$

2.17. Proposition. *Let C be a canonical algebra of Euclidean type and let*

$$\mathcal{T}^C = \{\mathcal{T}^C_\lambda\}_{\lambda \in \mathbb{P}_1(K)}$$

be the complete $\mathbb{P}_1(K)$-family of tubes in $\mathcal{R}(C)$, see (2.7) and (2.11). Let P be a postprojective C-module, Q a preinjective C-module, and $\lambda \in \mathbb{P}_1(K)$. Then for any homomorphism $f : P \to Q$ there exists a module $N_\lambda \in \text{add}\, \mathcal{T}^C_\lambda$ such that f factors through N_λ.

Proof. Let $e = e_0 + e_\omega$, let $A = eCe \cong \begin{bmatrix} K & 0 \\ K^2 & K \end{bmatrix}$ be the associated Kronecker algebra, and let $\text{res}_e : \text{mod}\, C \longrightarrow \text{mod}\, A$ be the restriction functor, see (2.1) and (I.6.6).

We fix $\lambda \in \mathbb{P}_1(K)$ and a homomorphism $f : P \to Q$. It follows from (2.16) that we may embed P into a module P' lying in $\text{add}\, \mathcal{P}(C)$ such that $P'/P \in \text{add}\, \mathcal{R}(C)$ and $\text{Hom}_C(P', M) = 0$, for any C-module M with $\text{res}_e(M) = 0$. We then get an exact sequence

$$\text{Hom}_C(P', Q) \longrightarrow \text{Hom}_C(P, Q) \longrightarrow \text{Ext}^1_C(P'/P, Q)$$

with $\text{Ext}^1_C(P'/P, Q) \cong D\overline{\text{Hom}}_C(Q, \tau_C(P'/P)) = 0$. Hence, the homomorphism $f : P \to Q$ can be extended to $f' : P' \to Q$. Therefore, without loss of generality, we may assume that $\text{Hom}_C(P, M) = 0$, for any C-module M with $\text{res}_e(M) = 0$.

We recall that the idempotent induced functor $T_e : \text{mod}\, A \longrightarrow \text{mod}\, C$ is left adjoint to $\text{res}_e : \text{mod}\, C \longrightarrow \text{mod}\, A$. Thus, for each C-module X, we have a functorial isomorphism

$$\text{Hom}_C(T_e(\text{res}_e(X)), X) \xrightarrow{\ \simeq\ } \text{Hom}_A(\text{res}_e(X), \text{res}_e(X))$$

of K-vector spaces. Denote by $\beta_X : T_e(\text{res}_e(X)) \longrightarrow X$ the homomorphism adjoint to the identity homomorphism on $\text{res}_e(X)$.

We claim that the homomorphism $\beta_P : T_e(\text{res}_e(P)) \longrightarrow P$ is a monomorphism. Indeed, because $\text{res}_e(\beta_P)$ is an isomorphism and res_e is an exact functor then $\text{res}_e(\text{Ker}\,\beta_P) = 0$. Moreover, by (2.1), $P \in \text{add}\, \mathcal{P}(C)$ implies

$$\text{res}_e(T_e(\text{res}_e(P))) \cong \text{res}_e(P) \in \text{add}\, \mathcal{P}(A),$$

and consequently $T_e(\text{res}_e(P)) \in \text{add}\, \mathcal{P}(C)$. Thus, $\text{Ker}\,\beta_P$ lies in $\text{add}\, \mathcal{P}(C)$. But then $\text{res}_e(\text{Ker}\,\beta_P) = 0$ implies $\text{Ker}\,\beta_P = 0$, and so β_P is a monomorphism.

Consider the induced exact sequence

$$0 \longrightarrow T_e(\text{res}_e(P)) \xrightarrow{\ \beta_P\ } P \xrightarrow{\ \gamma_P\ } Z \longrightarrow 0,$$

where Z is the cokernel of β_P. Clearly, then $\mathrm{res}_e(Z) = 0$. Let $Z = Z_\lambda \oplus Z'$ be a decomposition such that Z_λ is the maximal direct summand of Z lying in add $\mathcal{T}_\lambda^C$. Observe now that $\mathrm{res}_e(P) \in \mathrm{add}\,\mathcal{P}(A)$ and $\mathrm{res}_e(Q) \in \mathrm{add}\,\mathcal{Q}(A)$, by (2.1). Therefore, by (XI.4.7), the homomorphism $\mathrm{res}_e(f)$: $\mathrm{res}_e(P) \longrightarrow \mathrm{res}_e(Q)$ of Kronecker modules admits a factorisation through a module $R \in \mathrm{add}\,\mathcal{T}_{\xi_C(\lambda)}^A$, so $\mathrm{res}_e(f) = hg$, for some homomorphisms g : $\mathrm{res}_e(P) \longrightarrow R$ and $h : R \longrightarrow \mathrm{res}_e(Q)$.

Consider the commutative diagram with exact rows

$$
\begin{array}{ccccccccc}
0 & \longrightarrow & T_e(\mathrm{res}_e(P)) & \xrightarrow{\beta_P} & P & \xrightarrow{\gamma_P} & Z_\lambda \oplus Z' & \longrightarrow & 0 \\
& & \downarrow{\scriptstyle T_e(g)} & & \downarrow{\scriptstyle g'} & & \| & & \\
0 & \longrightarrow & T_e(R) & \longrightarrow & X & \longrightarrow & Z_\lambda \oplus Z' & \longrightarrow & 0.
\end{array}
$$

Because $\mathrm{res}_e(T_e(R)) = R$ lies in add $\mathcal{T}_{\xi_C(\lambda)}^A$ then, by (2.15), the module $T_e(R)$ lies in add $\mathcal{T}_\lambda^C$. It follows from our choice of Z_λ that Z' has no simple regular factors from $\mathcal{T}_\lambda^C$, and consequently,

$$\mathrm{Ext}_C^1(Z', T_e(R)) \cong D\mathrm{Hom}_C(\tau_C^{-1}T_e(R), Z') = 0.$$

Therefore, there is a decomposition $X = X_\lambda \oplus Z'$, where $X_\lambda \in \mathrm{add}\,\mathcal{T}_\lambda^C$. Because $\beta : T_e \mathrm{res}_e \longrightarrow \mathrm{id}$ is a functorial morphism, we have the equalities

$$f\beta_P = \beta_Q T_e(\mathrm{res}_e(f)) = \beta_Q T_e(hg) = \beta_Q T_e(h) T_e(g).$$

Invoking now the pushout property of the above commutative diagram, we conclude that there is a homomorphism $h' : X \to Q$ satisfying $h'g' = f$. Because $X = X_\lambda \oplus Z'$, we can write $f = h_1 g_1 + h_2 g_2$, where $g_1 : P \to X_\lambda$, $h_1 : X_\lambda \to Q$, $g_2 : P \to Z'$, $h_2 : Z' \to Q$. But $\mathrm{res}_e(Z') = 0$ implies $\mathrm{Hom}_C(P, Z') = 0$, by the property imposed on P. Therefore, $f = h_1 g_1$ is a required factorisation of f through the module $X_\lambda \in \mathrm{add}\,\mathcal{T}_\lambda^C$. This completes the proof. $\qquad\square$

XII.3. A separating family of tubes over a concealed algebra of Euclidean type

In this section we collect the basic properties of the module category over any concealed algebra of Euclidean type.

We recall that in Section XII.1 we have introduced the quivers

$\Delta(\widetilde{\mathbb{A}}_{p,q})$, with $1 \le p \le q$, $\Delta(\widetilde{\mathbb{D}}_m)$, with $m \ge 4$, $\Delta(\widetilde{\mathbb{E}}_6)$, $\Delta(\widetilde{\mathbb{E}}_7)$, and $\Delta(\widetilde{\mathbb{E}}_8)$,

whose underlying graphs are the Euclidean graphs $\widetilde{\mathbb{A}}_{p+q-1}$, $\widetilde{\mathbb{D}}_m$, $\widetilde{\mathbb{E}}_6$, $\widetilde{\mathbb{E}}_7$ and $\widetilde{\mathbb{E}}_8$, respectively. The following proposition shows that these special quivers are in fact the types of all concealed algebras of Euclidean types.

3.1. Proposition. *Let Q be an acyclic quiver whose underlying graph $\overline{Q}$ is Euclidean, and let B be a concealed algebra of type Q. Then B is a concealed algebra of type Δ, where Δ is a quiver such that:*

- *$\Delta \in \{\Delta(\widetilde{\mathbb{A}}_{p,q}), 1 \leq p \leq q, \Delta(\widetilde{\mathbb{D}}_m), m \geq 4, \Delta(\widetilde{\mathbb{E}}_6), \Delta(\widetilde{\mathbb{E}}_7), \Delta(\widetilde{\mathbb{E}}_8)\}$,*
- *the underlying graph $\overline{\Delta}$ of Δ equals $\overline{Q}$,*
- *if $\overline{Q} = \widetilde{\mathbb{A}}_m$ and $m \geq 1$, then $p = \min\{p', p''\}$ and $q = \max\{p', p''\}$, where p' and p'' is the number of counterclockwise-oriented arrows in Q and clockwise-oriented arrows in Q, respectively, and*
- *the quiver Δ is obtained from Q by a finite sequence of reflections.*

Proof. Let A be the path algebra KQ of Q, and let T be a postprojective tilting A-module such that $\operatorname{End}_A(T) = B$. Observe that then, for any positive integer $s \geq 1$, the module $\tau_A^{-s}T$ is also a postprojective tilting A-module and there is an isomorphism of algebras $\operatorname{End}_A(\tau_A^{-s}T) \cong \operatorname{End}_A(T) = B$. Because $\overline{Q}$ is a Euclidean graph, then there exists exactly one quiver Δ in the set $\{\Delta(\widetilde{\mathbb{A}}_{p,q}), 1 \leq p \leq q, \Delta(\widetilde{\mathbb{D}}_m), m \geq 4, \Delta(\widetilde{\mathbb{E}}_6), \Delta(\widetilde{\mathbb{E}}_7), \Delta(\widetilde{\mathbb{E}}_8)\}$ such that $\overline{\Delta} = \overline{Q}$.

If $\overline{Q}$ is one of the graphs $\widetilde{\mathbb{D}}_m$, $\widetilde{\mathbb{E}}_6$, $\widetilde{\mathbb{E}}_7$ or $\widetilde{\mathbb{E}}_8$ then, according to (VII.5.2), the quiver Q can be obtained from Δ by a finite sequence of reflections.

Assume that $\overline{Q} = \widetilde{\mathbb{A}}_m$ for some $m \geq 1$. Let $p = \min\{p', p''\}$ and $q = \max\{p', p''\}$, where p' and p'' is the number of counterclockwise-oriented arrows in Q and clockwise-oriented arrows in Q, respectively. Observe that $p + q = m + 1$ and $1 \leq p \leq q$. Because Q is acyclic, then Q can be obtained from $\Delta(\widetilde{\mathbb{A}}_{p,q})$ by a finite sequence of reflections (see (VIII.1.8)).

It follows that, in both cases, the postprojective component $\mathcal{P}(A)$ contains a cofinite full translation subquiver $\Gamma = (-\mathbb{N})\Delta^{\mathrm{op}}$ which is closed under successors in $\mathcal{P}(A)$. Moreover, the mesh-category $K(\Gamma)$ of Γ is equivalent to the mesh-category $K(\mathcal{P}(K\Delta))$ of the postprojective component $\mathcal{P}(K\Delta)$ of the path algebra $K\Delta$. It follows now from the first part of our proof that an isomorphism of algebras $B \cong \operatorname{End}_A(T')$, for a postprojective tilting A-module T' whose indecomposable direct summands lie in Γ, and, consequently, $B \cong \operatorname{End}_{K\Delta}(T'')$, for a postprojective tilting $K\Delta$-module T''. $\square$

To formulate the next result, we need some new concepts.

3.2. Definition. Let $\boldsymbol{\mathcal{T}} = \{\mathcal{T}_i\}_{i\in\Lambda}$ be a family of stable tubes and $(m_1, \ldots, m_s)$ a sequence of integers with $1 \leq m_1 \leq \ldots \leq m_s$. We say that $\boldsymbol{\mathcal{T}}$ is of **tubular type** $(m_1, \ldots, m_s)$ if $\boldsymbol{\mathcal{T}}$ admits s tubes $\mathcal{T}_{i_1}, \ldots, \mathcal{T}_{i_s}$ of ranks $m_1, \ldots, m_s$, respectively, and the remaining tubes $\mathcal{T}_i$ of $\boldsymbol{\mathcal{T}}$, with $i \notin \{i_1, \ldots, i_s\}$, are homogeneous.

3.3. Definition. A family $\boldsymbol{\mathcal{C}} = \{\mathcal{C}_i\}_{i\in\Lambda}$ of components in the Auslander–Reiten quiver $\Gamma(\operatorname{mod} A)$ of an algebra A is said to be **separating** [215] if the indecomposable A-modules outside the family $\mathcal{C}$ fall into two classes $\mathcal{P}$ and $\mathcal{Q}$ such that the following conditions are satisfied:

(a) The components in $\mathcal{C}$ are pairwise orthogonal and standard.

(b) $\mathrm{Hom}_A(\mathcal{Q},\mathcal{P}) = 0$, $\mathrm{Hom}_A(\mathcal{C},\mathcal{P}) = 0$ and $\mathrm{Hom}_A(\mathcal{Q},\mathcal{C}) = 0$.

(c) For each $i \in \Lambda$, any homomorphism $f : P \to Q$, with $P \in \mathrm{add}\,\mathcal{P}$ and $Q \in \mathrm{add}\,\mathcal{Q}$, factors through a module from $\mathrm{add}\,\mathcal{C}_i$.

In this case, we say that $\mathcal{C}$ **separates** $\mathcal{P}$ from $\mathcal{Q}$.

3.4. Theorem. *Let Q be an acyclic quiver whose underlying graph $\overline{Q}$ is Euclidean, and let B be a concealed algebra of type Q.*

(a) *The Auslander–Reiten quiver $\Gamma(\mathrm{mod}\,B)$ of B consists of the following three types of components:*

- *a postprojective component $\mathcal{P}(B)$ containing all the indecomposable projective modules,*
- *a preinjective component $\mathcal{Q}(B)$ containing all the indecomposable injective modules, and*
- *a unique tubular $\mathbb{P}_1(K)$-family $\mathcal{T}^B = \{\mathcal{T}^B_\lambda\}_{\lambda \in \mathbb{P}_1(K)}$ of stable tubes separating $\mathcal{P}(B)$ from $\mathcal{Q}(B)$.*

(b) *The tubular type $\mathbf{m}_B = (m_1,\dots,m_s)$ of the $\mathbb{P}_1(K)$-family $\mathcal{T}^B$ depends only on the Euclidean quiver Q and equals $\mathbf{m}_Q$, where*

- $\mathbf{m}_Q = (p,q)$, *if $\overline{Q} = \widetilde{\mathbb{A}}_m$, $m \geq 1$, $p = \min\{p',p''\}$ and $q = \max\{p',p''\}$, where p' and p'' is the number of counterclockwise-oriented arrows in Q and clockwise-oriented arrows in Q, respectively,*
- $\mathbf{m}_Q = (2,2,m-2)$, *if $\overline{Q} = \widetilde{\mathbb{D}}_m$ and $m \geq 4$,*
- $\mathbf{m}_Q = (2,3,3)$, *if $\overline{Q} = \widetilde{\mathbb{E}}_6$,*
- $\mathbf{m}_Q = (2,3,4)$, *if $\overline{Q} = \widetilde{\mathbb{E}}_7$, and*
- $\mathbf{m}_Q = (2,3,5)$, *if $\overline{Q} = \widetilde{\mathbb{E}}_8$.*

(c) *Every component of $\Gamma(\mathrm{mod}\,B)$ is generalised standard.*

Proof. Let Q be an acyclic quiver whose underlying graph $\overline{Q}$ is Euclidean and let $A = KQ$ be the path algebra of Q. Because B is a concealed algebra of type Q then, according to (3.1), there exists a quiver Δ in the set

$$\{\Delta(\widetilde{\mathbb{A}}_{p,q}),\ \text{with } 1 \leq p \leq q,\ \Delta(\widetilde{\mathbb{D}}_m),\ \text{with } m \geq 4,\ \Delta(\widetilde{\mathbb{E}}_6),\ \Delta(\widetilde{\mathbb{E}}_7),\ \Delta(\widetilde{\mathbb{E}}_8)\},$$

such that $\overline{\Delta} = \overline{Q}$, the quiver Q is obtained from Δ by a finite sequence of reflections and B is a concealed algebra of type Δ.

Let C be the canonical algebra of type Δ. It follows from (3.1), (1.5), (1.8), (1.11), and (1.14) that there are algebra isomorphisms

$$B \cong \mathrm{End}_{K\Delta}(T') \quad \text{and} \quad C \cong \mathrm{End}_{K\Delta}(T''),$$

for some postprojective tilting $K\Delta$-modules T' and T''. Because B is a concealed algebra of type Q, then $B \cong \operatorname{End}_A(T)$, for some postprojective tilting A-module T.

Applying now (VIII.4.5), (VIII.4.7), and (XI.5.2), we conclude that there exist

(i) a full translation subquiver $\mathcal{P}'(K\Delta) \cong (-\mathbb{N})\Delta^{\mathrm{op}}$ of $\mathcal{P}(K\Delta)$ closed under successors, and a full translation subquiver $\mathcal{Q}'(K\Delta) = \mathbb{N}\Delta^{\mathrm{op}}$ of $\mathcal{Q}(K\Delta)$ closed under predecessors,

(ii) a full translation subquiver $\mathcal{P}'(B) \cong (-\mathbb{N})\Delta^{\mathrm{op}}$ of $\mathcal{P}(B)$ closed under successors, and a full translation subquiver $\mathcal{Q}'(B) \cong \mathbb{N}\Delta^{\mathrm{op}}$ of $\mathcal{Q}(B)$ closed under predecessors,

(iii) a full translation subquiver $\mathcal{P}'(C) \cong (-\mathbb{N})\Delta^{\mathrm{op}}$ of $\mathcal{P}(C)$ closed under successors, and a full translation subquiver $\mathcal{Q}'(C) \cong \mathbb{N}\Delta^{\mathrm{op}}$ of $\mathcal{Q}(C)$ closed under predecessors, and

(iv) a full translation subquiver $\mathcal{P}'(A) \cong (-\mathbb{N})\Delta^{\mathrm{op}}$ of $\mathcal{P}(A)$ closed under successors, and a full translation subquiver $\mathcal{Q}'(A) = \mathbb{N}\Delta^{\mathrm{op}}$ of $\mathcal{Q}(A)$ closed under predecessors

such that the functor $\operatorname{Hom}_{K\Delta}(T', -)$ induces an equivalence of categories

$$\operatorname{add}(\mathcal{P}'(K\Delta) \cup \mathcal{R}(K\Delta) \cup \mathcal{Q}'(K\Delta)) \xrightarrow{\ \cong\ } \operatorname{add}(\mathcal{P}'(B) \cup \mathcal{R}(B) \cup \mathcal{Q}'(B)),$$

the functor $\operatorname{Hom}_{K\Delta}(T'', -)$ induces an equivalence of categories

$$\operatorname{add}(\mathcal{P}'(K\Delta) \cup \mathcal{R}(K\Delta) \cup \mathcal{Q}'(K\Delta)) \xrightarrow{\ \cong\ } \operatorname{add}(\mathcal{P}'(C) \cup \mathcal{R}(C) \cup \mathcal{Q}'(C)),$$

and the functor $\operatorname{Hom}_A(T, -)$ induces an equivalence of categories

$$\operatorname{add}(\mathcal{P}'(A) \cup \mathcal{R}(A) \cup \mathcal{Q}'(A)) \xrightarrow{\ \cong\ } \operatorname{add}(\mathcal{P}'(B) \cup \mathcal{R}(B) \cup \mathcal{Q}'(B)).$$

Let $\mathbf{m}_Q$ be the tubular type of the $\mathbb{P}_1(K)$-family

$$\boldsymbol{\mathcal{T}}^B = \{\mathcal{T}^B_\lambda\}_{\lambda \in \mathbb{P}_1(K)}$$

in (b). It follows from (2.13) that the regular part $\mathcal{R}(C)$ of $\Gamma(\operatorname{mod} C)$ consists of a $\mathbb{P}_1(K)$-family $\boldsymbol{\mathcal{T}}^C = \{\mathcal{T}^C_\lambda\}_{\lambda \in \mathbb{P}_1(K)}$ of stable tubes of tubular type $\mathbf{m}_Q$. Therefore, the regular part $\mathcal{R}(K\Delta)$ of $\Gamma(\operatorname{mod} K\Delta)$ consists of a $\mathbb{P}_1(K)$-family

$$\boldsymbol{\mathcal{T}}^{K\Delta} = \{\mathcal{T}^{K\Delta}_\lambda\}_{\lambda \in \mathbb{P}_1(K)}$$

of stable tubes of tubular type $\mathbf{m}_Q$, and $\mathcal{R}(B)$ consists of a $\mathbb{P}_1(K)$-family

$$\boldsymbol{\mathcal{T}}^B = \{\mathcal{T}^B_\lambda\}_{\lambda \in \mathbb{P}_1(K)}$$

of stable tubes $\mathcal{T}_\lambda^B$ of tubular type $\mathbf{m}_Q$. It then follows that $\mathcal{R}(A)$ consists of a $\mathbb{P}_1(K)$-family

$$\boldsymbol{\mathcal{T}}^A = \{\mathcal{T}_\lambda^A\}_{\lambda \in \mathbb{P}_1(K)}$$

of stable tubes of the same tubular type $\mathbf{m}_Q$. Moreover, by (XI.3.4), the tubes $\mathcal{T}_\lambda^B$, $\lambda \in \mathbb{P}_1(K)$, are pairwise orthogonal and standard. Then, according to (X.3.2) and (X.4.5), every component of $\Gamma(\mathrm{mod}\,B)$ is generalised standard and, hence, the statement (c) follows.

It remains to show that the tubular family $\boldsymbol{\mathcal{T}}^B = \{\mathcal{T}_\lambda^B\}_{\lambda \in \mathbb{P}_1(K)}$ separates $\mathcal{P}(B)$ from $\mathcal{Q}(B)$. For this purpose, we remark that, by (2.17), the tubular family

$$\boldsymbol{\mathcal{T}}^C = \{\mathcal{T}_\lambda^C\}_{\lambda \in \mathbb{P}_1(K)}$$

separates $\mathcal{P}(C)$ from $\mathcal{Q}(C)$. Let $P \in \mathrm{add}\,\mathcal{P}(B)$, $Q \in \mathrm{add}\,\mathcal{Q}(B)$ and $f : P \to Q$ be a non-zero homomorphism. We may clearly assume that P and Q are indecomposable. Because P and Q lie in different components of $\Gamma(\mathrm{mod}\,B)$, there is no finite path of irreducible morphisms from P to Q. Thus, applying (IV.5.1) we conclude that there exist modules $P' \in \mathrm{add}\,\mathcal{P}'(B)$, $Q' \in \mathcal{Q}'(B)$, and homomorphisms $u : P \to P'$, $f' : P' \to Q'$, and $v : Q' \to Q$ such that $f = vf'u$. Invoking the above equivalences of categories and the fact that the family $\boldsymbol{\mathcal{T}}^C = \{\mathcal{T}_\lambda^C\}_{\lambda \in \mathbb{P}_1(K)}$ separates $\mathcal{P}(C)$ from $\mathcal{Q}(C)$, we deduce that, for any $\lambda \in \mathbb{P}_1(K)$, there exists a module $R_\lambda \in \mathrm{add}\,\mathcal{T}_\lambda^B$ and homomorphisms $g'_\lambda : P' \to R_\lambda$, $h'_\lambda : R_\lambda \to Q'$ such that $f' = h'_\lambda g'_\lambda$. Taking $g_\lambda = g'_\lambda u$ and $h_\lambda = vh'_\lambda$ we obtain a factorisation $f = h_\lambda g_\lambda$ of f through a module $R_\lambda \in \mathrm{add}\,\mathcal{T}_\lambda^B$. This shows that the family $\boldsymbol{\mathcal{T}}^B = \{\mathcal{T}_\lambda^B\}_{\lambda \in \mathbb{P}_1(K)}$ separates $\mathcal{P}(B)$ from $\mathcal{Q}(B)$. $\qquad\square$

We may visualise the structure of the Auslander–Reiten quiver $\Gamma(\mathrm{mod}\,B)$ of a concealed algebra B of Euclidean type Q as follows

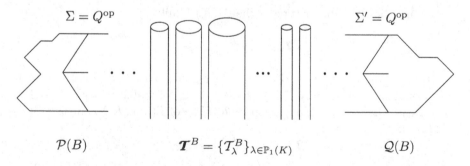

$$\Sigma = Q^{\mathrm{op}} \qquad\qquad\qquad\qquad\qquad\qquad \Sigma' = Q^{\mathrm{op}}$$

$$\mathcal{P}(B) \qquad\qquad \boldsymbol{\mathcal{T}}^B = \{\mathcal{T}_\lambda^B\}_{\lambda \in \mathbb{P}_1(K)} \qquad\qquad \mathcal{Q}(B)$$

3.5. Corollary. *Let B be a concealed algebra of a Euclidean type Q and let $\boldsymbol{\mathcal{T}}^B = \{\mathcal{T}_\lambda^B\}_{\lambda \in \mathbb{P}_1(K)}$ be the family of all stable tubes of $\Gamma(\mathrm{mod}\,B)$.*

(a) *If r_λ^B is the rank of the tube $\mathcal{T}_\lambda^B$, for $\lambda \in \mathbb{P}_1(K)$, then*

$$\sum_{\lambda \in \mathbb{P}_1(K)} (r_\lambda^B - 1) = n - 2,$$

where n is the rank of the Grothendieck group $K_0(B)$ of B.

(b) *The family $\mathcal{T}^B$ contains at most 3 non-homogeneous tubes $\mathcal{T}_\lambda^B$, and*

- $\mathcal{T}^B$ *contains exactly 3 non-homogeneous tubes if the underlying Euclidean graph $\overline{Q}$ of Q is one of the graphs $\widetilde{\mathbb{D}}_m$, with $m \geq 4$, $\widetilde{\mathbb{E}}_6$, $\widetilde{\mathbb{E}}_7$ or $\widetilde{\mathbb{E}}_8$,*
- $\mathcal{T}^B$ *contains exactly 2 non-homogeneous tubes if $\overline{Q} = \widetilde{\mathbb{A}}_m$, $m \geq 3$,*

$$p = \min\{p', p''\} \geq 2 \quad \text{and} \quad q = \max\{p', p''\} \geq 2,$$

  where p' and p'' is the number of counterclockwise-oriented arrows in Q and clockwise-oriented arrows in Q, respectively,
- $\mathcal{T}^B$ *contains exactly 1 non-homogeneous tube if $\overline{Q} = \widetilde{\mathbb{A}}_m$, $m \geq 2$, $p = \min\{p', p''\} = 1$ and $q = \max\{p', p''\} \geq 2$, and*
- $\mathcal{T}^B$ *consists of homogeneous tubes if and only if $\overline{Q} = \widetilde{\mathbb{A}}_1$, that is, Q is the Kronecker quiver $\circ \overset{\longleftarrow}{\underset{\longleftarrow}{}} \circ$.*

Proof. Apply (3.4). $\qquad\qquad\qquad\qquad\qquad\qquad\qquad\qquad\qquad\square$

3.6. Corollary. *Let B be a concealed algebra of a Euclidean type Q, let $\mathcal{T}^B = \{\mathcal{T}_\lambda^B\}_{\lambda \in \mathbb{P}_1(K)}$ be the family of all stable tubes of $\Gamma(\operatorname{mod} B)$, and M be an indecomposable B-module. Fix $\lambda \in \mathbb{P}_1(K)$ and denote by r_λ^B the rank of the tube $\mathcal{T}_\lambda^B$.*

(a) *If M is postprojective, then $\operatorname{Hom}_B(M, X) \neq 0$, for any indecomposable B-module X in the tube $\mathcal{T}_\lambda^B$ such that $r\ell(X) \geq r_\lambda^B$.*

(b) *If M is preinjective, then $\operatorname{Hom}_B(X, M) \neq 0$, for any indecomposable B-module X in the tube $\mathcal{T}_\lambda^B$ such that $r\ell(X) \geq r_\lambda^B$.*

Proof. We only prove the statement (a), because the proof of (b) is similar.

Assume that M is a postprojective indecomposable B-module and let $f : M \longrightarrow E$ be an injective envelope of M in $\operatorname{mod} B$. The injective module E is preinjective, because the preinjective component $\mathcal{Q}(B)$ of $\Gamma(\operatorname{mod} B)$ contains all the indecomposable injective B-modules, by (3.4)(a). It follows that the homomorphism f admits a factorisation through a module in $\operatorname{add} \mathcal{T}_\lambda^B$, because the family $\mathcal{T}^B$ separates the component $\mathcal{P}(B)$ from $\mathcal{Q}(B)$, again by (3.4)(a). Hence, there exists an indecomposable B-module L in the tube $\mathcal{T}_\lambda^B$ such that $\operatorname{Hom}_B(M, L) \neq 0$. Then, by applying (X.1.9)(b), we conclude that there exists an indecomposable module E' lying on the mouth of the tube $\mathcal{T}_\lambda^B$ such that $\operatorname{Hom}_B(M, E') \neq 0$.

Let $r_\lambda = r_\lambda^B$, and let $E_1, \dots, E_{r_\lambda}$ be pairwise non-isomorphic B-modules lying on the mouth of the tube $\mathcal{T}_\lambda^B$. Assume that

$$\tau E_1 \cong E_{r_\lambda}, \tau E_2 \cong E_1, \dots, \tau E_{r_\lambda} \cong E_{r_\lambda - 1}.$$

Without loss of generality, we may assume that $E' = E_{r_\lambda}$.

Because $\operatorname{Hom}_B(M, E_{r_\lambda}) \neq 0$ and the module M is not in the tube $\mathcal{T}_\lambda^B$ then, according to (X.1.8)(b), there exists a non-zero homomorphism $h_j : M \longrightarrow E_j[r_\lambda - j + 1]$, for any $j \in \{1, \dots, r_\lambda\}$. Take now an arbitrary indecomposable module X in the tube $\mathcal{T}_\lambda^B$ such that $r\ell(X) \geq r_\lambda$. Then there exist $j \in \{1, \dots, r_\lambda\}$ and a monomorphism $g_j : E_j[r_\lambda - j + 1] \longrightarrow X$. It follows that the composite homomorphism $g_j h_j : M \longrightarrow X$ is non-zero, that is, $\operatorname{Hom}_B(M, X) \neq 0$. This finishes the proof. $\qquad \square$

We finish this section by a theorem that establishes some properties of the separating family of stable tubes of any concealed algebra of Euclidean type that are stronger than the properties collected in (3.4). In the proof, we use the following terminology and notation introduced in Chapter VI.

Let A be an arbitrary algebra and M, N be modules in $\operatorname{mod} A$.

- The module M is said to be **generated by** N if there exist an integer $d \geq 1$ and an epimorphism $N^d \longrightarrow M$ of A-modules, where $N^d = N \oplus \dots \oplus N$ is the direct sum of d copies of N.
- Dually, the module M is **cogenerated by** N if there exist an integer $d \geq 1$ and a monomorphism $M \longrightarrow N^d$ of A-modules.
- The module N is **faithful** if its right annihilator

$$\operatorname{Ann} N = \{a \in A; \ Na = 0\}$$

  vanishes.

Obviously, $\operatorname{Ann} N = \{a \in A; \ Na = 0\}$ is a two-sided ideal of A. It is proved in (VI.2.2) that an A-module N is faithful if and only if the right module A_A is cogenerated by N, or equivalently, if and only if the right module $D(A)_A$ is generated by N.

3.7. Theorem. *Let B be a concealed algebra of a Euclidean type Q, let*

$$\boldsymbol{\mathcal{T}}^B = \{\mathcal{T}_\lambda^B\}_{\lambda \in \mathbb{P}_1(K)}$$

be the family of all stable tubes of $\Gamma(\operatorname{mod} B)$, and M be a B-module. Fix an element $\lambda \in \mathbb{P}_1(K)$.

(a) *If M is postprojective, then M is cogenerated by all but a finite number of indecomposable modules X of the stable tube $\mathcal{T}_\lambda^B$.*

(b) *If M is preinjective, then M is generated by all but a finite number of indecomposable modules X of the stable tube $\mathcal{T}_\lambda^B$.*

Proof. We only prove the statement (a), because the proof of (b) is similar.

Assume that M is a postprojective B-module. Let $f : M \longrightarrow E(M)$ be an injective envelope of M in mod B.

We recall from (3.4) that the family $\boldsymbol{T}^B$ separates the postprojective component $\mathcal{P}(B)$ from the preinjective component $\mathcal{Q}(B)$ of the Auslander–Reiten quiver $\Gamma(\text{mod }B)$ of B. Because $M \in \text{add}\,\mathcal{P}(B)$ and $E(M) \in \text{add}\,\mathcal{Q}(B)$ then, by (3.4), there exists a module

$$R_\lambda \in \text{add}\,\mathcal{T}_\lambda^B$$

and two homomorphisms of B-modules

$$M \xrightarrow{\ g\ } R_\lambda \xrightarrow{\ h\ } E(M)$$

such that $f = hg$. Moreover, g is a monomorphism, because f is a monomorphism.

Ler $r = r_\lambda^B \geq 1$ be the rank of the tube $\mathcal{T}_\lambda^B$ and let $E_1, E_2, \ldots, E_r$ be the pairwise non-isomorphic modules lying on the mouth of $\mathcal{T}_\lambda^B$. It follows from (XI.3.4) that add $\mathcal{T}_\lambda^B$ is a uniserial abelian subcategory of mod B with the indecomposable modules of the form $E_i[j]$, where $j \geq 1$, $i \in \{1, \ldots, r\}$ and $E_i[1] = E_i$. By (X.1.3), for each $i \in \{1, \ldots, r\}$, there exist

- an irreducible monomorphism $u_{ij} : E_i[j-1] \longrightarrow E_i[j]$ and an irreducible epimorphism $p_{ij} : E_i[j] \longrightarrow E_i[j-1]$, for $j \geq 2$,
- an almost split sequence

$$0 \longrightarrow E_i[1] \xrightarrow{\ u_{i2}\ } E_i[2] \longrightarrow E_{i+1}[1] \longrightarrow 0,$$

  in mod B, and
- an almost split sequence

$$0 \longrightarrow E_i[j-1] \xrightarrow{\begin{bmatrix} p_{i,j-1} \\ u_{ij} \end{bmatrix}} E_{i+1}[j-2] \oplus E_i[j] \xrightarrow{[\,u_{i+1,j-1}\ \ p_{ij}\,]} E_{i+1}[j-1] \longrightarrow 0$$

  in mod B, for $j \geq 3$, where we set $E_{r+1} = E_1$.

Because the tube $\mathcal{T}_\lambda^B$ is standard, any non-zero non-isomorphism between indecomposable modules in $\mathcal{T}_\lambda^B$ is a K-linear combination of the irreducible morphisms u_{ij} and p_{ij}. Observe also that, for each $i \in \{1, \ldots, r\}$, there is an infinite chain of irreducible monomorphisms

$$(*_i)\ \ E_i = E_i[1] \xrightarrow{\ u_{i1}\ } E_i[2] \xrightarrow{\ u_{i3}\ } E_i[3] \longrightarrow \cdots \longrightarrow E_i[j-1] \xrightarrow{\ u_{ij}\ } E_i[j] \longrightarrow \cdots$$

induced by the ray of the stable tube $\mathcal{T}_\lambda^B$ starting at E_i.

Given $i \in \{1, \ldots, r\}$, we consider the family $\mathcal{E}_i$ consisting of the modules $E_i[j]$, with $j \geq 1$. Fix a decomposition

$$R_\lambda = N_1 \oplus N_2 \oplus \ldots \oplus N_m$$

of the module R_λ into a direct sum of indecomposable modules $N_1, \ldots, N_m$ from the tube $\mathcal{T}_\lambda^B$. Then the homomorphism g has the form

$$g = \begin{bmatrix} g_1 \\ g_2 \\ \vdots \\ g_m \end{bmatrix} : M \longrightarrow N_1 \oplus N_2 \oplus \ldots \oplus N_m = R_\lambda,$$

where $g_s \in \mathrm{Hom}_B(M, N_s)$, for $s = 1, \ldots, m$.

Because the module M is postprojective, it has no non-zero direct summands lying in the tube $\mathcal{T}_\lambda^B$. Then, for any integer $k \geq 1$ and any $s \in \{1, \ldots, m\}$, the homomorphism $g_s : M \longrightarrow N_s$ has a factorisation

$$
\begin{array}{ccc}
M & \xrightarrow{\quad g_s \quad} & N_s \\
& {\scriptstyle g_{sk}} \searrow \quad \nearrow {\scriptstyle h_{sk}} & \\
& N_{sk} &
\end{array}
$$

with a module N_{sk} in $\mathrm{add}\, \mathcal{T}_\lambda^B$ such that the restriction of the homomorphism $h_{sk} : N_{sk} \longrightarrow N_s$ to any indecomposable direct summand of N_{sk} is a K-linear combination of k irreducible morphisms of the form u_{ij} and p_{ij}. Because there is a common bound of the regular lengths of the modules $N_1, N_2, \ldots, N_m$ then, invoking the standardness of the tube $\mathcal{T}_\lambda^B$ again, we infer that, given $i \in \{1, \ldots, r\}$ and $s \in \{1, \ldots, m\}$, there exist a module Z_{is} in $\mathrm{add}\, \mathcal{E}_i$ and two homomorphisms $v_{is} : M \longrightarrow Z_{is}$ and $w_{is} : Z_{is} \longrightarrow N_s$ of B-modules such that $g_s = w_{is} v_{is}$. If we set $Z_i = Z_{i1} \oplus \ldots \oplus Z_{im}$ and

$$
v_i = \begin{bmatrix} v_{i1} \\ v_{i2} \\ \vdots \\ v_{im} \end{bmatrix} : M \longrightarrow Z_i, \quad w_i = \begin{bmatrix} w_{i1} & 0 & \cdots & 0 \\ 0 & w_{i2} & \cdots & 0 \\ \vdots & & \ddots & \vdots \\ 0 & 0 & \cdots & w_{im} \end{bmatrix} : Z_i \longrightarrow R_\lambda,
$$

then $g : M \longrightarrow R_\lambda$ has the factorisation $g_s = w_i v_i$. It follows that $v_i : M \longrightarrow Z_i$ is a monomorphism. Because the indecomposable direct summands of Z_i lie on the ray $(*_i)$ starting from $E_i[1] = E_i$ and the irreducible morphisms $u_{ij} : E_i[j-1] \longrightarrow E_i[j]$ in $(*_i)$ are monomorphisms, then there are integers $q_i \geq 1$ and $j_i \geq 1$ such that there is a monomorphism $\varphi_{ij} : Z_i \longrightarrow E_i[j]^{q_i}$, for each $j \geq j_i$ and $i \in \{1, \ldots, r\}$, where $r = r_\lambda^B$.

This shows that the module M is cogenerated by each of the indecomposable modules $E_i[j]$, with $j \geq j_i$ and $i \in \{1, \ldots, r_\lambda^B\}$. It follows that all but a finite number of the indecomposable modules in the tube $\mathcal{T}_\lambda^B$ cogenerate the module M. The proof is complete. $\qquad\square$

The following important corollary is a consequence of (3.7).

3.8. Corollary. *Let B be a concealed algebra of a Euclidean type Q, and let*
$$\mathcal{T}^B = \{\mathcal{T}_\lambda^B\}_{\lambda \in \mathbb{P}_1(K)}$$
be the family of all stable tubes of $\Gamma(\operatorname{mod} B)$. For any fixed $\lambda \in \mathbb{P}_1(K)$, all but a finite number of indecomposable modules of the tube $\mathcal{T}_\lambda^B$ are faithful.

Proof. Fix a tube $\mathcal{T}_\lambda^B$ of the family $\mathcal{T}^B$. It follows from (3.7) that the postprojective B-module B_B is cogenerated by all but a finite number of the indecomposable modules of the tube $\mathcal{T}_\lambda^B$. It follows that all but a finite number of the indecomposable modules of the tube $\mathcal{T}_\lambda^B$ are faithful, by (VI.2.2).

XII.4. A controlled property of the Euler form of a concealed algebra of Euclidean type

We recall from (VII.5.10), (VIII.4.3) and (IX.3.3) that, if B is the path algebra of a Dynkin quiver, or more generally a representation-finite tilted algebra, then the map $X \mapsto \mathbf{dim}\, X$ induces a bijection between the set of isoclasses of indecomposable B-modules and the set of all positive roots of the Euler quadratic form $q_B : K_0(B) \longrightarrow \mathbb{Z}$ of B. On the other hand, we have seen in the previous sections that for concealed algebras C of Euclidean types we may have even infinitely many pairwise non-isomorphic indecomposable modules having the same dimension vector. However, the module category $\operatorname{mod} C$ of C is controlled by the Euler quadratic form $q_C : K_0(C) \longrightarrow \mathbb{Z}$ of C in the following sense.

4.1. Definition. Assume that A is a basic connected algebra of finite global dimension, $q_A : K_0(A) \longrightarrow \mathbb{Z}$ is the Euler quadratic form of A, and $Q = Q_A$ the ordinary quiver of A with $Q_0 = \{1, \ldots, n\}$, where $n \geq 1$ is the rank of the Grothendieck group $K_0(A)$ of A.

   (a) A vector $\mathbf{x} = [x_1 \ldots x_n]^t \in K_0(A) = \mathbb{Z}^n$ is defined to be **connected** if the full subquiver of Q, whose vertices are all $j \in Q_0$ such that $x_j \neq 0$, is connected.
   (b) The category $\operatorname{mod} A$ is defined to be **link controlled** by the quadratic form $q_A : K_0(A) \longrightarrow \mathbb{Z}$ if the following three conditions are satisfied.
      (b1) For any indecomposable A-module X, we have
         $q_A(\mathbf{dim}\, X) \in \{0, 1\}$.
      (b2) For any connected positive vector $\mathbf{x} \in K_0(A)$ with $q_A(\mathbf{x}) = 1$ there is precisely one indecomposable A-module X, up to isomorphism, such that $\mathbf{dim}\, X = \mathbf{x}$.

(b3) For any connected positive vector $\mathbf{x} \in K_0(A)$, with $q_A(\mathbf{x})=0$, there is an infinite family $\{X_\lambda\}_{\lambda \in \Lambda}$ of pairwise non-isomorphic indecomposable A-modules X_λ such that $\mathbf{dim}\, X_\lambda = \mathbf{x}$, for any $\lambda \in \Lambda$.

(c) The category $\mathrm{mod}\, A$ is defined to be **controlled** by the quadratic form $q_A : K_0(A) \longrightarrow \mathbb{Z}$ if the following three conditions are satisfied.

(c1) For any indecomposable A-module X in $\mathrm{mod}\, A$, we have
$$q_A(\mathbf{dim}\, X) \in \{0, 1\}.$$

(c2) For any positive vector $\mathbf{x} \in K_0(A)$, with $q_A(\mathbf{x}) = 1$, there is precisely one indecomposable A-module X, up to isomorphism, such that $\mathbf{dim}\, X = \mathbf{x}$.

(c3) For any positive vector $\mathbf{x} \in K_0(A)$, with $q_A(\mathbf{x}) = 0$, there is an infinite family $\{X_\lambda\}_{\lambda \in \Lambda}$ of pairwise non-isomorphic indecomposable A-modules X_λ such that $\mathbf{dim}\, X_\lambda = \mathbf{x}$, for any $\lambda \in \Lambda$.

We note that $\mathrm{mod}\, A$ is link controlled by q_A if it is controlled by q_A in the sense of Ringel [215, p.79]. Then our definition of a module category $\mathrm{mod}\, A$ controlled by q_A differs from that one introduced in [215]. In (6.2) of Chapter XVII, we give an example of an algebra B such that B is a tubular extension of the Kronecker algebra, the module category $\mathrm{mod}\, B$ is link controlled by q_B, but $\mathrm{mod}\, B$ is not controlled by q_B.

We recall from (VI.4.7), (VII.4.2), and (XI.3.7) that, if B is a concealed algebra of Euclidean type, then the Euler quadratic form $q_B : K_0(B) \longrightarrow \mathbb{Z}$ of B is positive semidefinite of corank 1 and there is a positive generator of the group $\mathrm{rad}\, q_B$, that is, a unique positive vector $\mathbf{h}_B \in K_0(B)$ such that $\mathrm{rad}\, q_B = \mathbb{Z} \cdot \mathbf{h}_B$. Recall also that, for a regular B-module X, we denote by $r\ell(X)$ the regular length of X, in the serial category $\mathrm{add}\,\mathcal{R}(B)$.

4.2. Theorem. *Let B be a concealed algebra of Euclidean type and let*

$$\boldsymbol{\mathcal{T}}^B = \{\mathcal{T}_\lambda^B\}_{\lambda \in \mathbb{P}_1(K)}$$

be the unique tubular $\mathbb{P}_1(K)$-family of stable tubes $\mathcal{T}_\lambda^B$ separating the post-projective component $\mathcal{P}(B)$ from the preinjective component $\mathcal{Q}(B)$. Let $\mathbf{h}_B$ be the positive generator of $\mathrm{rad}\, q_B$, where

$$q_B : K_0(B) \longrightarrow \mathbb{Z}$$

is the Euler quadratic form of B.

(a) *If X is an indecomposable B-module lying in a tube $\mathcal{T}_\lambda^B$ of rank $r_\lambda^B \geq 1$, where $\lambda \in \mathbb{P}_1(K)$, then*

$$\mathbf{dim}\, X = m \cdot \mathbf{h}_B \quad \text{if and only if} \quad r\ell(X) = m \cdot r_\lambda^B.$$

(b) *The category* $\mod B$ *is link controlled by the quadratic form* $q_B : K_0(C) \longrightarrow \mathbb{Z}$.

(c) *For any positive vector* $\mathbf{x} \in K_0(B)$ *with* $q_B(\mathbf{x}) = 0$, *there is a* $\mathbb{P}_1(K)$-*family* $\{X_\lambda\}_{\lambda \in \mathbb{P}_1(K)}$ *of pairwise non-isomorphic indecomposable* B-*modules* X_λ *such that* X_λ *lies in the tube* $\mathcal{T}_\lambda^B$ *and* $\dim X_\lambda = \mathbf{x}$, *for any* $\lambda \in \mathbb{P}_1(K)$.

Proof. Let X be an indecomposable B-module. If X is postprojective or preinjective then X is directing, and, consequently, $q_B(\dim X) = 1$, by (IX.1.5).

Assume that X is regular, say X belongs to a tube $\mathcal{T}_\lambda^B$ of rank r_λ^B. We conclude from (XI.3.6) that

$$q_B(\dim X) \in \{0, 1\}, \text{ and } q_B(\dim X) = 0$$

if and only if $r\ell(X)$ is divisible by r_λ^B. We claim now that the equality $r\ell(X) = r_\lambda^B$ implies $\dim X = \mathbf{h}_B$.

We know from (3.1) that $B \cong \operatorname{End}_A(T')$, for a postprojective tilting module T' over the path algebra $A = K\Delta$ of a quiver

$$\Delta \in \{\Delta(\widetilde{\mathbb{A}}_{p,q}), \text{with } 1 \le p \le q, \; \Delta(\widetilde{\mathbb{D}}_m), \text{with } m \ge 4, \; \Delta(\widetilde{\mathbb{E}}_6), \Delta(\widetilde{\mathbb{E}}_\lambda), \Delta(\widetilde{\mathbb{E}}_8)\}.$$

Let $C \cong \operatorname{End}_A(T'')$ be the associated canonical algebra of type Δ, where T'' is a postprojective tilting A-module. Observe that the category $\operatorname{add} \mathcal{R}(A)$ is contained in each of the torsion class $\mathcal{T}(T')$ and $\mathcal{T}(T'')$, that are determined by the postprojective tilting A-modules T' and T''.

Then, it follows from (VI.4.3) and (VI.4.5) that there are group isomorphisms $f : K_0(A) \longrightarrow K_0(B)$ and $g : K_0(A) \longrightarrow K_0(C)$ such that

$$f(\dim M) = \dim \operatorname{Hom}_A(T', M),$$
$$g(\dim M) = \dim \operatorname{Hom}_A(T'', M) \text{ and}$$
$$q_B(f(\dim M)) = q_A(\dim M) = q_C(g(\dim M)),$$

for any module M from $\operatorname{add} \mathcal{R}(A)$. Let

$$\boldsymbol{\mathcal{T}}^C = \{\mathcal{T}_\lambda^C\}_{\lambda \in \mathbb{P}_1(K)}$$

be the tubular family described in (2.7), for $C = C(p, q)$, and in (2.11), for $C = C(p, q, r)$, where (p, q, r) is any of the triples $(2, 2, m - 2)$, with $m \ge 4$, $(2, 3, 3)$, $(2, 3, 4)$, or $(2, 3, 5)$. We recall from (2.13) that the regular part $\mathcal{R}(C)$ of $\Gamma(\mod C)$ consists of the tubes $\mathcal{T}_\lambda^C$ from $\boldsymbol{\mathcal{T}}^C$.

By (3.4), there exists a tubular family

$$\boldsymbol{\mathcal{T}}^A = \{\mathcal{T}_\lambda^A\}_{\lambda \in \mathbb{P}_1(K)}$$

of stable tubes such that $\mathcal{R}(A)$ consists of the tubes $\mathcal{T}_\lambda^A$ from $\boldsymbol{\mathcal{T}}^A$. Similarly, there exists a tubular family

$$\boldsymbol{\mathcal{T}}^B = \{\mathcal{T}_\lambda^B\}_{\lambda \in \mathbb{P}_1(K)}$$

of stable tubes such that $\mathcal{R}(B)$ consists of the tubes $\mathcal{T}_\lambda^B$ from $\boldsymbol{\mathcal{T}}^B$. It follows from (VII.4.5) that, the tubular families $\boldsymbol{\mathcal{T}}^B$ and $\boldsymbol{\mathcal{T}}^C$ are the images of the tubular family $\boldsymbol{\mathcal{T}}^A$ under the functors $\mathrm{Hom}_A(T', -)$ and $\mathrm{Hom}_A(T'', -)$, respectively.

We recall from (1.18) and (2.14) that

$$\mathrm{rad}\, q_C = \mathbb{Z} \cdot \mathbf{h}_C$$

and, for each C-module Z of the regular length r_λ^B in the tubes $\mathcal{T}_\lambda^C$, we have $\mathbf{dim}\, Z = \mathbf{h}_C$.

It follows that the vector $\mathbf{h}' = g^{-1}(\mathbf{h}_C)$ generates the group $\mathrm{rad}\, q_A$ and the vector

$$\mathbf{h}'' = fg^{-1}(\mathbf{h}_C) = f(\mathbf{h}')$$

generates the group $\mathrm{rad}\, q_B$. Moreover, $\mathbf{h}'$ is the dimension vector of all A-modules Y of regular length r_λ^B in the tubes $\mathcal{T}_\lambda^A$ and $\mathbf{h}''$ is the dimension vector of all B-modules X of regular length r_λ^B in the tubes $\mathcal{T}_\lambda^B$. It follows that $\mathbf{h}'$ is a positive generator of the group $\mathrm{rad}\, q_A = \mathbb{Z} \cdot \mathbf{h}_Q$ (see (VII.4.2)) and $\mathbf{h}''$ is a positive generator of the group $\mathrm{rad}\, q_B = \mathbb{Z} \cdot \mathbf{h}_B$. Hence we conclude that

$$\mathbf{h}'' = \mathbf{h}_B, \quad \mathbf{h}' = \mathbf{h}_Q, \quad \text{and} \quad \mathbf{dim}\, X = \mathbf{h}_B,$$

for every B-module X of the regular length r_λ^B in a tube $\mathcal{T}_\lambda^B$. Furthermore, it follows from (XI.3.10) that, for each B-module X in $\mathcal{T}_\lambda^B$, we have $r\ell(X) = m \cdot r_\lambda^B$ if and only if $\mathbf{dim}\, X = m \cdot \mathbf{h}_B$. This proves the statement (a).

To prove (b), it remains to show that the conditions (b) and (c) of (4.1) are satisfied.

Let $\mathbf{x} \in K_0(B)$ be a positive vector such that $q_B(\mathbf{x}) = 1$. Clearly then $\mathbf{x} = \mathbf{dim}\, X$ for some B-module X. Choose such a module X with $\dim_K \mathrm{End}_B(X)$ minimal. Let

$$X = X_1 \oplus \ldots \oplus X_t$$

be a decomposition of X into a direct sum of indecomposable B-modules.

If $t = 1$, then $x = \mathbf{dim}\, X = \mathbf{dim}\, X_1$, and there is nothing to show.

Assume that $t \geq 2$. It follows from (VIII.2.8) and our minimality assumption that $\mathrm{Ext}_B^1(X_i, X_j) = 0$, for all $i \neq j$, see also (XIV.2.3) for more details. Because B is tilted, then gl.dim $B \leq 2$ and we have the following equalities

$$1 = q_B(\mathbf{x})$$
$$= q_B(\mathbf{dim}\, X)$$
$$= \chi_B(X, X)$$
$$= \dim_K \mathrm{End}_B(X) - \dim_K \mathrm{Ext}^1_B(X, X) + \dim_K \mathrm{Ext}^2_B(X, X)$$
$$= \sum_{i \neq j} \left[\dim_K \mathrm{Hom}_B(X_i, X_j) + \dim_K \mathrm{Ext}^2_B(X_i, X_j) \right]$$
$$+ \sum_{i=1}^{t} \left[\dim_K \mathrm{End}_B(X_i) - \dim_K \mathrm{Ext}^1_B(X_i, X_i) + \dim_K \mathrm{Ext}^2_B(X_i, X_i) \right]$$
$$= \sum_{i \neq j} \left[\dim_K \mathrm{Hom}_B(X_i, X_j) + \dim_K \mathrm{Ext}^2_B(X_i, X_j) \right] + \sum_{i=1}^{t} \chi_B(X_i, X_i)$$
$$= \sum_{i \neq j} \left[\dim_K \mathrm{Hom}_B(X_i, X_j) + \dim_K \mathrm{Ext}^2_B(X_i, X_j) \right] + \sum_{i=1}^{t} q_B(\mathbf{dim}\, X_i).$$

Now we show that there exists i such that $1 \leq i \leq t$ and $q_B(\mathbf{dim}\, X_i) = 1$. Suppose to the contrary that $q_B(\mathbf{dim}\, X_i) = 0$, for any $i \in \{1, \ldots, t\}$. Then $X_1, \ldots, X_t$ are indecomposable regular modules, say X_i belongs to $\mathcal{T}^B_{\lambda_i}$, where $i \in \{1, \ldots, t\}$. Moreover, $q_B(\mathbf{dim}\, X_i) = 0$ implies that the regular length $r\ell(X_i)$ of X_i is greater or equal than the rank of $\mathcal{T}^B_{\lambda_i}$. Hence, if $\lambda_i = \lambda_j$, for some $i \neq j$, then $\mathrm{Hom}_B(X_i, X_j) \neq 0$ and $\mathrm{Hom}_B(X_i, X_j) \neq 0$. It follows that

$$\sum_{i \neq j} \dim_K \mathrm{Hom}_B(X_i, X_j) \geq 2,$$

and we get a contradiction. On the other hand, if the modules $X_1, \ldots, X_t$ lie in pairwise different tubes, then

$$\sum_{i \neq j} \dim_K \mathrm{Hom}_B(X_i, X_j) = 0.$$

Moreover, because pd $Z \leq 1$ and id $Z \leq 1$, for any indecomposable regular B-module Z, see (VIII.4.5) and (XI.3.3)(e), then

$$\sum_{i \neq j} \dim_K \mathrm{Ext}^2_B(X_i, X_j) = 0,$$

and again we get a contradiction. This shows that there exists $i \in \{1, \ldots, t\}$ such that $q_B(\mathbf{dim}\, X_i) = 1$.

Without lost of generality, we may assume that $i = 1$. Consequently, we have

- $q_B(\mathbf{dim}\, X_i) = 0$, for $i \in \{2, \ldots, t\}$, and
- $\sum_{i \neq j} \left[\dim_K \mathrm{Hom}_B(X_i, X_j) + \dim_K \mathrm{Ext}^2_B(X_i, X_j) \right] = 0.$

In particular, again, if $2 \leq i \leq t$, then $X_i \in \mathcal{T}^B_{\lambda_i}$, for some $\lambda_i \in \mathbb{P}_1(K)$, and the regular length $r\ell(X_i)$ of X_i is greater than or equal to the rank of $\mathcal{T}^B_{\lambda_i}$.

Invoking now (3.6), we conclude that the module X_1 also belongs to a tube $\mathcal{T}^B_{\lambda_1}$, for some $\lambda_1 \in \mathbb{P}_1(K) \setminus \{\lambda_2, \dots, \lambda_r\}$. We know also from the first part of our proof, (XI.3.6), and (XI.3.7) that, for each $i \in \{2, \dots, t\}$, we have $\mathbf{dim}\, X_i = m_i \cdot \mathbf{h}_B$, where $m_i \cdot r_{\lambda_i} = r\ell(X_i)$ is a positive integer. Then

$$\mathbf{x} = \mathbf{dim}\, X = \sum_{1 \leq i \leq t} \mathbf{dim}\, X_i = \mathbf{dim}\, X_1 + m \cdot \mathbf{h}_B,$$

where $m = m_2 + \dots + m_t$. Let $r = r^B_{\lambda_1}$ be the rank of the tube $\mathcal{T}^B_{\lambda_1}$ containing the module X_1. Consider now the ray

$$E = E[1] \longrightarrow E[2] \longrightarrow \cdots \longrightarrow E[l] \longrightarrow E[l+1] \longrightarrow \cdots$$

in the tube $\mathcal{T}^B_{\lambda_1}$, with E lying on the mouth of $\mathcal{T}_{\lambda_1}$, containing the module X_1, say $X_1 = E[l]$, for some $l \geq 1$.

We set $M = E[l + mr]$. Because $\mathbf{dim}\, E[sr] = s \cdot \mathbf{h}_B$, for any $s \geq 1$, then

$$
\begin{aligned}
\mathbf{dim}\, M &= \mathbf{dim}\, E[l+mr] \\
&= \mathbf{dim}\, E[l] + \mathbf{dim}\, E[mr] \\
&= \mathbf{dim}\, X_1 + m \cdot \mathbf{h}_B \\
&= \mathbf{x}.
\end{aligned}
$$

It follows that M is a required indecomposable B-module such that $\mathbf{dim}\, M = \mathbf{x}$.

It remains to prove that, if X is an indecomposable B-module such that $q_B(\mathbf{dim}\, X) = 1$, then X is uniquely determined (up to isomorphism) by $\mathbf{dim}\, X$. If X is postprojective or preinjective, then X is directing and (IX.3.1) applies; but if X is regular, then (XI.3.12) applies. This completes the proof of (b).

(c) Let $\mathbf{x} \in K_0(B)$ be a positive vector such that $q_B(\mathbf{x}) = 0$. Then $\mathbf{x} \in \mathrm{rad}\, q_B = \mathbb{Z} \cdot \mathbf{h}_B$, that is, $\mathbf{x} = m \cdot \mathbf{h}_B$, for some integer $m \geq 1$.

Given $\lambda \in \mathbb{P}_1(K)$, choose an indecomposable module X_λ in the tube $\mathcal{T}^B_\lambda$ of regular length $m \cdot r^B_\lambda$. Then it follows from (a) that

$$\mathbf{dim}\, X_\lambda = m \cdot \mathbf{h}_B.$$

Obviously, $X_\lambda \not\cong X_\mu$, for $\lambda \neq \mu$, because the tubes $\mathcal{T}^B_\lambda$ and $\mathcal{T}^B_\mu$ are different. This finishes the proof of (c) and completes the proof of the theorem. $\qquad \square$

XII.5. Exercises

1. Let (p, q, r) be a triple of positive integers such that $r \geq q \geq p \geq 2$ and let $C(p, q, r) = K\Delta(p, q, r)/I(p, q, r)$ be the algebra defined in (1.2) Show that

(a) pd $S(j) \leq 1$, for any $j \neq w$;

(b) gl.dim $C(p, q, r) = 2$.

Hint: Apply the arguments used in the proof of (1.4).

2. Let $C = C(p, q, r)$ be one of the canonical K-algebras $C(2, 2, m-2)$, with $m \geq 4$, $C(2, 3, 3)$, $C(2, 3, 4)$, $C(2, 3, 5)$ of Euclidean type.

(a) Compute the Coxeter matrix of the algebra C and the Coxeter transformation $\Phi_C : K_0(C) \longrightarrow K_0(C)$ of C.

(b) Show that $\Phi_C(\mathbf{h}_C) = \mathbf{h}_C$, where $\mathbf{h}_C \in K_0(C)$ is the canonical radical vector with the identity coordinate over each vertex of the quiver $\Delta(p, q, r)$ of C.

3. Let $C = C(p, q)$, where $q \geq p \geq 1$.

(a) Under the notation of (2.8), prove that, for each $i \in \{1, \dots, p\}$, there exists an isomorphism $\tau_C E_{i+1}^{(\infty)} \cong E_i^{(\infty)}$ of right C-modules.

(b) Show that the regular C-modules $E_p^{(\infty)}$, $E_q^{(0)}$ and $E^{(\lambda)}$, with $\lambda \in K \setminus \{0\}$, are simple regular. **Hint:** Follow the proof of (2.8).

4. Let $C = C(p, q, r)$, where (p, q, r) is any of the triples $(2, 2, m-2)$, with $m \geq 4$, $(2, 3, 3)$, $(2, 3, 4)$, or $(2, 3, 5)$. Under the notation of (2.12), prove that there exist isomorphisms of right C-modules:

(a) $\tau_C E_2^{(\infty)} \cong E_1^{(\infty)}$ and $\tau_C E_1^{(\infty)} \cong E_2^{(\infty)}$,

(b) $\tau_C E_{i+1}^{(1)} \cong E_i^{(1)}$, for each $i \in \{1, \dots, p\}$,

(c) $\tau_C E_{j+1}^{(0)} \cong E_j^{(0)}$, for each $j \in \{1, \dots, q\}$,

(d) $\tau_C E_{t+1}^{(1)} \cong E_t^{(1)}$, for each $t \in \{1, \dots, r\}$, and

(e) $\tau_C E^{(\lambda)} \cong E^{(\lambda)}$, for each $\lambda \in K \setminus \{0, 1\}$.

Show that each of the modules $E_1^{(\infty)}$, $E_2^{(\infty)}$, $E_i^{(1)}$, with $i \in \{1, \dots, p\}$, $E_j^{(0)}$, with $j \in \{1, \dots, q\}$, and $E^{(\lambda)}$, with $\lambda \in K \setminus \{0, 1\}$, is simple regular.

Hint: Follow the proof of (2.12).

5. Let $C = C(p, q)$, where $q \geq p \geq 1$. Describe the dimension vectors of the indecomposable regular C-modules.

Hint: Apply (XI.3.10) and (2.8).

6. Let $C = C(p, q, r)$, where (p, q, r) is any of the triples $(2, 2, m-2)$, with $m \geq 4$, $(2, 3, 3)$, $(2, 3, 4)$, or $(2, 3, 5)$. Describe the dimension vectors of the indecomposable regular C-modules.

Hint: Apply (XI.3.10) and (2.12).

7. Let B be the algebra given by the quiver

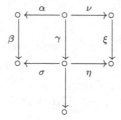

bound by two commutativity relations $\alpha\beta = \gamma\sigma\delta$ and $\xi\eta = \rho\nu$.

(a) Show that B is a concealed algebra of the Euclidean type $\widetilde{\mathbb{D}}_9$.

(b) Describe the dimension vectors $\mathbf{dim}\, M \in \mathbb{Z}^{10}$ of the indecomposable regular B-modules M.

Hint: Apply the criterion (XI.5.1).

8. Let B be the algebra given by the quiver

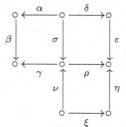

bound by two commutativity relations $\alpha\beta = \gamma\sigma$ and $\nu\gamma = \xi\eta$.

(a) Show that B is a concealed algebra of the Euclidean type $\widetilde{\mathbb{E}}_6$.

(b) Describe the dimension vectors $\mathbf{dim}\, M \in \mathbb{Z}^7$ of the indecomposable regular B-modules M.

9. Let B be the algebra given by the quiver

bound by two commutativity relations $\alpha\beta = \sigma\gamma$, $\sigma\delta = \delta\varepsilon$, and $\nu\rho\xi\eta$.

(a) Show that B is a concealed algebra of the Euclidean type $\widetilde{\mathbb{E}}_7$.

(b) Describe the dimension vectors $\mathbf{dim}\, M \in \mathbb{Z}^8$ of the indecomposable regular B-modules M.

10. Let B be the algebra given by the quiver

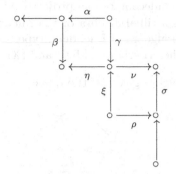

bound by the relations $\xi\eta = 0$, $\alpha\beta = \gamma\eta$, and $\xi\nu = \rho\sigma$.

(a) Show that B is a concealed algebra of the Euclidean type $\widetilde{\mathbb{E}}_8$.

(b) Describe the dimension vectors $\mathbf{dim}\, M \in \mathbb{Z}^9$ of the indecomposable regular B-modules M.

11. Let B be a concealed algebra of the Euclidean type and $\mathcal{T}$ a stable tube in the Auslander–Reiten quiver $\Gamma(\mathrm{mod}\, B)$ of B. Prove that all but finitely many indecomposable B-modules in $\mathcal{T}$ are faithful.

Hint: Apply (VI.2.2) and (3.4).

12. Let $A = \begin{bmatrix} K & 0 \\ K^2 & K \end{bmatrix}$ be the Kronecker algebra and let

$$\boldsymbol{\mathcal{T}}^A = \{\mathcal{T}_\lambda^A\}_{\lambda \in \mathbb{P}_1(K)}$$

be the $\mathbb{P}_1(K)$-family of pairwise orthogonal standard stable tubes $\mathcal{T}_\lambda^B$ described in (XI.4). Show that all modules lying on the mouth of the rank one tubes of $\boldsymbol{\mathcal{T}}^A$ are not faithful.

13. Let B be the algebra given by the quiver

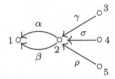

bound by the three relations $\gamma\alpha = 0$, $\rho\beta = 0$, and $\sigma\alpha = \sigma\beta$.

(a) Show that $\Gamma(\mathrm{mod}\, B)$ admits a preinjective component $\mathcal{Q}$ containing all the indecomposable injective modules.

(b) Show that $\Gamma(\mathrm{mod}\,B)$ admits a postprojective component $\mathcal{P}$ containing exactly two indecomposable projective modules.

(c) Show that B is a tilted algebra of the Euclidean type $\widetilde{\mathbb{D}}_4$, but it is not a concealed algebra of Euclidean type.
   Hint: Apply the criteria (VIII.5.6) and (XI.5.1).

14. Let B be the algebra given by the quiver

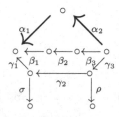

bound by the relations $\alpha_2\alpha_1 + \beta_3\beta_2\beta_1 + \gamma_3\gamma_2\gamma_1$, $\gamma_2\sigma = 0$, and $\gamma_3\rho = 0$.

(a) Show that $\Gamma(\mathrm{mod}\,B)$ admits a postprojective component $\mathcal{P}$ containing all the indecomposable projective modules.

(b) Show that $\Gamma(\mathrm{mod}\,B)$ admits a preinjective component $\mathcal{Q}$ containing all the indecomposable injective modules except two of them.

(c) Show that B is a tilted algebra of the Euclidean type $\widetilde{\mathbb{E}}_8$, but it is not a concealed algebra of Euclidean type.
   Hint: Apply the criteria (VIII.5.6) and (XI.5.1).

Chapter XIII

Indecomposable modules and tubes over hereditary algebras of Euclidean type

The aim of this chapter is to present a classification of indecomposable A-modules and a detailed description of the Auslander–Reiten quiver $\Gamma(\mathrm{mod}\, A)$ of any hereditary path algebra

$$A = KQ$$

of a finite acyclic quiver Q whose underlying graph $\overline{Q}$ is Euclidean. It follows from the structure theorem (XII.3.4) that the quiver $\Gamma(\mathrm{mod}\, A)$ has the disjoint union form

$$\Gamma(\mathrm{mod}\, A) = \mathcal{P}(A) \cup \mathcal{R}(A) \cup \mathcal{Q}(A),$$

and can be visualised as follows

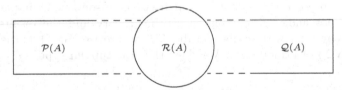

where $\mathcal{P}(A)$ is the unique postprojective component containing all the indecomposable projective A-modules, $\mathcal{Q}(A)$ is the unique preinjective component containing all the indecomposable injective A-modules, and the regular part $\mathcal{R}(A)$ is a $\mathbb{P}_1(K)$-family

$$\boldsymbol{T}^A = \{\mathcal{T}_\lambda^A\}_{\lambda \in \mathbb{P}_1(K)}$$

of stable tubes separating $\mathcal{P}(A)$ from $\mathcal{Q}(A)$. At most 3 of the tubes in $\boldsymbol{T}^A$ are of rank ≥ 2, the remaining ones are of rank 1.

The procedure presented in the proof of (IX.4.5) provides us with an efficient tool for a description of the components $\mathcal{P}(A)$, $\mathcal{Q}(A)$ and their indecomposable modules. Furthermore, it was shown in Section X.2 that the regular A-modules in a tube $\mathcal{T}_\lambda^A$ can be constructed by means of the simple regular modules in $\mathcal{T}_\lambda^A$. It follows that a classification of tubes and their indecomposable modules reduces to the classification of simple regular A-modules. Moreover, we know from (2.4)(a) below that, by applying the reflection functors, the problem reduces to the case when Q is a Euclidean graph Δ equipped with a particular admissible orientation of its edges. Therefore, we restrict our considerations to a quiver Δ which is one of the canonically oriented Euclidean quivers

$$\Delta(\widetilde{\mathbb{A}}_{p,q}), \text{ with } 1 \leq p \leq q, \ \Delta(\widetilde{\mathbb{D}}_m), \text{ with } m \geq 4, \ \Delta(\widetilde{\mathbb{E}}_6), \ \Delta(\widetilde{\mathbb{E}}_7), \ \Delta(\widetilde{\mathbb{E}}_8)$$

presented in Table 1.1 below, compare with Section XII.1.

In the first two sections, for each such a quiver $\Delta = (\Delta_0, \Delta_1)$, we compute the Coxeter matrix

$$\Phi_{K\Delta} \in \mathbb{M}_n(\mathbb{Z})$$

of the path algebra $K\Delta$ and the defect

$$\partial_{K\Delta} : K_0(K\Delta) \longrightarrow \mathbb{Z}$$

of $K\Delta$, where $K_0(K\Delta) \cong \mathbb{Z}^n$ is the Grothendieck of the algebra $K\Delta$ and $n = |\Delta_0|$.

We also present a complete list of pairwise non-isomorphic simple regular $K\Delta$-modules and, by applying the tilting theory and the results of the previous chapters, we describe all tubes $\mathcal{T}_\lambda^A$, where $\lambda \in \mathbb{P}_1(K)$, in the regular part $\mathcal{R}(K\Delta)$ of $\Gamma(\operatorname{mod} K\Delta)$. We show that the tubular type

$$\mathbf{m}_\Delta = (m_1, \ldots, m_s)$$

of the family $\mathbf{\mathcal{T}}^A = \{\mathcal{T}_\lambda^A\}_{\lambda \in \mathbb{P}_1(K)}$ equals (p,q), $(2,2,m-2)$, $(2,3,3)$, $(2,3,4)$, and $(2,3,5)$, if Δ is the quiver $\Delta(\widetilde{\mathbb{A}}_{p,q})$, with $1 \leq p \leq q$, $\Delta(\widetilde{\mathbb{D}}_m)$, with $m \geq 4$, $\Delta(\widetilde{\mathbb{E}}_6)$, $\Delta(\widetilde{\mathbb{E}}_7)$, and $\Delta(\widetilde{\mathbb{E}}_8)$, respectively.

In the third section we consider in details the case, when Q is the quiver

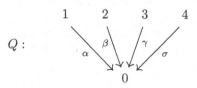

of the Euclidean type $\widetilde{\mathbb{D}}_4$. In this case, we give a complete description of the indecomposable modules over the path algebra KQ and we construct all the standard stable tubes in the $\mathbb{P}_1(K)$-family

$$\boldsymbol{\mathcal{T}}^{KQ} = \{\mathcal{T}_\lambda^{KQ}\}_{\lambda \in \mathbb{P}_1(K)}.$$

We show that three of the tubes in $\boldsymbol{\mathcal{T}}^{KQ}$ are of rank 2, and the remaining ones are of rank 1.

In other words, we give a solution of the problem of classifying the indecomposable K-linear representations of the four arrows quiver Q shown above, known as a **four subspace problem**, and solved by Gelfand and Ponomariev in [97].

XIII.1. Canonically oriented Euclidean quivers, their Coxeter matrices and the defect

Throughout this section, K is an algebraically closed field, Δ is any of the canonically oriented Euclidean quivers $\Delta(\widetilde{\mathbb{A}}_{p,q})$, with $q \geq p \geq 1$, $\Delta(\widetilde{\mathbb{D}}_m)$, with $m \geq 4$, $\Delta(\widetilde{\mathbb{E}}_6)$, $\Delta(\widetilde{\mathbb{E}}_7)$, and $\Delta(\widetilde{\mathbb{E}}_8)$ presented in Table 1.1 below, and

$$A = K\Delta$$

is the path algebra of the quiver Δ. We also agree, as in previous chapters, that

$$\mathbb{P}_1(K) = K \cup \{\infty\}$$

is a projective line over K.

Our main aim of this section is to present, for each Δ, a complete list of simple regular A-modules, and, as a consequence of our results in Chapter XII, a description of the tubular $\mathbb{P}_1(K)$-family

$$\boldsymbol{\mathcal{T}}^A = \{\mathcal{T}_\lambda^A\}_{\lambda \in \mathbb{P}_1(K)}$$

of all regular components of the Auslander–Reiten quiver $\Gamma(\operatorname{mod} A)$ of A.

We start by presenting in (1.1) below a list of particularly oriented Euclidean quivers $\Delta(\widetilde{\mathbb{A}}_{p,q})$, $\Delta(\widetilde{\mathbb{D}}_m)$, $\Delta(\widetilde{\mathbb{E}}_6)$, $\Delta(\widetilde{\mathbb{E}}_7)$, and $\Delta(\widetilde{\mathbb{E}}_8)$, called **canonically oriented Euclidean quivers**. They have appeared already in Chapter XII as a tool in our study of canonical algebras of Euclidean type and their regular components.

1.1. Table. Canonically oriented Euclidean quivers

$$\Delta(\widetilde{\mathbb{A}}_{p,q}):0\quad\begin{array}{c}1\longleftarrow 2\longleftarrow\cdots\longleftarrow p-1\\ \\ p\longleftarrow p+1\longleftarrow\cdots\longleftarrow p+q-2\end{array}\quad p+q-1,\quad 1\le p\le q.$$

$$\Delta(\widetilde{\mathbb{D}}_m):\quad\begin{array}{c}1\\ \\ 3\longleftarrow 4\longleftarrow\cdots\longleftarrow m-1\\ \\ 2\end{array}\quad\begin{array}{c}m\\ \\ \\ m+1\end{array}\quad,\ m\ge 4.$$

$$\Delta(\widetilde{\mathbb{E}}_6):\qquad\begin{array}{c}5\\ \downarrow\\ 4\\ \downarrow\\ 3\longrightarrow 2\longrightarrow 1\longleftarrow 6\longleftarrow 7.\end{array}$$

$$\Delta(\widetilde{\mathbb{E}}_7):\qquad\begin{array}{c}5\\ \downarrow\\ 4\longrightarrow 3\longrightarrow 2\longrightarrow 1\longleftarrow 6\longleftarrow 7\longleftarrow 8.\end{array}$$

$$\Delta(\widetilde{\mathbb{E}}_8):\qquad\begin{array}{c}4\\ \downarrow\\ 3\longrightarrow 2\longrightarrow 1\longleftarrow 5\longleftarrow 6\longleftarrow 7\longleftarrow 8\longleftarrow 9.\end{array}$$

Now we prove two important technical lemmata.

1.2. Lemma. *Let Δ be a canonically oriented Euclidean quiver, $n = |\Delta_0|$, $A = K\Delta$ and let $\Phi_A \in \mathbb{M}_n(\mathbb{Z})$ be the Coxeter matrix of A.*

(a) *If $\Delta = \Delta(\widetilde{\mathbb{A}}_{p,q})$ and $1 \leq p \leq q$, then $n = p + q$ and*

$$
\Phi_A = \begin{bmatrix}
-1 & 1 & 0 & 0 & \cdots & 0 & 0 & 1 & 0 & \cdots & 0 & 0 \\
-1 & 0 & 1 & 0 & \cdots & 0 & 0 & 1 & 0 & \cdots & 0 & 0 \\
-1 & 0 & 0 & 1 & \ddots & 0 & 0 & 1 & 0 & \cdots & 0 & 0 \\
\vdots & \vdots & & \ddots & \ddots & & & \vdots & \vdots & & \vdots & \vdots \\
-1 & 0 & 0 & 0 & \ddots & 1 & 0 & 1 & 0 & \cdots & 0 & 0 \\
-1 & 0 & 0 & 0 & \ddots & 0 & 1 & 1 & 0 & \cdots & 0 & 0 \\
-1 & 0 & 0 & 0 & \cdots & 0 & 0 & 1 & 0 & \cdots & 0 & 1 \\
-1 & 1 & 0 & 0 & \cdots & 0 & 0 & 0 & 1 & \ddots & 0 & 0 \\
\vdots & \vdots & \vdots & \vdots & & \vdots & \vdots & \vdots & & \ddots & \vdots & \vdots \\
-1 & 1 & 0 & 0 & \cdots & 0 & 0 & 0 & 0 & \ddots & 1 & 0 \\
-1 & 1 & 0 & 0 & \cdots & 0 & 0 & 0 & 0 & \ddots & 0 & 1 \\
-2 & 1 & 0 & 0 & \cdots & 0 & 0 & 1 & 0 & \cdots & 0 & 1
\end{bmatrix} \in \mathbb{M}_{p+q}(\mathbb{Z}).
$$

(b) *If $\Delta = \Delta(\widetilde{\mathbb{D}}_m)$ and $m \geq 4$, then $n = m + 1$ and*

$$
\Phi_A = \begin{bmatrix}
-1 & 0 & 1 & 0 & 0 & \cdots & 0 & 0 & 0 & 0 \\
0 & -1 & 1 & 0 & 0 & \cdots & 0 & 0 & 0 & 0 \\
-1 & -1 & 1 & 1 & 0 & \cdots & 0 & 0 & 0 & 0 \\
-1 & -1 & 1 & 0 & 1 & \ddots & 0 & 0 & 0 & 0 \\
\vdots & \vdots & \vdots & \vdots & & \ddots & \ddots & & \vdots & \vdots \\
-1 & -1 & 1 & 0 & 0 & \ddots & 1 & 0 & 0 & 0 \\
-1 & -1 & 1 & 0 & 0 & \cdots & 0 & 1 & 0 & 0 \\
-1 & -1 & 1 & 0 & 0 & \cdots & 0 & 0 & 1 & 1 \\
-1 & -1 & 1 & 0 & 0 & \cdots & 0 & 0 & 0 & 1 \\
-1 & -1 & 1 & 0 & 0 & \cdots & 0 & 0 & 1 & 0
\end{bmatrix} \in \mathbb{M}_{m+1}(\mathbb{Z}).
$$

(c) *If $\Delta = \Delta(\widetilde{\mathbb{E}}_6)$, then $n = 7$ and*

$$
\Phi_A = \begin{bmatrix}
-1 & 1 & 0 & 1 & 0 & 1 & 0 \\
-1 & 0 & 1 & 1 & 0 & 1 & 0 \\
-1 & 0 & 0 & 1 & 0 & 1 & 0 \\
-1 & 1 & 0 & 0 & 1 & 1 & 0 \\
-1 & 1 & 0 & 0 & 0 & 1 & 0 \\
-1 & 1 & 0 & 1 & 0 & 0 & 1 \\
-1 & 1 & 0 & 1 & 0 & 0 & 0
\end{bmatrix} \in \mathbb{M}_7(\mathbb{Z}).
$$

(d) *If $\Delta = \Delta(\widetilde{\mathbb{E}}_7)$, then $n = 8$ and*

$$
\Phi_A = \begin{bmatrix}
-1 & 1 & 0 & 0 & 1 & 1 & 0 & 0 \\
-1 & 0 & 1 & 0 & 1 & 1 & 0 & 0 \\
-1 & 0 & 0 & 1 & 1 & 1 & 0 & 0 \\
-1 & 0 & 0 & 0 & 1 & 1 & 0 & 0 \\
-1 & 1 & 0 & 0 & 0 & 1 & 0 & 0 \\
-1 & 1 & 0 & 0 & 1 & 0 & 1 & 0 \\
-1 & 1 & 0 & 0 & 1 & 0 & 0 & 1 \\
-1 & 1 & 0 & 0 & 1 & 0 & 0 & 0
\end{bmatrix} \in \mathbb{M}_8(\mathbb{Z}).
$$

(e) *If $\Delta = \Delta(\widetilde{\mathbb{E}}_8)$, then $n = 9$ and*

$$\Phi_A = \begin{bmatrix} -1 & 1 & 0 & 1 & 1 & 0 & 0 & 0 & 0 \\ -1 & 0 & 1 & 1 & 1 & 0 & 0 & 0 & 0 \\ -1 & 0 & 0 & 1 & 1 & 0 & 0 & 0 & 0 \\ -1 & 1 & 0 & 0 & 1 & 0 & 0 & 0 & 0 \\ -1 & 1 & 0 & 1 & 0 & 1 & 0 & 0 & 0 \\ -1 & 1 & 0 & 1 & 0 & 0 & 1 & 0 & 0 \\ -1 & 1 & 0 & 1 & 0 & 0 & 0 & 1 & 0 \\ -1 & 1 & 0 & 1 & 0 & 0 & 0 & 0 & 1 \\ -1 & 1 & 0 & 1 & 0 & 0 & 0 & 0 & 0 \end{bmatrix} \in \mathbb{M}_9(\mathbb{Z}).$$

Proof. We recall from (III.3.14) that

$$\Phi_A = -\mathbf{C}_A^t \mathbf{C}_A^{-1} \in \mathbb{M}_n(\mathbb{Z}),$$

where $n = |\Delta_0|$ and $\mathbf{C}_A$ is the Cartan matrix of the path algebra $A = K\Delta$.

(a) Assume that $\Delta = \Delta(\widetilde{\mathbb{A}}_{p,q})$ and $1 \le p \le q$. Then the Cartan matrix $\mathbf{C}_A$ of $A = K\Delta$ and its inverse are the square $(p+q) \times (p+q)$ matrices

$$\mathbf{C}_A = \begin{bmatrix} 1 & 1 & 1 & \cdots & 1 & 1 & 1 & \cdots & 1 & 2 \\ 0 & 1 & 1 & \cdots & 1 & 0 & 0 & \cdots & 0 & 1 \\ 0 & 0 & 1 & \cdots & 1 & 0 & 0 & \cdots & 0 & 1 \\ \vdots & & & \ddots & \vdots & \vdots & \vdots & & \vdots & \vdots \\ 0 & 0 & 0 & \cdots & 1 & 0 & 0 & \ddots & 0 & 1 \\ 0 & 0 & 0 & \cdots & 0 & 1 & 1 & \cdots & 1 & 1 \\ 0 & 0 & 0 & \cdots & 0 & 0 & 1 & \cdots & 1 & 1 \\ \vdots & & & \ddots & & & \ddots & & \vdots & \vdots \\ 0 & 0 & 0 & \cdots & 0 & 0 & 0 & \cdots & 1 & 1 \\ 0 & 0 & 0 & \cdots & 0 & 0 & 0 & \cdots & 0 & 1 \end{bmatrix} \in \mathbb{M}_{p+q}(\mathbb{Z}),$$

$$\mathbf{C}_A^{-1} = \begin{bmatrix} 1 & -1 & 0 & 0 & \cdots & 0 & 0 & -1 & 0 & \cdots & 0 & 0 \\ 0 & 1 & -1 & 0 & \cdots & 0 & 0 & 0 & 0 & \cdots & 0 & 0 \\ 0 & 0 & 1 & -1 & \ddots & 0 & 0 & 0 & 0 & \cdots & 0 & 0 \\ \vdots & \vdots & & \ddots & \ddots & \vdots & \vdots & \vdots & & & \vdots & \vdots \\ 0 & 0 & 0 & 0 & \ddots & -1 & 0 & 0 & 0 & \cdots & 0 & 0 \\ 0 & 0 & 0 & 0 & \ddots & 1 & -1 & 0 & 0 & \cdots & 0 & 0 \\ 0 & 0 & 0 & 0 & \cdots & 0 & 1 & 0 & 0 & \cdots & 0 & -1 \\ 0 & 0 & 0 & 0 & \cdots & 0 & 0 & 1 & -1 & \ddots & 0 & 0 \\ \vdots & \vdots & \vdots & \vdots & & \vdots & \vdots & & \ddots & \ddots & \vdots & \vdots \\ 0 & 0 & 0 & 0 & \cdots & 0 & 0 & 0 & 0 & \ddots & -1 & 0 \\ 0 & 0 & 0 & 0 & \cdots & 0 & 0 & 0 & 0 & \ddots & 1 & -1 \\ 0 & 0 & 0 & 0 & \cdots & 0 & 0 & 0 & 0 & \cdots & 0 & 1 \end{bmatrix} \in \mathbb{M}_{p+q}(\mathbb{Z}).$$

Consequently, the Coxeter matrix $\Phi_A \in \mathbb{M}_{p+q}(\mathbb{Z})$ is of the form presented in (a).

(b) Assume that $\Delta = \Delta(\widetilde{\mathbb{D}}_m)$ and $m \ge 4$. Then the Cartan matrix $\mathbf{C}_A$

of $A = K\Delta$ and its inverse are the square $(m+1) \times (m+1)$ matrices

$$
\mathbf{C}_A = \begin{bmatrix}
1 & 0 & 1 & 1 & 1 & \ldots & 1 & 1 & 1 & 1 \\
0 & 1 & 1 & 1 & 1 & \ldots & 1 & 1 & 1 & 1 \\
0 & 0 & 1 & 1 & 1 & \ldots & 1 & 1 & 1 & 1 \\
0 & 0 & 0 & 1 & 1 & \ldots & 1 & 1 & 1 & 1 \\
0 & 0 & 0 & 0 & 1 & \ldots & 1 & 1 & 1 & 1 \\
\vdots & \vdots & \vdots & \vdots & \vdots & \ddots & \vdots & \vdots & \vdots & \vdots \\
0 & 0 & 0 & 0 & 0 & \ddots & 1 & 1 & 1 & 1 \\
0 & 0 & 0 & 0 & 0 & \ldots & 0 & 1 & 1 & 1 \\
0 & 0 & 0 & 0 & 0 & \ldots & 0 & 0 & 1 & 0 \\
0 & 0 & 0 & 0 & 0 & \ldots & 0 & 0 & 0 & 1
\end{bmatrix}
\quad \text{and} \quad
\mathbf{C}_A^{-1} = \begin{bmatrix}
1 & 0 & -1 & 0 & 0 & \ldots & 0 & 0 & 0 & 0 \\
0 & 1 & -1 & 0 & 0 & \ldots & 0 & 0 & 0 & 0 \\
0 & 0 & 1 & -1 & 0 & \ldots & 0 & 0 & 0 & 0 \\
0 & 0 & 0 & 1 & -1 & \ddots & 0 & 0 & 0 & 0 \\
\vdots & & & & & \ddots & & & & \vdots \\
0 & 0 & 0 & 0 & 0 & \ddots & -1 & 0 & 0 & 0 \\
0 & 0 & 0 & 0 & 0 & \ddots & 1 & -1 & -1 & 0 \\
0 & 0 & 0 & 0 & 0 & \ldots & 0 & 1 & -1 & -1 \\
0 & 0 & 0 & 0 & 0 & \ldots & 0 & 0 & 1 & 0 \\
0 & 0 & 0 & 0 & 0 & \ldots & 0 & 0 & 0 & 1
\end{bmatrix}.
$$

Thus, the Coxeter matrix $\mathbf{\Phi}_A \in \mathbb{M}_{m+1}(\mathbb{Z})$ has the form required in (b).

(c) Assume that $\Delta = \Delta(\widetilde{\mathbb{E}}_6)$. Then the Cartan matrix $\mathbf{C}_A$ of $A = K\Delta$ and its inverse $\mathbf{C}_A^{-1}$ are the square 7×7 matrices

$$
\mathbf{C}_A = \begin{bmatrix}
1 & 1 & 1 & 1 & 1 & 1 & 1 \\
0 & 1 & 1 & 0 & 0 & 0 & 0 \\
0 & 0 & 1 & 0 & 0 & 0 & 0 \\
0 & 0 & 0 & 1 & 1 & 0 & 0 \\
0 & 0 & 0 & 0 & 1 & 0 & 0 \\
0 & 0 & 0 & 0 & 0 & 1 & 1 \\
0 & 0 & 0 & 0 & 0 & 0 & 1
\end{bmatrix} \in \mathbb{M}_7(\mathbb{Z}) \quad \text{and} \quad
\mathbf{C}_A^{-1} = \begin{bmatrix}
1 & -1 & 0 & -1 & 0 & -1 & 0 \\
0 & 1 & -1 & 0 & 0 & 0 & 0 \\
0 & 0 & 1 & 0 & 0 & 0 & 0 \\
0 & 0 & 0 & 1 & -1 & 0 & 0 \\
0 & 0 & 0 & 0 & 1 & 0 & 0 \\
0 & 0 & 0 & 0 & 0 & 1 & -1 \\
0 & 0 & 0 & 0 & 0 & 0 & 1
\end{bmatrix} \in \mathbb{M}_7(\mathbb{Z}),
$$

respectively. Consequently, the Coxeter matrix $\mathbf{\Phi}_A \in \mathbb{M}_7(\mathbb{Z})$ has the forms required in (c).

(d) Assume that $\Delta = \Delta(\widetilde{\mathbb{E}}_7)$. Then the Cartan matrix $\mathbf{C}_A$ of $A = K\Delta$ and its inverse $\mathbf{C}_A^{-1}$ are the square 8×8 matrices

$$
\mathbf{C}_A = \begin{bmatrix}
1 & 1 & 1 & 1 & 1 & 1 & 1 & 1 \\
0 & 1 & 1 & 1 & 0 & 0 & 0 & 0 \\
0 & 0 & 1 & 1 & 0 & 0 & 0 & 0 \\
0 & 0 & 0 & 1 & 0 & 0 & 0 & 0 \\
0 & 0 & 0 & 0 & 1 & 0 & 0 & 0 \\
0 & 0 & 0 & 0 & 0 & 1 & 1 & 1 \\
0 & 0 & 0 & 0 & 0 & 0 & 1 & 1 \\
0 & 0 & 0 & 0 & 0 & 0 & 0 & 1
\end{bmatrix} \in \mathbb{M}_8(\mathbb{Z}) \quad \text{and} \quad
\mathbf{C}_A^{-1} = \begin{bmatrix}
1 & -1 & 0 & 0 & -1 & -1 & 0 & 0 \\
0 & 1 & -1 & 0 & 0 & 0 & 0 & 0 \\
0 & 0 & 1 & -1 & 0 & 0 & 0 & 0 \\
0 & 0 & 0 & 1 & 0 & 0 & 0 & 0 \\
0 & 0 & 0 & 0 & 1 & 0 & 0 & 0 \\
0 & 0 & 0 & 0 & 0 & 1 & -1 & 0 \\
0 & 0 & 0 & 0 & 0 & 0 & 1 & -1 \\
0 & 0 & 0 & 0 & 0 & 0 & 0 & 1
\end{bmatrix} \in \mathbb{M}_8(\mathbb{Z}).
$$

Thus, the Coxeter matrix $\mathbf{\Phi}_A \in \mathbb{M}_8(\mathbb{Z})$ has the form required in (d).

(e) Assume that $\Delta = \Delta(\widetilde{\mathbb{E}}_8)$. Then the Cartan matrix $\mathbf{C}_A$ of $A = K\Delta$ and its inverse $\mathbf{C}_A^{-1}$ are the square 9×9 matrices

$$
\mathbf{C}_A = \begin{bmatrix}
1 & 1 & 1 & 1 & 1 & 1 & 1 & 1 & 1 \\
0 & 1 & 1 & 0 & 0 & 0 & 0 & 0 & 0 \\
0 & 0 & 1 & 0 & 0 & 0 & 0 & 0 & 0 \\
0 & 0 & 0 & 1 & 0 & 0 & 0 & 0 & 0 \\
0 & 0 & 0 & 0 & 1 & 1 & 1 & 1 & 1 \\
0 & 0 & 0 & 0 & 0 & 1 & 1 & 1 & 1 \\
0 & 0 & 0 & 0 & 0 & 0 & 1 & 1 & 1 \\
0 & 0 & 0 & 0 & 0 & 0 & 0 & 1 & 1 \\
0 & 0 & 0 & 0 & 0 & 0 & 0 & 0 & 1
\end{bmatrix} \quad \text{and} \quad
\mathbf{C}_A^{-1} = \begin{bmatrix}
1 & -1 & 0 & -1 & -1 & 0 & 0 & 0 & 0 \\
0 & 1 & -1 & 0 & 0 & 0 & 0 & 0 & 0 \\
0 & 0 & 1 & 0 & 0 & 0 & 0 & 0 & 0 \\
0 & 0 & 0 & 1 & 0 & 0 & 0 & 0 & 0 \\
0 & 0 & 0 & 0 & 1 & -1 & 0 & 0 & 0 \\
0 & 0 & 0 & 0 & 0 & 1 & -1 & 0 & 0 \\
0 & 0 & 0 & 0 & 0 & 0 & 1 & -1 & 0 \\
0 & 0 & 0 & 0 & 0 & 0 & 0 & 1 & -1 \\
0 & 0 & 0 & 0 & 0 & 0 & 0 & 0 & 1
\end{bmatrix} \in \mathbb{M}_9(\mathbb{Z}).
$$

Then, the Coxeter matrix $\mathbf{\Phi}_A \in \mathbb{M}_9(\mathbb{Z})$ has the form required in (e), and the proof is complete. $\square$

1.3. Lemma. *Let Δ be a canonically oriented Euclidean quiver, $A = K\Delta$ be the path algebra of Δ and let $\operatorname{rad} q_A$ be the radical of the Euler form $q_A : \mathbb{Z}^n \longrightarrow \mathbb{Z}$ of A, where $n = |\Delta_0|$.*

(a) The cyclic group $\operatorname{rad} q_A$ admits a unique positive generator $\mathbf{h}_\Delta$ of the form

$$1{\overset{1\ldots1}{\underset{1\ldots1}{}}}1, \quad 1{\overset{1}{\underset{2}{}}}2\cdots2{\overset{1}{\underset{1}{}}}, \quad 1\,2{\overset{1}{\underset{\overset{2}{3}}{}}}2\,1, \quad 1\,2\,3\,{\overset{2}{4}}\,3\,2\,1 \quad \text{and} \quad 2\,4\,6\,{\overset{3}{5}}\,4\,3\,2\,1$$

if Δ is the quiver $\Delta(\widetilde{\mathbb{A}}_{p,q})$, $\Delta(\widetilde{\mathbb{D}}_m)$, $\Delta(\widetilde{\mathbb{E}}_6)$, $\Delta(\widetilde{\mathbb{E}}_7)$ and $\Delta(\widetilde{\mathbb{E}}_8)$, respectively.

(b) The defect number d_Δ of Δ and the defect $\partial_A : \mathbb{Z}^n \longrightarrow \mathbb{Z}$ of A, with $n = |\Delta_0|$, are given by the formulae

- $d_\Delta = \operatorname{lcm}(p,q)$, *and* $\partial_A(\mathbf{x}) = \frac{p+q}{\gcd(p,q)}\cdot(x_{p+q-1} - x_0)$,
  if $\Delta = \Delta(\widetilde{\mathbb{A}}_{p,q})$ *and* $1 \leq p \leq q$,
- $d_\Delta = m - 2$ *and* $\partial_A(\mathbf{x}) = x_{m+1} + x_m - x_2 - x_1$,
  if $\Delta = \Delta(\widetilde{\mathbb{D}}_m)$ *and* $m \geq 4$ *is even,*
- $d_\Delta = 2(m-2)$ *and* $\partial_A(\mathbf{x}) = 2(x_{m+1} + x_m - x_2 - x_1)$,
  if $\Delta = \Delta(\widetilde{\mathbb{D}}_m)$ *and* $m \geq 5$ *is odd,*
- $d_\Delta = 6$ *and* $\partial_A(\mathbf{x}) = -3x_1 + x_2 + x_3 + x_4 + x_5 + x_6 + x_7$,
  if $\Delta = \Delta(\widetilde{\mathbb{E}}_6)$,
- $d_\Delta = 12$ *and* $\partial_A(\mathbf{x}) = -4x_1 + x_2 + x_3 + x_4 + 2x_5 + x_6 + x_7 + x_8$,
  if $\Delta = \Delta(\widetilde{\mathbb{E}}_7)$,
- $d_\Delta = 30$ *and* $\partial_A(\mathbf{x}) = -6x_1 + 2x_2 + 2x_3 + 3x_4 + x_5 + x_6 + x_7 + x_8 + x_9$,
  if $\Delta = \Delta(\widetilde{\mathbb{E}}_8)$.

Proof. Assume that $A = K\Delta$, where Δ is one of the quivers $\Delta(\widetilde{\mathbb{A}}_{p,q})$, $\Delta(\widetilde{\mathbb{D}}_m)$, $\Delta(\widetilde{\mathbb{E}}_6)$, $\Delta(\widetilde{\mathbb{E}}_7)$, and $\Delta(\widetilde{\mathbb{E}}_8)$.

(a) Because the quiver Δ is acyclic then, by (VII.4.1), the Euler quadratic form q_A of $A = K\Delta$ equals the quadratic form q_Δ and depends only on the underlying graph $\overline{\Delta}$ of Δ. Then (a) is a consequence of (VII.4.2) applied to $Q = \Delta$.

(b) We recall from (XI.1.4) that the defect of $A = K\Delta$ is the group homomorphism $\partial_A : \mathbb{Z}^n \longrightarrow \mathbb{Z}$ such that

$$\Phi_A^{d_\Delta}(\mathbf{x}) = \mathbf{x} + \partial_A(\mathbf{x})\cdot\mathbf{h}_\Delta,$$

for all $\mathbf{x} \in \mathbb{Z}^n$, where $\Phi_A : \mathbb{Z}^n \longrightarrow \mathbb{Z}^n$ is the Coxeter transformation of A, $n = |\Delta_0|$, and the defect number $d_\Delta = d_A$ is a positive integer depending on Δ and minimal with respect to these properties. Now, for each canonically oriented quiver Δ, we determine the defect number d_Δ and the defect $\partial_A : \mathbb{Z}^n \longrightarrow \mathbb{Z}$, for $A = K\Delta$.

1° Assume that $\Delta = \Delta(\widetilde{\mathbb{A}}_{p,q})$ and $1 \leq p \leq q$. Because the Coxeter matrix $\mathbf{\Phi}_A \in \mathbb{M}_{p+q}(\mathbb{Z})$ of $A = K\Delta$ is of the form shown in (1.2)(a) and

$$\mathbf{h}_\Delta = 1{\overset{1\ldots 1}{\underset{1\ldots 1}{}}}1,$$

then a direct calculation shows that, for $d_\Delta = \operatorname{lcm}(p,q)$ and $p \uplus q = \frac{p+q}{\gcd(p,q)}$, we have

$$\mathbf{\Phi}_A^{d_\Delta} = \begin{bmatrix} -p\uplus q+1 & 0 & 0 & \ldots & 0 & 0 & p\uplus q \\ -p\uplus q & 1 & 0 & \ldots & 0 & 0 & p\uplus q \\ -p\uplus q & 0 & 1 & & 0 & 0 & p\uplus q \\ \vdots & \vdots & & \ddots & \ddots & \vdots & \vdots \\ -p\uplus q & 0 & 0 & & 1 & 0 & p\uplus q \\ -p\uplus q & 0 & 0 & \ldots & 0 & 1 & p\uplus q \\ -p\uplus q & 0 & 0 & \ldots & 0 & 0 & p\uplus q+1 \end{bmatrix} \in \mathbb{M}_{p+q}(\mathbb{Z}).$$

Moreover, for each vector $\mathbf{x} = [x_0 \ x_1 \ \ldots \ x_{p+q-1}]^t \in \mathbb{Z}^{p+q}$, we have

$$\mathbf{\Phi}_A^{d_\Delta}(\mathbf{x}) = \mathbf{x} + \frac{p+q}{\gcd(p,q)}(x_{p+q-1} - x_0) \cdot \mathbf{h}_\Delta,$$

and $d_\Delta = \operatorname{lcm}(p,q)$ is the minimal positive integer with this property. This shows that $\partial_A(\mathbf{x}) = \frac{p+q}{\gcd(p,q)}(x_{p+q-1} - x_0)$, and (b) follows in case $\Delta = \Delta(\widetilde{\mathbb{A}}_{p,q})$.

2° Assume that $\Delta = \Delta(\widetilde{\mathbb{D}}_m)$ and $m \geq 4$. We set

$$d_\Delta = \begin{cases} 2(m-2), & \text{if } m \text{ is odd, and} \\ m-2, & \text{if } m \text{ is even.} \end{cases}$$

Because the Coxeter matrix $\mathbf{\Phi}_A \in \mathbb{M}_{m+1}(\mathbb{Z})$ of $A = K\Delta$ is of the form shown in (1.2)(b) and $\mathbf{h}_\Delta = {\overset{1}{\underset{1}{}}}2\ldots 2{\overset{1}{\underset{1}{}}}$, then

$$\mathbf{\Phi}_A^{d_\Delta} = \begin{bmatrix} -1 & -2 & 0 & 0 & 0 & \ldots & 0 & 0 & 2 & 2 \\ -2 & -1 & 0 & 0 & 0 & \ldots & 0 & 0 & 2 & 2 \\ -4 & -4 & 1 & 0 & 0 & \ldots & 0 & 0 & 4 & 4 \\ -4 & -4 & 0 & 1 & 0 & \ddots & 0 & 0 & 4 & 4 \\ -4 & -4 & 0 & 0 & 1 & \ddots & 0 & 0 & 4 & 4 \\ \vdots & \vdots & \vdots & & \ddots & \ddots & \ddots & & \vdots & \vdots \\ -4 & -4 & 0 & 0 & 0 & \ddots & 1 & 0 & 4 & 4 \\ -4 & -4 & 0 & 0 & 0 & \ddots & 0 & 1 & 4 & 4 \\ -2 & -2 & 0 & 0 & 0 & \ldots & 0 & 0 & 3 & 2 \\ -2 & -2 & 0 & 0 & 0 & \ldots & 0 & 0 & 2 & 3 \end{bmatrix} \in \mathbb{M}_{m+1}(\mathbb{Z}),$$

if m is odd, and

$$\mathbf{\Phi}_A^{d_\Delta} = \begin{bmatrix} 0 & -1 & 0 & 0 & 0 & \ldots & 0 & 0 & 1 & 1 \\ -1 & 0 & 0 & 0 & 0 & \ldots & 0 & 0 & 1 & 1 \\ -2 & -2 & 1 & 0 & 0 & \ldots & 0 & 0 & 2 & 2 \\ -2 & -2 & 0 & 1 & 0 & \ddots & 0 & 0 & 2 & 2 \\ -2 & -2 & 0 & 0 & 1 & \ddots & 0 & 0 & 2 & 2 \\ \vdots & \vdots & \vdots & & \ddots & \ddots & \ddots & & \vdots & \vdots \\ -2 & -2 & 0 & 0 & 0 & \ddots & 1 & 0 & 2 & 2 \\ -2 & -2 & 0 & 0 & 0 & \ddots & 0 & 1 & 2 & 2 \\ -1 & -1 & 0 & 0 & 0 & \ldots & 0 & 0 & 2 & 1 \\ -1 & -1 & 0 & 0 & 0 & \ldots & 0 & 0 & 1 & 2 \end{bmatrix} \in \mathbb{M}_{m+1}(\mathbb{Z}),$$

if m is even. A direct calculation shows that, for each $\mathbf{x} \in \mathbb{Z}^{m+1}$, we have

$$\Phi_A^{d_\Delta}(\mathbf{x}) = \begin{cases} \mathbf{x} + 2(x_{m+1} + x_m - x_2 - x_1) \cdot \mathbf{h}_\Delta, & \text{if } m \text{ is odd,} \\ \mathbf{x} + (x_{m+1} + x_m - x_2 - x_1) \cdot \mathbf{h}_\Delta, & \text{if } m \text{ is even,} \end{cases}$$

and that d_Δ is the minimal positive integer with this property. Hence, we get

$$\partial_A(\mathbf{x}) = \begin{cases} 2(x_{m+1} + x_m - x_2 - x_1), & \text{if } m \text{ is odd, and} \\ \partial_A(\mathbf{x}) = x_{m+1} + x_m - x_2 - x_1, & \text{if } m \text{ is even.} \end{cases}$$

$3°$ Assume that $\Delta = \Delta(\widetilde{\mathbb{E}}_6)$. Because the Coxeter matrix $\Phi_A \in \mathbb{M}_7(\mathbb{Z})$ of $A = K\Delta$ is of the form shown in (1.2)(c) and $\mathbf{h}_\Delta = {}_{1\,2\,3\,2\,1}^{\quad 1 \atop 2}$ then, for $d_\Delta = 6$, we get

$$\Phi_A^d = \begin{bmatrix} -8 & 3 & 3 & 3 & 3 & 3 & 3 \\ -6 & 3 & 2 & 2 & 2 & 2 & 2 \\ -3 & 1 & 2 & 1 & 1 & 1 & 1 \\ -6 & 2 & 2 & 3 & 2 & 2 & 2 \\ -3 & 1 & 1 & 1 & 2 & 1 & 1 \\ -6 & 2 & 2 & 2 & 2 & 3 & 2 \\ -3 & 1 & 1 & 1 & 1 & 1 & 2 \end{bmatrix} \in \mathbb{M}_7(\mathbb{Z}).$$

A simple calculation shows that, for each $\mathbf{x} \in \mathbb{Z}^7$, we have

$$\Phi_A^6(\mathbf{x}) = \mathbf{x} + (-3x_1 + x_2 + x_3 + x_4 + x_5 + x_6 + x_7) \cdot \mathbf{h}_\Delta,$$

and $d_\Delta = 6$ is the minimal positive integer with this property. This shows that

$$\partial_A(\mathbf{x}) = -3x_1 + x_2 + x_3 + x_4 + x_5 + x_6 + x_7, \text{ and}$$
$$d_\Delta = 6.$$

$4°$ Assume that $\Delta = \Delta(\widetilde{\mathbb{E}}_7)$. Because the Coxeter matrix $\Phi_A \in \mathbb{M}_8(\mathbb{Z})$ of $A = K\Delta$ is of the form shown in (1.2)(d) and $\mathbf{h}_\Delta = {}_{1\,2\,3\,4\,3\,2\,1}^{\qquad 2}$ then, for $d_\Delta = 12$, we get

$$\Phi_A^{d_\Delta} = \begin{bmatrix} -15 & 4 & 4 & 4 & 8 & 4 & 4 & 4 \\ -12 & 4 & 3 & 3 & 6 & 3 & 3 & 3 \\ -8 & 2 & 3 & 2 & 4 & 2 & 2 & 2 \\ -4 & 1 & 1 & 2 & 2 & 1 & 1 & 1 \\ -8 & 2 & 2 & 2 & 5 & 2 & 2 & 2 \\ -12 & 3 & 3 & 3 & 6 & 4 & 3 & 3 \\ -8 & 2 & 2 & 2 & 4 & 2 & 3 & 2 \\ -4 & 1 & 1 & 1 & 2 & 1 & 1 & 2 \end{bmatrix} \in \mathbb{M}_8(\mathbb{Z}).$$

A straightforward calculation shows that, for each vector $\mathbf{x} \in \mathbb{Z}^8$, we have

$$\Phi_A^{12}(\mathbf{x}) = \mathbf{x} + (-4x_1 + x_2 + x_3 + x_4 + 2x_5 + x_6 + x_7 + x_8) \cdot \mathbf{h}_\Delta,$$

and $d_\Delta = 12$ is the minimal positive integer with this property. This shows that

$$\partial_A(\mathbf{x}) = -4x_1 + x_2 + x_3 + x_4 + 2x_5 + x_6 + x_7 + x_8, \text{ and}$$
$$d_\Delta = 12.$$

5° Finally, we assume that $\Delta = \Delta(\widetilde{\mathbb{E}}_8)$. Because the Coxeter matrix $\mathbf{\Phi}_A \in \mathbb{M}_9(\mathbb{Z})$ of $A = K\Delta$ is of the form shown in $(1.2)(e)$ and $\mathbf{h}_\Delta = {}_2 4 \overset{3}{6} 5 4 3 2 1$ then, for $d_\Delta = 30$, we get

$$\mathbf{\Phi}_A^{d_\Delta} = \begin{bmatrix} -35 & 12 & 12 & 18 & 6 & 6 & 6 & 6 & 6 \\ -24 & 9 & 8 & 12 & 4 & 4 & 4 & 4 & 4 \\ -12 & 4 & 5 & 6 & 2 & 2 & 2 & 2 & 2 \\ -18 & 6 & 6 & 10 & 3 & 3 & 3 & 3 & 3 \\ -30 & 10 & 10 & 15 & 6 & 5 & 5 & 5 & 5 \\ -24 & 8 & 8 & 12 & 4 & 5 & 4 & 4 & 4 \\ -18 & 6 & 6 & 9 & 3 & 3 & 4 & 3 & 3 \\ -12 & 4 & 4 & 6 & 2 & 2 & 2 & 3 & 2 \\ -6 & 2 & 2 & 3 & 1 & 1 & 1 & 1 & 2 \end{bmatrix} \in \mathbb{M}_9(\mathbb{Z}).$$

A straightforward calculation shows that, for each vector $\mathbf{x} \in \mathbb{Z}^9$, we have

$$\mathbf{\Phi}_A^{d_\Delta}(\mathbf{x}) = \mathbf{x} + (-6x_1 + 2x_2 + 2x_3 + 3x_4 + x_5 + x_6 + x_7 + x_8 + x_9) \cdot \mathbf{h}_\Delta,$$

and $d_\Delta = 30$ is the minimal positive integer with this property. This shows that

$$\partial_A(\mathbf{x}) = -6x_1 + 2x_2 + 2x_3 + 3x_4 + x_5 + x_6 + x_7 + x_8 + x_9, \text{ and}$$
$$d_\Delta = 30.$$

and completes the proof. $\square$

XIII.2. Tubes and simple regular modules over hereditary algebras of Euclidean type

Our main aim in this section is a description of the tubes $\mathcal{T}_\lambda^\Delta$ and their indecomposable $K\Delta$-modules, and, in particular, a description of the simple regular $K\Delta$-modules, where Δ is one of the canonically oriented Euclidean quivers $\Delta(\widetilde{\mathbb{A}}_{p,q})$, $\Delta(\widetilde{\mathbb{D}}_m)$, $\Delta(\widetilde{\mathbb{E}}_6)$, $\Delta(\widetilde{\mathbb{E}}_7)$, and $\Delta(\widetilde{\mathbb{E}}_8)$ of Table 1.1.

We start with the following structure theorem presenting a description of the Auslander–Reiten quiver $\Gamma(\mathrm{mod}\, A)$ of any hereditary algebra $A = KQ$ of Euclidean type.

2.1. Theorem. *Let Q be an acyclic quiver whose underlying graph $\overline{Q}$ is Euclidean, and let $A = KQ$ be the path algebra of Q.*

(a) *There exists a canonically oriented Euclidean quiver Δ such that Q is obtained from Δ by a finite sequence of reflections and A is a tilted algebra of type Δ.*

(b) *The Auslander–Reiten quiver $\Gamma(\bmod A)$ of A consists of the following three types of components:*

- *a postprojective component $\mathcal{P}(A)$ containing all indecomposable projective modules,*
- *a preinjective component $\mathcal{Q}(A)$ containing all indecomposable injective modules, and*
- *a unique $\mathbb{P}_1(K)$-family*

$$\boldsymbol{T}^Q = \{\mathcal{T}_\lambda^Q\}_{\lambda \in \mathbb{P}_1(K)}$$

of stable tubes, in the regular part $\mathcal{R}(A)$ of $\Gamma(\bmod A)$, separating $\mathcal{P}(A)$ from $\mathcal{Q}(A)$.

(c) *Let $\mathbf{m}_Q = (m_1, \dots, m_s)$ be the tubular type of the $\mathbb{P}_1(K)$-family $\boldsymbol{T}^Q$. Then*

- $\mathbf{m}_Q = (p, q)$, *if $\overline{Q} = \tilde{\mathbb{A}}_m$, $m \geq 1$, $p = \min\{p', p''\}$, and $q = \max\{p', p''\}$, where p' and p'' are the numbers of counterclockwise-oriented arrows in Q and clockwise-oriented arrows in Q, respectively,*
- $\mathbf{m}_Q = (2, 2, m - 2)$, *if $\overline{Q} = \tilde{\mathbb{D}}_m$ and $m \geq 4$,*
- $\mathbf{m}_Q = (2, 3, 3)$, *if $\overline{Q} = \tilde{\mathbb{E}}_6$,*
- $\mathbf{m}_Q = (2, 3, 4)$, *if $\overline{Q} = \tilde{\mathbb{E}}_7$, and*
- $\mathbf{m}_Q = (2, 3, 5)$, *if $\overline{Q} = \tilde{\mathbb{E}}_8$.*

Proof. Because the hereditary algebra $A = KQ$ is obviously concealed of Euclidean type Q then the part (b) of the theorem is a direct consequence of (XII.3.4) with B and KQ interchanged. The statement (a) follows from (VII.5.2) and (VIII.1.8). $\qquad\square$

It follows from (2.1) that the regular part $\mathcal{R}(A)$ of $\Gamma(\bmod A)$ is just the $\mathbb{P}_1(K)$-family $\boldsymbol{T}^Q = \{\mathcal{T}_\lambda^Q\}_{\lambda \in \mathbb{P}_1(K)}$ of stable tubes.

By (2.1) and (XII.3.5), applied to the hereditary algebra $A = KQ$, we get the following useful facts.

2.2. Corollary. *Let Q be an acyclic quiver whose underlying graph $\overline{Q}$ is Euclidean, and let $A = KQ$ be the path algebra of Q.*

(a) *The unique $\mathbb{P}_1(K)$-family $\boldsymbol{T}^Q = \{\mathcal{T}_\lambda^Q\}_{\lambda \in \mathbb{P}_1(K)}$ of stable tubes of (2.1) contains at most 3 non-homogeneous tubes $\mathcal{T}_\lambda^Q$.*

(b) *The family* $\boldsymbol{T}^Q$ *contains exactly* 3 *non-homogeneous tubes, if the Euclidean graph* $\overline{Q}$ *of* Q *is one of the graphs* $\widetilde{\mathbb{D}}_m$, *with* $m \geq 4$, $\widetilde{\mathbb{E}}_6$, $\widetilde{\mathbb{E}}_7$, *or* $\widetilde{\mathbb{E}}_8$.

(c) *The family* $\boldsymbol{T}^Q$ *contains exactly* 2 *non-homogeneous tubes, if* $\overline{Q} = \widetilde{\mathbb{A}}_m$, $m \geq 3$, $p = \min\{p', p''\} \geq 2$, *and* $q = \max\{p', p''\} \geq 2$, *where* p' *and* p'' *are the numbers of counterclockwise-oriented arrows in* Q *and clockwise-oriented arrows in* Q, *respectively.*

(d) *The family* $\boldsymbol{T}^Q$ *contains exactly* 1 *non-homogeneous tube, if* $\overline{Q} = \widetilde{\mathbb{A}}_m$, $m \geq 2$, $p = \min\{p', p''\} = 1$, *and* $q = \max\{p', p''\} \geq 2$, *and*

(e) $\boldsymbol{T}^Q$ *consists of homogeneous tubes if and only if* Q *is the Kronecker quiver* $\circ \leftleftarrows \circ$. $\qquad\qquad\square$

We remark that we have described above only the shape of the quiver $\Gamma(\mathrm{mod}\, A)$ and of the tubes in $\mathcal{R}(A)$. However, this is still far from a complete description of the indecomposable A-modules and the homomorphisms between them.

However, it follows from (2.1) that any indecomposable regular A-module lies in a stable tube $\mathcal{T}_\lambda^Q$. Hence, in view of our results of Chapter X, to get a description of the indecomposable regular A-modules lying in the tube $\mathcal{T}_\lambda^Q$ it is sufficient to describe the simple regular modules in $\mathcal{T}_\lambda^Q$. Indeed, an arbitrary indecomposable regular A-module in a tube $\mathcal{T}_\lambda^Q$ can be constructed from the simple regular ones by the procedure explained in Section X.2 (below the formulation of Theorem X.2.2).

In view of (2.1)(a), any Euclidean quiver Q is obtained from a canonically oriented Euclidean quiver Δ by a finite sequence of reflections and the indecomposable KQ-modules can be obtained from the indecomposable $K\Delta$-modules by applying the reflection functors defined in Section VII.5.

Therefore, up to applying the reflection functors, the description of the tubes $\mathcal{T}_\lambda^Q$ and their indecomposable KQ-modules reduces to a description of the tubes $\mathcal{T}_\lambda^\Delta$ and, in particular, to a description of the simple regular $K\Delta$-modules, where Δ is one of the canonically oriented Euclidean quivers $\Delta(\widetilde{\mathbb{A}}_{p,q})$, $\Delta(\widetilde{\mathbb{D}}_m)$, $\Delta(\widetilde{\mathbb{E}}_6)$, $\Delta(\widetilde{\mathbb{E}}_7)$, and $\Delta(\widetilde{\mathbb{E}}_8)$ of Table 1.1.

For each such a quiver Δ, we present in (2.4), (2.5), (2.6), (2.7), (2.12), (2.13), (2.16), (2.17), (2.20), and (2.21) below a list of simple regular $K\Delta$-modules and a $\mathbb{P}_1(K)$-family

$$\boldsymbol{T}^\Delta = \{\mathcal{T}_\lambda^\Delta\}_{\lambda \in \mathbb{P}_1(K)} \tag{2.3}$$

of stable tubes in $\mathcal{R}(K\Delta)$. We then show that every simple regular $K\Delta$-module is isomorphic to one of the modules in the list and every stable tube in $\mathcal{R}(K\Delta)$ has the form $\mathcal{T}_\lambda^\Delta$ for some $\lambda \in \mathbb{P}_1(K)$.

We do it in Theorems (2.5), (2.9), (2.15), (2.19), and (2.23) below by a case by case inspection of the canonically oriented Euclidean quivers presented in Table 1.1.

We start from the case when Δ is the quiver $\Delta(\widetilde{\mathbb{A}}_{p,q})$, where $1 \leq p \leq q$. In this case the algebra $K\Delta$ is the canonical algebra $C = C(p,q)$. Then the description of the tubes $\mathcal{T}_\lambda^\Delta$ in $\mathcal{R}(K\Delta) = \mathcal{R}(C)$ and a complete list of the simple regular $K\Delta$-modules, in each of the tubes $\mathcal{T}_\lambda^\Delta$, follow from (XII.2.8) and (XII.2.13). Now, for the convenience of the reader, we recall the list.

2.4. Table. Simple regular representations of $\Delta(\widetilde{\mathbb{A}}_{p,q})$

Assume that p and q are integers such that $q \geq p \geq 1$. Let $\Delta = \Delta(\widetilde{\mathbb{A}}_{p,q})$ be the canonically oriented Euclidean quiver

$$\Delta(\widetilde{\mathbb{A}}_{p,q}):$$

$$\begin{array}{c} 1 \leftarrow 2 \leftarrow \cdots \leftarrow p-1 \\ 0 \qquad\qquad p+q-1 \\ p \leftarrow p+1 \leftarrow \cdots \leftarrow p+q-2 \end{array}$$

presented in Table 1.1. Let $A = K\Delta$.

Consider the following family of indecomposable regular A-modules, viewed as K-linear representations of Δ. It is shown in (XII.2.8) that they are simple regular.

(a) **Simple regular representations in the tube** $\mathcal{T}_\infty^{\Delta(\widetilde{\mathbb{A}}_{p,q})}$

$$E_i^{(\infty)} = S(i), \text{ with } 1 \leq i \leq p-1, \quad E_p^{(\infty)}: \begin{array}{c} 0 \leftarrow 0 \leftarrow \cdots \leftarrow 0 \\ K \qquad\qquad K, \\ K \leftarrow K \leftarrow \cdots \leftarrow K \end{array}$$

(b) **Simple regular representations in the tube** $\mathcal{T}_0^{\Delta(\widetilde{\mathbb{A}}_{p,q})}$

$$E_j^{(0)} = S(p+j-1), \text{ with } 1 \leq j \leq q-1, \quad E_q^{(0)}: \begin{array}{c} K \leftarrow K \leftarrow \cdots \leftarrow K \\ K \qquad\qquad K, \\ 0 \leftarrow 0 \leftarrow \cdots \leftarrow 0 \end{array}$$

(c) Simple regular representations in the tube $\mathcal{T}_\lambda^{\Delta(\widetilde{\mathbb{A}}_{p,q})}$, with $\lambda \in K \setminus \{0\}$

$$
E^{(\lambda)} : \quad
\begin{array}{c}
K \xleftarrow{\;1\;} K \xleftarrow{\;1\;} \cdots \xleftarrow{\;1\;} K \\[-2pt]
\end{array}
$$

$E^{(\lambda)}$:
$$
\begin{array}{ccc}
 & K \xleftarrow{1} K \xleftarrow{1} \cdots \xleftarrow{1} K & \\
 {}^{\lambda}\nearrow & & \searrow^{1} \\
 K & & K. \\
 {}^{1}\nwarrow & & \nearrow^{1} \\
 & K \xleftarrow{1} K \xleftarrow{} \cdots \xleftarrow{1} K &
\end{array}
$$

We now recall from (XII.2.6) and (XII.2.8) the main properties of the modules listed in (2.4).

2.5. Theorem. *Assume that $A = K\Delta$, where Δ is the canonically oriented Euclidean quiver $\Delta(\widetilde{\mathbb{A}}_{p,q})$, with $q \geq p \geq 1$, shown in Table 2.4.*

(a) *Every simple regular A-module is isomorphic to one of the modules of (2.4).*

(b) *Every component in the regular part $\mathcal{R}(A)$ of $\Gamma(\operatorname{mod} A)$ is one of the following stable tubes:*

- *the tube $\mathcal{T}_\infty^{\Delta(\widetilde{\mathbb{A}}_{p,q})}$ of rank p containing the A-modules $E_1^{(\infty)}, \ldots, E_p^{(\infty)}$,*
- *the tube $\mathcal{T}_0^{\Delta(\widetilde{\mathbb{A}}_{p,q})}$ of rank q containing the A-modules $E_1^{(0)}, \ldots, E_q^{(0)}$, and*
- *the tube $\mathcal{T}_\lambda^{\Delta(\widetilde{\mathbb{A}}_{p,q})}$ of rank 1 containing the A-module $E^{(\lambda)}$, for $\lambda \in K \setminus \{0\}$, where $E_j^{(\infty)}$, $E_i^{(0)}$, and $E^{(\lambda)}$ are the simple regular A-modules of (2.4).*

(c) *The Auslander–Reiten quiver $\Gamma(\operatorname{mod} A)$ of A consists of a postprojective component $\mathcal{P}(A)$, a preinjective component $\mathcal{Q}(A)$ and the $\mathbb{P}_1(K)$-family $\boldsymbol{\mathcal{T}}^{\Delta(\widetilde{\mathbb{A}}_{p,q})} = \{\mathcal{T}_\lambda^{\Delta(\widetilde{\mathbb{A}}_{p,q})}\}_{\lambda \in \mathbb{P}_1(K)}$ of stable tubes separating $\mathcal{P}(A)$ from $\mathcal{Q}(A)$.*

2.6. Table. Simple regular representations of $\Delta(\widetilde{\mathbb{D}}_m)$

Assume that $m \geq 4$ and $\Delta = \Delta(\widetilde{\mathbb{D}}_m)$ is the canonically oriented Euclidean quiver

$$
\Delta(\widetilde{\mathbb{D}}_m) : \quad
\begin{array}{ccccccccc}
1 & & & & & & & & m \\
 & \nwarrow & & & & & & \swarrow & \\
 & & 3 & \longleftarrow & 4 & \longleftarrow \cdots \longleftarrow & m-1 & & \\
 & \swarrow & & & & & & \nwarrow & \\
2 & & & & & & & & m+1.
\end{array}
\quad , \; m \geq 4,
$$

presented in Table 1.1. Let $A = K\Delta$.

Consider the following family of indecomposable regular A-modules, viewed as K-linear representations of Δ. We show below that they are simple regular.

(a) Simple regular representations in the tube $\mathcal{T}_\infty^{\Delta(\widetilde{\mathbb{D}}_m)}$

$$
F_1^{(\infty)} :
\begin{array}{c}
K \\
\nwarrow{\scriptstyle 1} \\
K \xleftarrow{1} K \xleftarrow{1} \cdots \xleftarrow{1} K \\
\swarrow \qquad\qquad \nwarrow{\scriptstyle 1} \\
0 \qquad\qquad K
\end{array}
\qquad
F_2^{(\infty)} :
\begin{array}{c}
0 \qquad\qquad 0 \\
\nwarrow \qquad\qquad \swarrow \\
K \xleftarrow{1} K \xleftarrow{1} \cdots \xleftarrow{1} K \\
{\scriptstyle 1}\swarrow \qquad\qquad \nwarrow \\
K \qquad\qquad 0
\end{array}
\qquad K
$$

(b) Simple regular representations in the tube $\mathcal{T}_0^{\Delta(\widetilde{\mathbb{D}}_m)}$

$$
F_1^{(0)} :
\begin{array}{c}
0 \qquad\qquad 0 \\
\nwarrow \qquad\qquad \swarrow \\
K \xleftarrow{1} K \xleftarrow{1} \cdots \xleftarrow{1} K \\
{\scriptstyle 1}\swarrow \qquad\qquad \nwarrow{\scriptstyle 1} \\
K \qquad\qquad K
\end{array}
\qquad
F_2^{(0)} :
\begin{array}{c}
K \qquad\qquad K \\
\nwarrow{\scriptstyle 1} \qquad\qquad {\scriptstyle 1}\nearrow \\
K \xleftarrow{1} K \xleftarrow{1} \cdots \xleftarrow{1} K \\
\swarrow \qquad\qquad \nwarrow \\
0 \qquad\qquad 0
\end{array}
$$

(c) Simple regular representations in the tube $\mathcal{T}_1^{\Delta(\widetilde{\mathbb{D}}_m)}$

$$
F_1^{(1)} :
\begin{array}{c}
0 \qquad\qquad 0 \\
\nwarrow \qquad\qquad \swarrow \\
K \xleftarrow{} 0 \xleftarrow{} 0 \xleftarrow{} \cdots \xleftarrow{} 0 \\
\swarrow \qquad\qquad \nwarrow \\
0 \qquad\qquad 0
\end{array}
\qquad
F_2^{(1)} :
\begin{array}{c}
0 \qquad\qquad 0 \\
\nwarrow \qquad\qquad \swarrow \\
0 \xleftarrow{} K \xleftarrow{} 0 \xleftarrow{} \cdots \xleftarrow{} 0 \\
\swarrow \qquad\qquad \nwarrow \\
0 \qquad\qquad 0
\end{array}
$$

$$\vdots \qquad\qquad\qquad\qquad \vdots$$

$$
F_{m-3}^{(1)} :
\begin{array}{c}
0 \qquad\qquad 0 \\
\nwarrow \qquad\qquad \swarrow \\
0 \xleftarrow{} 0 \xleftarrow{} \cdots \xleftarrow{} 0 \xleftarrow{} K \\
\swarrow \qquad\qquad \nwarrow \\
0 \qquad\qquad 0
\end{array}
\qquad
F_{m-2}^{(1)} :
\begin{array}{c}
K \qquad\qquad K \\
\nwarrow{\scriptstyle 1} \qquad\qquad {\scriptstyle 1}\nearrow \\
K \xleftarrow{1} K \xleftarrow{1} K \xleftarrow{1} \cdots \xleftarrow{1} K \\
{\scriptstyle 1}\swarrow \qquad\qquad \nwarrow{\scriptstyle 1} \\
K \qquad\qquad K
\end{array}
$$

(d) Simple regular representations in the tube $\mathcal{T}_\lambda^{\Delta(\widetilde{\mathbb{D}}_m)}$

$$
F^{(\lambda)} :
\begin{array}{c}
K \qquad\qquad\qquad\qquad \begin{bmatrix}1\\0\end{bmatrix} K \\
\nwarrow{\scriptstyle [1\,1]} \qquad\qquad\qquad\qquad \swarrow \\
K^2 \xleftarrow{\begin{bmatrix}1&0\\0&1\end{bmatrix}} K^2 \xleftarrow{} \cdots \xleftarrow{} K^2 \xleftarrow{\begin{bmatrix}1&0\\0&1\end{bmatrix}} K^2 \qquad \lambda \in K \setminus \{0,1\}. \\
\swarrow{\scriptstyle [\lambda\,1]} \qquad\qquad\qquad\qquad \nwarrow \\
K \qquad\qquad\qquad\qquad \begin{bmatrix}0\\1\end{bmatrix} K
\end{array}
$$

2.7. Definition. Assume that $A = K\Delta$, where Δ is the canonically oriented Euclidean quiver $\Delta(\widetilde{\mathbb{D}}_m)$, with $m \geq 4$, shown in (2.6). We define the $\mathbb{P}_1(K)$-family (2.3)

$$\mathcal{T}^{\Delta(\widetilde{\mathbb{D}}_m)} = \{\mathcal{T}_\lambda^{\Delta(\widetilde{\mathbb{D}}_m)}\}_{\lambda \in \mathbb{P}_1(K)}$$

of connected components $\mathcal{T}_\lambda^{\Delta(\widetilde{\mathbb{D}}_m)}$ of the Auslander–Reiten quiver $\Gamma(\text{mod } A)$ of A as follows:

- $\mathcal{T}_\infty^{\Delta(\widetilde{\mathbb{D}}_m)}$ is the component containing the A-module $F_1^{(\infty)}$,
- $\mathcal{T}_0^{\Delta(\widetilde{\mathbb{D}}_m)}$ is the component containing the A-module $F_1^{(0)}$,
- $\mathcal{T}_1^{\Delta(\widetilde{\mathbb{D}}_m)}$ is the component containing the A-module $F_1^{(1)}$, and
- for each $\lambda \in K \setminus \{0,1\}$, $\mathcal{T}_\lambda^{\Delta(\widetilde{\mathbb{D}}_m)}$ is the component containing the A-module $F^{(\lambda)}$, where $F_1^{(\infty)}$, $F_1^{(0)}$, $F_1^{(1)}$, and $F^{(\lambda)}$ are the modules listed in Table 2.6.

It follows from (2.9) below that the components $\mathcal{T}_\lambda^{\Delta(\widetilde{\mathbb{D}}_m)}$ of the $\mathbb{P}_1(K)$-family $\mathcal{T}^{\Delta(\widetilde{\mathbb{D}}_m)}$ are stable tubes in $\mathcal{R}(A)$.

We now collect the main properties of the modules listed in (2.6).

2.8. Lemma. *Let $m \geq 4$ and Δ be the canonically oriented Euclidean quiver $\Delta(\widetilde{\mathbb{D}}_m)$ shown in Table 2.6. Let $A = K\Delta$ be the path algebra of Δ. The A-modules*

$$F_1^{(\infty)}, F_2^{(\infty)}, F_1^{(0)}, F_2^{(0)}, F_1^{(1)}, F_2^{(1)}, \ldots, F_{m-3}^{(1)}, F_{m-2}^{(1)}, \text{ and } F^{(\lambda)},$$

with $\lambda \in K \setminus \{0,1\}$, presented in Table 2.6 are pairwise orthogonal bricks, and each of them is simple regular.

Proof. It is obvious that the A-modules

$$F_1^{(\infty)}, F_2^{(\infty)}, \quad F_1^{(0)}, F_2^{(0)}, \quad F_1^{(1)}, F_2^{(1)}, \ldots, F_{m-3}^{(1)}, F_{m-2}^{(1)}$$

are bricks. Now we show that, for each $\lambda \in K \setminus \{0,1\}$, there is an isomorphism $\text{End}_A F^{(\lambda)} \cong K$. Let $f = (f_1, f_2 \ldots, f_{m+1}) : F^{(\lambda)} \longrightarrow F^{(\lambda)}$ be an endomorphism of the A-module $F^{(\lambda)}$, viewed as a K-linear representation of $\Delta = \Delta(\mathbb{D}_m)$, and assume that the K-linear map $f_3 : K^2 \longrightarrow K^2$ is defined by the matrix

$$f_3 = \begin{bmatrix} a & b \\ c & d \end{bmatrix}$$

in the standard basis of K^2, where $a, b, c, d \in K$. It follows that $f_3 = f_4 = \ldots = f_{m-1}$ and therefore

$$f_3 \cdot \begin{bmatrix} 0 \\ 1 \end{bmatrix} = \begin{bmatrix} 0 \\ 1 \end{bmatrix} \cdot f_{m+1} \quad \text{and} \quad f_3 \cdot \begin{bmatrix} 1 \\ 0 \end{bmatrix} = \begin{bmatrix} 1 \\ 0 \end{bmatrix} \cdot f_m.$$

Hence we get $b = 0$ and $c = 0$. Similarly the equality $\begin{bmatrix} 1 \\ 1 \end{bmatrix} \cdot f_3 = f_1 \cdot \begin{bmatrix} 1 \\ 1 \end{bmatrix}$

yields $a = d$ and, consequently, $f_3 = \begin{bmatrix} a & 0 \\ 0 & a \end{bmatrix}$. It follows that the map

$\mathrm{End}_A F^{(\lambda)} \longrightarrow K$, assigning to the endomorphism $f : F^{(\lambda)} \longrightarrow F^{(\lambda)}$ of $F^{(\lambda)}$ the scalar a on the diagonal of the matrix f_3, is a K-algebra isomorphism.

It follows that the modules of Table 2.6 are bricks. Now we show that they are pairwise orthogonal, that is, that

- $\mathrm{Hom}_A(F_i^{(s)}, F_j^{(t)}) = 0,$ $\mathrm{Hom}_A(F_j^{(\lambda)}, F_j^{(t)}) = 0,$ and
- $\mathrm{Hom}_A(F_j^{(t)}, F^{(\lambda)}) = 0,$ for all $i \neq j$, $s, t \in \{0, 1, \infty\}$, and $\lambda \in \mathbb{P}_1(K) \setminus \{0, 1, \infty\}$.

We only prove that $\mathrm{Hom}_A(F^{(\lambda)}, F_{m-2}^{(1)}) = 0$ and $\mathrm{Hom}_A(F_{m-2}^{(1)}, F^{(\lambda)}) = 0$, because the remaining equalities follow easily in a similar way.

Let $f = (f_1, \dots, f_{m+1}) : F^{(\lambda)} \longrightarrow F_{m-2}^{(1)}$ be a morphism of representations of the quiver $\Delta = \Delta(\widetilde{\mathbb{D}}_m)$. The equalities

$$f_{m-1} \cdot \begin{bmatrix} 0 \\ 1 \end{bmatrix} = f_{m+1} \quad \text{and} \quad f_{m-1} \cdot \begin{bmatrix} 1 \\ 0 \end{bmatrix} = f_m$$

yield $f_3 = \dots = f_{m-1} = \begin{bmatrix} f_m \\ f_{m+1} \end{bmatrix}$. On the other hand, the equalities $f_3 = f_1[1\ 1]$ and $f_3 = f_2[\lambda\ 1]$ yield $f_m = f_{m+1} = f_1 = f_2 = f_2\lambda$. Because $\lambda \neq 1$, then we get $f_2 = 0$ and, consequently, $f = 0$.

Let $g = (g_1, \dots, g_{m+1}) : F_{m-2}^{(1)} \longrightarrow F^{(\lambda)}$ be a morphism of representations of the quiver Δ. The equalities $\begin{bmatrix} 1 \\ 0 \end{bmatrix} \cdot g_m = g_{m-1} = \begin{bmatrix} 0 \\ 1 \end{bmatrix} \cdot g_{m+1}$ yield $g_m = 0$, $g_{m+1} = 0$, and $g_3 = \dots = g_{m-1} = 0$. It follows that $g_1 = 0$, $g_2 = 0$ and, consequently, $g = 0$.

We recall from (1.3)(b) that the defect $\partial_A : \mathbb{Z}^{m+1} \longrightarrow \mathbb{Z}$ of A is given by the formula

- $\partial_A(\mathbf{x}) = x_{m+1} + x_m - x_2 - x_1,$ if m is even, and
- $\partial_A(\mathbf{x}) = 2(x_{m+1} + x_m - x_2 - x_1),$ if m is odd.

Hence easily follows that $\partial_A(\mathbf{dim}\, F) = 0$, for any module F of Table 2.6. It follows from (XI.2.8) that the modules of Table 2.6 are regular.

Now we show that they are simple regular. For this purpose, we note that the modules $F_1^{(1)}$, $F_2^{(1)}$, $\dots, F_{m-3}^{(1)}$ are simple. Suppose that F is one of the modules

$$F_1^{(\infty)}, F_2^{(\infty)}, F_1^{(0)}, F_2^{(0)}, F_{m-2}^{(1)}, \text{ and } F^{(\lambda)}, \text{ with } \lambda \in K \setminus \{0, 1\},$$

and let X be an indecomposable regular submodule of F. Because X is a subrepresentation of F then $X_1 = F_1$ and $X_2 = F_2$. The regularity of X

yields $\partial_A(\mathbf{dim}\,X) = 0$ and we get

$$
\begin{aligned}
\dim{}_K F_2 + \dim{}_K F_1 &= \dim{}_K X_2 + \dim{}_K X_1 \\
&= \dim{}_K X_{m+1} + \dim{}_K X_m \\
&\leq \dim{}_K F_{m+1} + \dim{}_K F_m \\
&= \dim{}_K F_2 + \dim{}_K F_1.
\end{aligned}
$$

It follows that $X_{m+1} = F_{m+1}$, $X_m = F_m$. This implies the equality $X = F$ and the fact that F is simple regular. Indeed, this is obvious when $F \neq F^{(\lambda)}$. In the case $F = F^{(\lambda)}$, we have

$$
X_3 = \ldots = X_{m-1} = K[1,0] + K[0,1] = K^2 = F_3,
$$

because X is a subrepresentation of F. It follows that $X = F$ and, consequently, the module $F = F^{(\lambda)}$ is simple regular. This completes the proof. $\square$

Next, we collect the main properties of the family $\mathcal{T}^{\Delta(\widetilde{\mathbb{D}}_m)}$ of components defined in (2.7).

2.9. Theorem. *Let $m \geq 4$ and Δ be the canonically oriented Euclidean quiver $\Delta(\widetilde{\mathbb{D}}_m)$ shown in Table 1.1. Let $A = K\Delta$ be the path algebra of Δ and*

$$
F_1^{(\infty)},\ F_2^{(\infty)},\ F_1^{(0)},\ F_2^{(0)},\ F_1^{(1)},\ F_2^{(1)}, \ldots, F_{m-3}^{(1)}, F_{m-2}^{(1)},\ F^{(\lambda)},
$$

with $\lambda \in K \setminus \{0,1\}$, be the A-modules presented in Table 2.6. The family

$$
\mathcal{T}^{\Delta(\widetilde{\mathbb{D}}_m)} = \{\mathcal{T}_\lambda^{\Delta(\widetilde{\mathbb{D}}_m)}\}_{\lambda \in \mathbb{P}_1(K)}
$$

defined in (2.7) has the following properties.

(a) *The component $\mathcal{T}_\infty^{\Delta(\widetilde{\mathbb{D}}_m)}$ is a stable tube of rank 2 containing the modules $F_1^{(\infty)}$ and $F_2^{(\infty)}$, and there are isomorphisms $\tau_A F_2^{(\infty)} \cong F_1^{(\infty)}$ and $\tau_A F_1^{(\infty)} \cong F_2^{(\infty)}$.*

(b) *The component $\mathcal{T}_0^{\Delta(\widetilde{\mathbb{D}}_m)}$ is a stable tube of rank 2 containing the modules $F_1^{(0)}$ and $F_2^{(0)}$, and there are isomorphisms $\tau_A F_2^{(0)} \cong F_1^{(0)}$, $\tau_A F_1^{(0)} \cong F_2^{(0)}$.*

(c) *The component $\mathcal{T}_1^{\Delta(\widetilde{\mathbb{D}}_m)}$ is a stable tube of rank $m - 2$ containing the modules $F_1^{(1)}, \ldots, F_{m-2}^{(1)}$ and, for each $s \in \{1, \ldots, m - 2\}$, there is an isomorphism $\tau_A F_{s+1}^{(1)} \cong F_s^{(1)}$, where we set $F_{m-1}^{(1)} = F_1^{(1)}$.*

(d) *For each $\lambda \in K \setminus \{0,1\}$, the component $\mathcal{T}_\lambda^{\Delta(\widetilde{\mathbb{D}}_m)}$ is a stable tube of rank 1 containing the module $F^{(\lambda)}$, and there is an isomorphism $\tau_A F^{(\lambda)} \cong F^{(\lambda)}$.*

(e) *The Auslander–Reiten quiver* $\Gamma(\mathrm{mod}\,A)$ *of* A *consists of a postprojective component* $\mathcal{P}(A)$, *a preinjective component* $\mathcal{Q}(A)$ *and the* $\mathbb{P}_1(K)$-*family* $\boldsymbol{\mathcal{T}}^{\Delta(\widetilde{\mathbb{D}}_m)} = \{\mathcal{T}_\lambda^{\Delta(\widetilde{\mathbb{D}}_m)}\}_{\lambda\in\mathbb{P}_1(K)}$ *of stable tubes separating* $\mathcal{P}(A)$ *from* $\mathcal{Q}(A)$.

Proof. Let $\Delta = \Delta(\widetilde{\mathbb{D}}_m)$, where $m \geq 4$, and let $A = K\Delta$. We consider the canonical algebra $C = C(2,2,m-2)$ of type $\Delta(\widetilde{\mathbb{D}}_m)$ defined in (XII.1.3), and the family of C-modules:

$$E_1^{(\infty)},\, E_2^{(\infty)},\, E_1^{(0)},\, E_2^{(0)},\, E_1^{(1)},\ldots,E_{m-2}^{(1)},\ \text{and}\ E^{(\lambda)},\ \text{with}\ \lambda \in K \setminus \{0,1\},$$

defined in Table XII.2.9. It follows from (XII.2.12) that they are simple regular.

By (XII.1.4), the postprojective component $\mathcal{P}(C)$ of $\Gamma(\mathrm{mod}\,C)$ contains the section

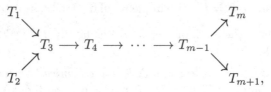

where

$$T_1 = P(a_1),\ T_2 = P(b_1),$$
$$T_3 = \tau_C^{-1}P(0),\ T_4 = \tau_C^{-1}P(c_1),\ldots,T_{m-1} = \tau_C^{-1}P(c_{m-3}),$$
$$T_m = P(\omega),\ \text{and}\ T_{m+1} = \tau_C^{-1}P(c_{m-2}).$$

Here we use the notation of (XII.1.2). Moreover,

$$T = T_1 \oplus \ldots \oplus T_m \oplus T_{m+1}$$

is a postprojective tilting C-module and there is a K-algebra isomorphism

$$A \cong \mathrm{End}\,T_C.$$

Then, according to (VIII.4.5) and (XI.3.3), the K-linear functor $\mathrm{Hom}_C(T,-) : \mathrm{mod}\,C \longrightarrow \mathrm{mod}\,A$ restricts to the equivalence

$$\mathrm{Hom}_C(T,-) : \mathrm{add}\,\mathcal{R}(C) \xrightarrow{\ \cong\ } \mathrm{add}\,\mathcal{R}(A) \qquad (2.10)$$

of abelian and serial categories $\mathrm{add}\,\mathcal{R}(C)$ and $\mathrm{add}\,\mathcal{R}(A)$ of regular modules such that the following diagram is commutative

$$
\begin{array}{ccc}
\mathrm{add}\,\mathcal{R}(C) & \overset{\tau_C}{\underset{\tau_C^{-1}}{\rightleftarrows}} & \mathrm{add}\,\mathcal{R}(C) \\
{\scriptstyle\mathrm{Hom}_C(T,-)}\downarrow{\scriptstyle\cong} & & {\scriptstyle\cong}\downarrow{\scriptstyle\mathrm{Hom}_C(T,-)} \\
\mathrm{add}\,\mathcal{R}(A) & \overset{\tau_A}{\underset{\tau_A^{-1}}{\rightleftarrows}} & \mathrm{add}\,\mathcal{R}(A),
\end{array}
\qquad (2.11)
$$

where the horizontal functors are mutually inverse equivalences of categories induced by the Auslander–Reiten translates in $\operatorname{mod} A$ and in $\operatorname{mod} C$, respectively.

It follows that the functor $\operatorname{Hom}_C(T, -) : \operatorname{add} \mathcal{R}(C) \xrightarrow{\cong} \operatorname{add} \mathcal{R}(A)$ carries irreducible morphisms to irreducible ones, and the image $\operatorname{Hom}_C(T, \mathcal{T}_\lambda^C)$ of any stable tube $\mathcal{T}_\lambda^C$ in $\mathcal{R}(C)$ of rank $r_\lambda \geq 1$ is a stable tube in $\mathcal{R}(A)$ of the same rank r_λ. Moreover, any stable tube in $\mathcal{R}(A)$ is of the form $\operatorname{Hom}_C(T, \mathcal{T}_\lambda^C)$.

We claim that, for each $\lambda \in \mathbb{P}_1(K)$, the tubes $\operatorname{Hom}_C(T, \mathcal{T}_\lambda^C)$ and $\mathcal{T}_\lambda^{\Delta(\widetilde{\mathbb{D}}_m)}$ coincide. For this purpose, it suffices to show that there are isomorphisms

$$\operatorname{Hom}_C(T, E_j^{(s)}) \cong F_j^{(s)} \quad \text{and} \quad \operatorname{Hom}_C(T, E^{(\lambda)}) \cong F^{(\lambda)},$$

for all $j, s \in \{0, 1, \infty\}$ and $\lambda \in \mathbb{P}_1(K) \setminus \{0, 1, \infty\}$.

To prove this, we recall that $A \cong \operatorname{End} T_C$ and, as in the proof of (XII.1.5), the left hand part of the postprojective component $\mathcal{P}(C)$ of $\Gamma(\operatorname{mod} C)$ containing all the indecomposable projective C-modules $P(0), \dots, P(\omega)$ looks as follows

where

$$P(0) = 1 \underset{00\dots00}{\overset{0}{0}} 0, \qquad P(\omega) = 2 \underset{11\dots11}{\overset{1}{1}} 1,$$

and the indecomposable modules are represented by their dimension vectors.

It follows from (VI.3.10) that $e_iA \cong \mathrm{Hom}_C(T, T_i)$ and, for each regular C-module X and, for each $i \in \{1, \dots, m+1\}$, we have

$$
\begin{aligned}
\mathrm{Hom}_C(T, X)_i &= \mathrm{Hom}_C(T, X)e_i \\
&\cong \mathrm{Hom}_A(e_iA, \mathrm{Hom}_C(T, X)) \\
&\cong \mathrm{Hom}_A(\mathrm{Hom}_C(T, T_i), \mathrm{Hom}_C(T, X)) \\
&\cong \mathrm{Hom}_C(T_i, X),
\end{aligned}
$$

and therefore $(\mathbf{dim}\,\mathrm{Hom}_C(T, X))_i = \dim_K \mathrm{Hom}_C(T_i, X)$. Note also that we have

- $\mathrm{Hom}_C(T_1, X) = \mathrm{Hom}_C(P(a_1), X) \cong X_{a_1}$,
- $\mathrm{Hom}_C(T_2, X) = \mathrm{Hom}_C(P(a_2), X) \cong X_{a_2}$,
- $\mathrm{Hom}_C(T_m, X) = \mathrm{Hom}_C(P(\omega), X) \cong X_\omega$, and
- $\mathrm{Hom}_C(T_j, X) = \mathrm{Hom}_C(\tau_C^{-1}P(c_{j-3}), X)$, for $j \in \{3, 4, \dots, m-1, m+1\}$,

where we set $c_0 = 0$. By consulting the beginning part of the component $\mathcal{P}(C)$ shown above we see that the ω-th coordinate of the dimension vector of each of the C-modules $T_1, \dots T_{m-1}, T_{m+1}$ equals zero. Let

$$
Q: \quad 0 \xleftarrow[\gamma_1]{\;\beta_1\;}
\begin{array}{c}
a_1 \\
\nwarrow^{\alpha_1} \\
b_1 \\
\swarrow \\
c_1
\end{array}
$$

be the full subquiver of $\Delta(2, 2, m-2)$, of the Dynkin type $\mathbb{D}_4$, defined by the points $0, a_1, b_1, c_1$, in the notation of (XII.1.2). Note that there is an isomorphism $KQ \cong e_Q C e_Q$ of algebras and the equivalence of categories $\mathrm{mod}\,KQ \cong \mathrm{rep}_K(Q)$, where e_Q is the idempotent

$$
e_Q = e_0 + e_{a_1} + e_{b_1} + e_{c_1}
$$

of the algebra C. Consider the restriction functor

$$
\mathrm{res}_{e_Q} : \mathrm{mod}\,C \longrightarrow \mathrm{mod}\,KQ \cong \mathrm{rep}_K(Q),
$$

see (I.6.6). It follows from the shape of the C-modules $T_1, \dots T_{m-1}, T_{m+1}$, shown in $\mathcal{P}(C)$ above, and the modules

$$
E_1^{(\infty)}, E_2^{(\infty)}, E_1^{(0)}, E_2^{(0)}, E_1^{(1)}, \dots, E_{m-2}^{(1)}, F^{(\lambda)}, \text{ with } \lambda \in K \setminus \{0, 1\},
$$

presented in (XII.2.9), that the restriction functor res_{e_Q} defines a functorial isomorphism

$$
\mathrm{Hom}_C(T_j, E) \cong \mathrm{Hom}_{KQ}(\mathrm{res}_{e_Q} T_j, \mathrm{res}_{e_Q} E),
$$

for $j = 1, \dots, m-1, m+1$ and for any module E in the family

$$
E_1^{(\infty)}, E_2^{(\infty)}, E_1^{(0)}, E_2^{(0)}, E_1^{(1)}, \dots, E_{m-2}^{(1)}, \text{ and } E^{(\lambda)}, \text{ with } \lambda \in K \setminus \{0, 1\}.
$$

By applying the above formulae to the C-module $E = E^{(\lambda)}$, with $\lambda \in K \setminus \{0,1\}$, we get

- $\mathrm{Hom}_C(T_1, E^{(\lambda)}) \cong F_1^{(\lambda)} = K$,
- $\mathrm{Hom}_C(T_2, E^{(\lambda)}) \cong F_2^{(\lambda)} = K$,
- $\mathrm{Hom}_C(T_j, E^{(\lambda)}) \cong \mathrm{Hom}_{KQ}(\mathrm{res}_{e_Q} T_j, \mathrm{res}_{e_Q} F^{(\lambda)}) \cong K^2$, for $j = 3, 4, \ldots, m-1$,
- $\mathrm{Hom}_C(T_m, E^{(\lambda)}) \cong \mathrm{Hom}_{KQ}(\mathrm{res}_{e_Q} T_m, \mathrm{res}_{e_Q} E^{(\lambda)}) \cong K$,
- $\mathrm{Hom}_C(T_{m+1}, E^{(\lambda)}) \cong F_{m+1}^{(\lambda)} = K$.

In particular, we get $\dim \mathrm{Hom}_C(T, E^{(\lambda)}) = \begin{smallmatrix}1\\2\,2\ldots2\,2\\1\end{smallmatrix}\begin{smallmatrix}1\\ \\1\end{smallmatrix}$. Moreover, it follows also that the chain of the irreducible monomorphisms

$$T_3 \hookrightarrow T_4 \hookrightarrow \ldots \hookrightarrow T_{m-1}$$

induces the chain of the identity homomorphisms

$$\mathrm{Hom}_{KQ}(\mathrm{res}_{e_Q} T_3, \mathrm{res}_{e_Q} F^{(\lambda)}) \xleftarrow[\cong]{1} \ldots \xleftarrow[\cong]{1} \mathrm{Hom}_{KQ}(\mathrm{res}_{e_Q} T_{m-1}, \mathrm{res}_{e_Q} F^{(\lambda)}).$$

Finally, by applying the above formulae, one shows that there is an A-module isomorphism $\mathrm{Hom}_C(T, E^{(\lambda)}) \cong F^{(\lambda)}$, and hence

$$\mathrm{Hom}_C(T, \mathcal{T}_\lambda^C) = \mathcal{T}_\lambda^{(\widetilde{\mathbb{D}}_m)},$$

for any $\lambda \in K \setminus \{0,1\}$. Because, by (XII.2.12), the C-module $E^{(\lambda)}$ is simple regular, then the A-module $\mathrm{Hom}_C(T, E^{(\lambda)}) \cong F^{(\lambda)}$ is simple regular. Furthermore, in view of the isomorphism $\tau_C E^{(\lambda)} \cong E^{(\lambda)}$ in (XII.2.12) and the commutativity of the diagram (2.11), we get the isomorphism $\tau_A F^{(\lambda)} \cong F^{(\lambda)}$. This finishes the proof of (d).

Similarly, by applying the above formulae to any module E in the family $E_1^{(\infty)}, E_2^{(\infty)}, E_1^{(0)}, E_2^{(0)}, E_1^{(1)}, \ldots, E_{m-2}^{(1)}$, we get

(a) $\dim \mathrm{Hom}_C(T, E_1^{(\infty)}) = \begin{smallmatrix}1\\1\,1\ldots1\\0\end{smallmatrix}\begin{smallmatrix}0\\ \\1\end{smallmatrix}$, $\dim \mathrm{Hom}_C(T, E_2^{(\infty)}) = \begin{smallmatrix}0\\1\,1\ldots1\\1\end{smallmatrix}\begin{smallmatrix}1\\ \\0\end{smallmatrix}$,

(b) $\dim \mathrm{Hom}_C(T, E_1^{(0)}) = \begin{smallmatrix}0\\1\,1\ldots1\\1\end{smallmatrix}\begin{smallmatrix}0\\ \\1\end{smallmatrix}$, $\dim \mathrm{Hom}_C(T, E_2^{(0)}) = \begin{smallmatrix}1\\1\,1\ldots1\\0\end{smallmatrix}\begin{smallmatrix}1\\ \\0\end{smallmatrix}$,

(c) $\dim \mathrm{Hom}_C(T, E_1^{(1)}) = \begin{smallmatrix}0\\1\,0\ldots0\\0\end{smallmatrix}\begin{smallmatrix}0\\ \\0\end{smallmatrix}$, $\dim \mathrm{Hom}_C(T, E_2^{(1)}) = \begin{smallmatrix}0\\0\,1\,0\ldots0\\0\end{smallmatrix}\begin{smallmatrix}0\\ \\0\end{smallmatrix}$,

$$\vdots \qquad\qquad\qquad\qquad \vdots$$

$\dim \mathrm{Hom}_C(T, E_{m-3}^{(1)}) = \begin{smallmatrix}0\\0\,0\ldots0\,1\\0\end{smallmatrix}\begin{smallmatrix}0\\ \\0\end{smallmatrix}$, $\dim \mathrm{Hom}_C(T, E_{m-2}^{(1)}) = \begin{smallmatrix}1\\1\,1\ldots1\\1\end{smallmatrix}\begin{smallmatrix}1\\ \\1\end{smallmatrix}$.

In particular, $\dim \mathrm{Hom}_C(T, E_j^{(s)}) = \dim F_j^{(s)}$, for all j and $s \in \{0,1,\infty\}$.

Because, by (XII.2.12), E is a simple regular C-module, then obviously the A-module $\mathrm{Hom}_C(T, E)$ is simple regular.

We recall from (XI.3.3) and (XII.4.2) that the simple regular A-modules lying in the non-homogeneous tubes of $\mathcal{R}(A)$ are uniquely determined by their dimension vectors. Because, by (2.8), the modules presented in Table 2.6 are simple regular, then there is an isomorphism

$$\mathrm{Hom}_C(T, E_j^{(s)}) \cong F_j^{(s)},$$

for all j, and $s \in \{0, 1, \infty\}$.

Therefore, by applying (XII.2.12) and the commutative diagram (2.11), we obtain the statements (a)–(c), and we conclude that the tubes

$$\mathcal{T}_\infty^{(\widetilde{\mathbb{D}}_m)} = \mathrm{Hom}_C(T, \mathcal{T}_\infty^C), \;\; \mathcal{T}_0^{(\widetilde{\mathbb{D}}_m)} = \mathrm{Hom}_C(T, \mathcal{T}_0^C), \;\; \mathcal{T}_1^{(\widetilde{\mathbb{D}}_m)} = \mathrm{Hom}_C(T, \mathcal{T}_1^C)$$

exhaust all the non-homogeneous tubes of $\Gamma(\mathrm{mod}\, A)$. Consequently, the tubes $\mathcal{T}_\lambda^{(\widetilde{\mathbb{D}}_m)} = \mathrm{Hom}_C(T, \mathcal{T}_\lambda^C)$, with $\lambda \in K \setminus \{0, 1\}$, exhaust all the homogeneous tubes of the regular part $\mathcal{R}(A)$ of $\Gamma(\mathrm{mod}\, A)$. Now the statement (e) follows from (2.1). This completes the proof. $\qquad\square$

2.12. Table. Simple regular representations of $\Delta(\widetilde{\mathbb{E}}_6)$

Assume that $\Delta = \Delta(\widetilde{\mathbb{E}}_6)$ is the canonically oriented Euclidean quiver

$$\Delta(\widetilde{\mathbb{E}}_6): \qquad \begin{array}{c} 5 \\ \downarrow \\ 4 \\ \downarrow \\ 3 \longrightarrow 2 \longrightarrow 1 \longleftarrow 6 \longleftarrow 7. \end{array}$$

presented in Table 1.1, and let $A = K\Delta$. Consider the following family of indecomposable regular A-modules, viewed as K-linear representations of Δ. We show below that they are simple regular.

(a) **Simple regular representations in the tube $\mathcal{T}_\infty^{\Delta(\widetilde{\mathbb{E}}_6)}$**

$$F_1^{(\infty)}: \qquad \begin{array}{c} 0 \\ \downarrow \\ K \\ \downarrow 1 \\ 0 \longrightarrow K \xrightarrow{1} K \xleftarrow{1} K \longleftarrow 0 \end{array}$$

$$F_2^{(\infty)}: \qquad \begin{array}{c} K \\ \downarrow 1 \\ K \\ \downarrow \left[\begin{smallmatrix}1\\1\end{smallmatrix}\right] \\ K \xrightarrow{1} K \xrightarrow{\left[\begin{smallmatrix}1\\0\end{smallmatrix}\right]} K^2 \xleftarrow{\left[\begin{smallmatrix}0\\1\end{smallmatrix}\right]} K \xleftarrow{1} K \end{array}$$

(b) Simple regular representations in the tube $\mathcal{T}_0^{\Delta(\widetilde{\mathbb{E}}_6)}$

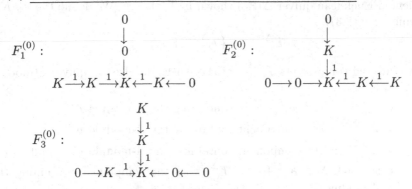

(c) Simple regular representations in the tube $\mathcal{T}_1^{\Delta(\widetilde{\mathbb{E}}_6)}$

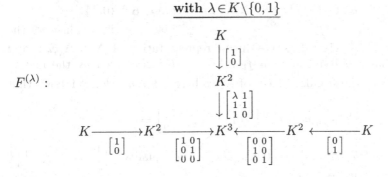

(d) Simple regular representations in the tube $\mathcal{T}_\lambda^{\Delta(\widetilde{\mathbb{E}}_6)}$, with $\lambda \in K \setminus \{0,1\}$

$$F^{(\lambda)} :$$

$$K$$
$$\downarrow \begin{bmatrix} 1 \\ 0 \end{bmatrix}$$
$$K^2$$
$$\downarrow \begin{bmatrix} \lambda & 1 \\ 1 & 1 \\ 1 & 0 \end{bmatrix}$$
$$K \xrightarrow{\begin{bmatrix} 1 \\ 0 \end{bmatrix}} K^2 \xrightarrow{\begin{bmatrix} 1 & 0 \\ 0 & 1 \\ 0 & 0 \end{bmatrix}} K^3 \xleftarrow{\begin{bmatrix} 0 & 0 \\ 1 & 0 \\ 0 & 1 \end{bmatrix}} K^2 \xleftarrow{\begin{bmatrix} 0 \\ 1 \end{bmatrix}} K$$

2.13. Definition. Assume that $A = K\Delta$, where Δ is the canonically oriented Euclidean quiver $\Delta(\widetilde{\mathbb{E}}_6)$ shown in Table 1.1. We define the $\mathbb{P}_1(K)$-family (see (2.3))

$$\mathcal{T}^{\Delta(\widetilde{\mathbb{E}}_6)} = \{\mathcal{T}_\lambda^{\Delta(\widetilde{\mathbb{E}}_6)}\}_{\lambda \in \mathbb{P}_1(K)}$$

of connected components $\mathcal{T}_\lambda^{\Delta(\widetilde{\mathbb{E}}_6)}$ of the Auslander–Reiten quiver $\Gamma(\mathrm{mod}\, A)$ of A as follows:

- $\mathcal{T}_\infty^{\Delta(\widetilde{\mathbb{E}}_6)}$ is the component containing the A-module $F_1^{(\infty)}$,
- $\mathcal{T}_0^{\Delta(\widetilde{\mathbb{E}}_6)}$ is the component containing the A-module $F_1^{(0)}$,
- $\mathcal{T}_1^{\Delta(\widetilde{\mathbb{E}}_6)}$ is the component containing the A-module $F_1^{(1)}$, and
- for each $\lambda \in K \setminus \{0,1\}$, $\mathcal{T}_\lambda^{\Delta(\widetilde{\mathbb{E}}_6)}$ is the component containing the A-module $F^{(\lambda)}$,

where $F_1^{(\infty)}$, $F_1^{(0)}$, $F_1^{(1)}$, and $F^{(\lambda)}$ are the modules shown in Table 2.12.

It follows from (2.15) below that the components $\mathcal{T}_\lambda^{\Delta(\widetilde{\mathbb{E}}_6)}$ of the $\mathbb{P}_1(K)$-family $\mathcal{T}^\Delta(\widetilde{\mathbb{E}}_6)$ are stable tubes in $\mathcal{R}(A)$.

We now collect the main properties of the modules listed in (2.12).

2.14. Lemma. *Let Δ be the the canonically oriented Euclidean quiver $\Delta(\widetilde{\mathbb{E}}_6)$ shown in Table 1.1. Let $A = K\Delta$ be the path algebra of Δ. The A-modules*

$$F_1^{(\infty)}, F_2^{(\infty)}, F_1^{(0)}, F_2^{(0)}, F_3^{(0)}, F_1^{(1)}, F_2^{(1)}, F_3^{(1)}, \text{ and } F^{(\lambda)}, \text{ with } \lambda \in K\backslash\{0,1\},$$

presented in Table 2.12 are pairwise orthogonal bricks, and each of them is simple regular.

Proof. It is obvious that the A-modules $F_1^{(\infty)}$, $F_1^{(0)}$, $F_2^{(0)}$, $F_3^{(0)}$, $F_1^{(1)}$, $F_2^{(1)}$, $F_3^{(1)}$ are bricks. Now we show that there are algebra isomorphisms $\mathrm{End}_A\, F_2^{(\infty)} \cong K$, and $\mathrm{End}_A\, F^{(\lambda)} \cong K$, for each $\lambda \in K \setminus \{0,1\}$.

Let $f = (f_1, \ldots, f_7) : F_2^{(\infty)} \longrightarrow F_2^{(\infty)}$ be an endomorphism of the A-module $F_2^{(\infty)}$, viewed as a K-linear representation of $\Delta = \Delta(\widetilde{\mathbb{E}}_6)$, and assume that the K-linear map $f_1 : K^2 \longrightarrow K^2$ is defined by the matrix $f_1 = \begin{bmatrix} a & b \\ c & d \end{bmatrix}$ in the standard basis of K^2, where $a, b, c, d \in K$. It follows that

$$\begin{bmatrix} 0 \\ 1 \end{bmatrix} \cdot f_6 = f_1 \cdot \begin{bmatrix} 0 \\ 1 \end{bmatrix} \quad \text{and} \quad f_1 \cdot \begin{bmatrix} 1 \\ 0 \end{bmatrix} = \begin{bmatrix} 1 \\ 0 \end{bmatrix} \cdot f_2.$$

Hence we get $b = 0$ and $c = 0$. Similarly, the equality $\begin{bmatrix} 1 \\ 1 \end{bmatrix} \cdot f_4 = f_1 \cdot \begin{bmatrix} 1 \\ 1 \end{bmatrix}$ yields $a = d$. Hence we get

$$f_1 = \begin{bmatrix} a & 0 \\ 0 & a \end{bmatrix},$$

and obviously the map $\mathrm{End}\, F_2^{(\infty)} \longrightarrow K$ that assigns to the endomorphism $f : F_2^{(\infty)} \longrightarrow F_2^{(\infty)}$ of $F_2^{(\infty)}$ the scalar a on the diagonal of the matrix f_1, is a K-algebra isomorphism.

Let $g = (g_1, \dots, g_7) : F^{(\lambda)} \longrightarrow F^{(\lambda)}$ be an endomorphism of the A-module $F^{(\lambda)}$, viewed as a K-linear representation of $\Delta = \Delta(\widetilde{\mathbb{E}}_6)$, and assume that the K-linear map $g_1 : K^3 \longrightarrow K^3$ is defined by the matrix

$$g_1 = \begin{bmatrix} a_{11} & a_{12} & a_{13} \\ a_{21} & a_{22} & a_{23} \\ a_{31} & a_{32} & a_{33} \end{bmatrix}$$

in the standard basis of K^3, where $a_{ij} \in K$. It follows that

$$\begin{bmatrix} 1 \\ 0 \\ 0 \end{bmatrix} \cdot g_3 = g_1 \cdot \begin{bmatrix} 1 \\ 0 \\ 0 \end{bmatrix} \quad \text{and} \quad g_1 \cdot \begin{bmatrix} 0 \\ 0 \\ 1 \end{bmatrix} = \begin{bmatrix} 0 \\ 0 \\ 1 \end{bmatrix} \cdot g_7.$$

Hence we get $a_{21} = 0$, $a_{31} = 0$, $a_{13} = 0$, and $a_{23} = 0$.

Similarly, the equality $\begin{bmatrix} 0 & 0 \\ 1 & 0 \\ 0 & 1 \end{bmatrix} \cdot g_6 = g_1 \cdot \begin{bmatrix} 0 & 0 \\ 1 & 0 \\ 0 & 1 \end{bmatrix}$ yields $a_{12} = 0$, and finally, from the equalities

$$\begin{bmatrix} \lambda & 1 \\ 1 & 1 \\ 1 & 0 \end{bmatrix} \cdot g_4 = g_1 \cdot \begin{bmatrix} \lambda & 1 \\ 1 & 1 \\ 1 & 0 \end{bmatrix} \quad \text{and} \quad \begin{bmatrix} 1 & 0 \\ 0 & 1 \\ 0 & 0 \end{bmatrix} \cdot g_2 = g_1 \cdot \begin{bmatrix} 1 & 0 \\ 0 & 1 \\ 0 & 0 \end{bmatrix}$$

we conclude that $a_{32} = 0$ and $a_{33} = a_{22} = a_{11}$. Consequently, g_1 has the diagonal form

$$g_1 = \begin{bmatrix} a_{11} & 0 & 0 \\ 0 & a_{11} & 0 \\ 0 & 0 & a_{11} \end{bmatrix},$$

and obviously the map $\mathrm{End}_A\, F^{(\lambda)} \longrightarrow K$, assigning to the endomorphism g of $F^{(\lambda)}$ the scalar a_{11} on the diagonal of the matrix g_1, is a K-algebra isomorphism.

It follows that the modules of Table 2.12 are bricks. Now we show that they are pairwise orthogonal. It is easy to check that $\mathrm{Hom}_A(F_i^{(s)}, F_j^{(t)}) = 0$, for all $i \neq j$ and $s, t \in \{0, 1, \infty\}$. Then, it remains to show that $\mathrm{Hom}_A(F^{(\lambda)}, F_j^{(t)}) = 0$ and $\mathrm{Hom}_A(F_j^{(t)}, F^{(\lambda)}) = 0$, for all $t \in \{0, 1, \infty\}$, all j, and $\lambda \in \mathbb{P}_1(K) \setminus \{0, 1, \infty\}$.

We only prove that $\mathrm{Hom}_A(F_3^{(1)}, F^{(\lambda)}) = 0$ and $\mathrm{Hom}_A(F^{(\lambda)}, F_3^{(1)}) = 0$, because the remaining equalities follow in a similar way.

Let $f = (f_1, \dots, f_7) : F_3^{(1)} \longrightarrow F^{(\lambda)}$ be a morphism of representations of the quiver $\Delta = \Lambda(\widetilde{\mathbb{F}}_6)$. It follows that

$$f_1 \cdot 1 = \begin{bmatrix} \lambda & 1 \\ 1 & 1 \\ 1 & 0 \end{bmatrix} \cdot \begin{bmatrix} 1 \\ 0 \end{bmatrix} \cdot f_5 = \begin{bmatrix} \lambda \\ 1 \\ 1 \end{bmatrix} \cdot f_5 \quad \text{and} \quad f_1 \cdot 1 = \begin{bmatrix} 0 & 0 \\ 1 & 0 \\ 0 & 1 \end{bmatrix} \cdot f_6 = \begin{bmatrix} 0 \\ a_1 \\ a_2 \end{bmatrix},$$

where $f_6 = \begin{bmatrix} a_1 \\ a_2 \end{bmatrix} : K \longrightarrow K^2$. Hence $f_5\lambda = 0$, and, because $\lambda \neq 0$, then $f_5 = 0$. It follows that $f_1 = 0$ and $f_6 = 0$. Moreover, $f_4 = \begin{bmatrix} 1 \\ 0 \end{bmatrix} \cdot f_1 = 0$ and, hence, $f = 0$.

Let $g = (g_1, \ldots, g_7) : F^{(\lambda)} \longrightarrow F_3^{(1)}$ be a morphism of representations of the quiver Δ and let $g_1 = [a_1\ a_2\ a_3] : K^3 \longrightarrow K$. The equalities

$$0 = g_1 \begin{bmatrix} 1 & 0 \\ 0 & 1 \\ 0 & 0 \end{bmatrix} = [a_1\ a_2\ a_3] \cdot \begin{bmatrix} 1 & 0 \\ 0 & 1 \\ 0 & 0 \end{bmatrix} = [a_1\ a_2]$$

yield $a_1 = a_2 = 0$. On the other hand, the equality

$$0 = g_1 \cdot \begin{bmatrix} 0 & 0 \\ 1 & 0 \\ 0 & 1 \end{bmatrix} \cdot \begin{bmatrix} 0 \\ 1 \end{bmatrix}$$

yields $a_3 = 0$. Hence, $g_1 = 0$ and, consequently, $g = 0$.

We recall from (1.3)(b) that the defect $\partial_A : \mathbb{Z}^7 \longrightarrow \mathbb{Z}$ of A is given by the formula $\partial_A(\mathbf{x}) = -3x_1 + x_2 + x_3 + x_4 + x_5 + x_6 + x_7$. It follows that $\partial_A(\mathbf{dim}\,F) = 0$ for any module F of Table 2.12. Hence, by (XI.2.8), the indecomposable modules of Table 2.12 are regular.

Now we show that they are simple regular. For this purpose, we suppose that F is one of the modules
$F_1^{(\infty)}, F_2^{(\infty)}, F_1^{(0)}, F_2^{(0)}, F_3^{(0)}, F_1^{(1)}, F_2^{(1)}, F_3^{(1)}$, and $F^{(\lambda)}$, with $\lambda \in K \setminus \{0, 1\}$, listed in (2.12), and let X be an indecomposable regular submodule of F. By (XI.2.8), we have $\partial_A(\mathbf{dim}\,F) = 0$, and a case by case inspection shows that $X = F$ (apply the arguments given in the proof of (2.8)). This is easily seen if $F \neq F^{(\lambda)}$. For $F = F^{(\lambda)}$ it is not immediate. However, it follows from the proof of the following theorem that $F^{(\lambda)} \cong \mathrm{Hom}_C(T, E^{(\lambda)})$ and, hence, $F^{(\lambda)}$ is simple regular. This completes the proof.   □

Next, we collect the main properties of the family $\mathcal{T}^{\Delta(\widetilde{\mathbb{E}}_6)}$ of components defined in (2.13).

2.15. Theorem. *Let Δ be the Euclidean quiver $\Delta(\widetilde{\mathbb{E}}_6)$. Let $A = K\Delta$ be the path algebra of Δ. Let*

$F_1^{(\infty)}, F_2^{(\infty)}, F_1^{(0)}, F_2^{(0)}, F_3^{(0)}, F_1^{(1)}, F_2^{(1)}, F_3^{(1)},$ *and $F^{(\lambda)}$, with $\lambda \in K \setminus \{0, 1\}$,*

be the A-modules presented in Table 2.12. The family

$$\mathcal{T}^{\Delta(\widetilde{\mathbb{E}}_6)} = \{\mathcal{T}_\lambda^{\Delta(\widetilde{\mathbb{E}}_6)}\}_{\lambda \in \mathbb{P}_1(K)}$$

defined in (2.13) has the following properties.

(a) *The component $\mathcal{T}_\infty^{\Delta(\widetilde{\mathbb{E}}_6)}$ is a stable tube of rank 2 containing the modules $F_1^{(\infty)}$ and $F_2^{(\infty)}$, and there exist isomorphisms*

$$\tau_A F_2^{(\infty)} \cong F_1^{(\infty)} \quad and \quad \tau_A F_1^{(\infty)} \cong F_2^{(\infty)}.$$

(b) *The component* $\mathcal{T}_0^{\Delta(\widetilde{\mathbb{E}}_6)}$ *is a stable tube of rank* 3 *containing the modules* $F_1^{(0)}, F_2^{(0)}, F_3^{(0)},$ *and there are isomorphisms*

$$\tau_A F_3^{(0)} \cong F_2^{(0)}, \quad \tau_A F_2^{(0)} \cong F_1^{(0)}, \quad \tau_A F_1^{(0)} \cong F_3^{(0)}.$$

(c) *The component* $\mathcal{T}_1^{\Delta(\widetilde{\mathbb{E}}_6)}$ *is a stable tube of rank* 3 *containing the modules* $F_1^{(1)}, F_2^{(1)}$ *and* $F_3^{(1)},$ *and there are isomorphisms*

$$\tau_A F_3^{(1)} \cong F_2^{(1)}, \quad \tau_A F_2^{(1)} \cong F_1^{(1)}, \quad \tau_A F_1^{(1)} \cong F_3^{(1)}.$$

(d) *For each* $\lambda \in K \setminus \{0, 1\},$ *the component* $\mathcal{T}_\lambda^{\Delta(\widetilde{\mathbb{E}}_6)}$ *is a stable tube of rank* 1 *containing the module* $F^{(\lambda)},$ *and there is an isomorphism* $\tau_A F^{(\lambda)} \cong F^{(\lambda)}.$

(e) *The Auslander–Reiten quiver* $\Gamma(\mathrm{mod}\,A)$ *of* A *consists of a postprojective component* $\mathcal{P}(A),$ *a preinjective component* $\mathcal{Q}(A)$ *and the* $\mathbb{P}_1(K)$-*family* $\boldsymbol{\mathcal{T}}^{\Delta(\widetilde{\mathbb{E}}_6)} = \{\mathcal{T}_\lambda^{\Delta(\widetilde{\mathbb{E}}_6)}\}_{\lambda \in \mathbb{P}_1(K)}$ *of stable tubes separating the component* $\mathcal{P}(A)$ *from* $\mathcal{Q}(A).$

Proof. Let $\Delta = \Delta(\widetilde{\mathbb{E}}_6)$ and let $A = K\Delta.$ We consider the canonical algebra $C = C(2, 3, 3)$ of type $\Delta(\widetilde{\mathbb{E}}_6)$ defined in (XII.1.3), and the family of C-modules

$$E_1^{(\infty)}, E_2^{(\infty)}, E_1^{(0)}, E_2^{(0)}, E_3^{(0)}, E_1^{(1)}, E_2^{(1)}, E_3^{(1)}, and\ E^{(\lambda)}, with\ \lambda \in K \setminus \{0, 1\},$$

defined in Table XII.2.9. It follows from (XII.2.12) that they are simple regular.

By (XII.1.10), the postprojective component $\mathcal{P}(C)$ of $\Gamma(\mathrm{mod}\,C)$ contains the section

$$T_5$$
$$\uparrow$$
$$T_4$$
$$\uparrow$$
$$T_3 \longleftarrow T_2 \longleftarrow T_1 \longrightarrow T_6 \longrightarrow T_7,$$

where

$$T_1 = \tau^{-2} P(0), \ T_2 = \tau^{-2} P(b_1), \ T_3 = \tau^{-2} P(b_2),$$
$$T_4 = \tau^{-2} P(a_1), \ T_5 = P(\omega), \ T_6 = \tau^{-2} P(c_1), \ T_7 = \tau^{-2} P(c_2).$$

Here we use the notation of (XII.1.2). Moreover,

$$T = T_1 \oplus T_2 \oplus T_3 \oplus T_4 \oplus T_5 \oplus T_6 \oplus T_7$$

is a postprojective tilting C-module and there is a K-algebra isomorphism

$$A \cong \mathrm{End}\,T_C.$$

Then, by (VIII.4.5), the functor $\mathrm{Hom}_C(T, -) : \mathrm{mod}\, C \longrightarrow \mathrm{mod}\, A$ restricts to the equivalence

$$\mathrm{Hom}_C(T, -) : \mathrm{add}\,\mathcal{R}(C) \xrightarrow{\;\cong\;} \mathrm{add}\,\mathcal{R}(A)$$

of abelian and serial categories $\mathrm{add}\,\mathcal{R}(C)$ and $\mathrm{add}\,\mathcal{R}(A)$ of regular modules such that the diagram (2.11) is commutative.

It follows that the functor $\mathrm{Hom}_C(T, -) : \mathrm{add}\,\mathcal{R}(C) \xrightarrow{\cong} \mathrm{add}\,\mathcal{R}(A)$ carries irreducible morphisms to irreducible ones, and the image $\mathrm{Hom}_C(T, \mathcal{T}_\lambda^C)$ of any stable tube $\mathcal{T}_\lambda^C$ in $\mathcal{R}(C)$ of rank $r_\lambda \geq 1$ is a stable tube in $\mathcal{R}(A)$ of the same rank r_λ. Moreover, any stable tube in $\mathcal{R}(A)$ is of the form $\mathrm{Hom}_C(T, \mathcal{T}_\lambda^C)$.

We claim that, for each $\lambda \in \mathbb{P}_1(K)$, the tubes $\mathrm{Hom}_C(T, \mathcal{T}_\lambda^C)$ and $\mathcal{T}_\lambda^{\Delta(\widetilde{\mathbb{E}}_6)}$ coincide. For this purpose, it suffices to show that there are isomorphisms

$$\mathrm{Hom}_C(T, E_j^{(s)}) \cong F_j^{(s)} \quad \text{and} \quad \mathrm{Hom}_C(T, E^{(\lambda)}) \cong F^{(\lambda)},$$

for all $j, s \in \{0, 1, \infty\}$, and all $\lambda \in \mathbb{P}_1(K) \setminus \{0, 1, \infty\}$.

To prove this, we recall that $A \cong \mathrm{End}\, T_C$ and, as in the proof of (XII.1.8), the left hand part of the postprojective component $\mathcal{P}(C)$ of $\Gamma(\mathrm{mod}\, C)$ containing all the indecomposable projective C-modules $P(0), \ldots, P(\omega)$ looks as follows

where

$$P(a_1) = 1 \underset{0\;0}{\overset{1}{0}} 0\, 0$$

and the indecomposable modules are represented by their dimension vectors.

It follows from (VI.3.10) that $e_i A \cong \operatorname{Hom}_C(T, T_i)$ and, for each regular C-module X and for each $i \in \{1, \ldots, 7\}$, we have

$$
\begin{aligned}
\operatorname{Hom}_C(T, X)_i = \operatorname{Hom}_C(T, X)e_i \\
\cong \operatorname{Hom}_A(e_i A, \operatorname{Hom}_C(T, X)) \\
\cong \operatorname{Hom}_A(\operatorname{Hom}_C(T, T_i), \operatorname{Hom}_C(T, X)) \\
\cong \operatorname{Hom}_C(T_i, X),
\end{aligned}
$$

and therefore $(\mathbf{dim}\,\operatorname{Hom}_C(T, X))_i = \dim_K \operatorname{Hom}_C(T_i, X)$. Note also that we have

$$\operatorname{Hom}_C(T_5, X) = \operatorname{Hom}_C(P(\omega), X) \cong X_\omega.$$

By consulting the beginning part of the component $\mathcal{P}(C)$ shown above we see that the ω-th coordinate of the dimension vector of each of the C-modules $T_1, T_2, T_3, T_4, T_6, T_7$ equals zero. Let

$$
Q : \quad 0 \xleftarrow{\beta_1} b_1 \xleftarrow{\beta_2} b_2
$$
with α_1 to a_1 and γ_1 to $c_1 \xleftarrow{\gamma_2} c_2$

be the full subquiver of $\Delta(2, 3, 3)$, of the Dynkin type $\mathbb{E}_6$, defined by the points $0, a_1, b_1, b_2, c_1, c_2$, in the notation of (XII.1.2).

Note that there is an isomorphism $KQ \cong e_Q C e_Q$ of algebras and the equivalence of categories $\operatorname{mod} KQ \cong \operatorname{rep}_K(Q)$, where e_Q is the idempotent

$$e_Q = e_0 + e_{a_1} + e_{b_1} + e_{b_2} + e_{c_1} + e_{c_2}$$

of the algebra C. It follows that the restriction functor

$$\operatorname{res}_{e_Q} : \operatorname{mod} C \longrightarrow \operatorname{mod} KQ \cong \operatorname{rep}_K(Q),$$

see (I.6.6), defines a functorial isomorphism

$$\operatorname{Hom}_C(T_j, E) \cong \operatorname{Hom}_{KQ}(\operatorname{res}_{e_Q} T_j, \operatorname{res}_{e_Q} E),$$

for $T_j \neq T_5 = P(\omega)$ and for any module E in the family
$E_1^{(\infty)}, E_2^{(\infty)}, E_1^{(0)}, E_2^{(0)}, E_3^{(0)}, E_1^{(1)}, E_2^{(1)}, E_3^{(1)}$, and $E^{(\lambda)}$, with $\lambda \in K \backslash \{0, 1\}$.
Because $\operatorname{res}_{e_Q} E^{(\lambda)} = I(0)$ is the injective envelope of the simple representation $S(0)$ in $\operatorname{rep}_K(Q)$ then, for each $j \neq 5$, there exist isomorphisms

$$\operatorname{Hom}_C(T_j, E^{(\lambda)}) \cong \operatorname{Hom}_{KQ}(\operatorname{res}_{e_Q} T_j, \operatorname{res}_{e_Q} E^{(\lambda)}) \cong D(T_j)_0,$$

that are functorial at T_j. By applying the above formulae to the module $E = F^{(\lambda)}$, with $\lambda \in K \backslash \{0, 1\}$, we get

$$\mathrm{Hom}_C(T_1, E^{(\lambda)}) \cong D(T_1)_0 = K^3,$$
$$\mathrm{Hom}_C(T_2, E^{(\lambda)}) \cong D(T_2)_0 = K^2,$$
$$\mathrm{Hom}_C(T_3, E^{(\lambda)}) \cong D(T_3)_0 = K,$$
$$\mathrm{Hom}_C(T_4, E^{(\lambda)}) \cong D(T_4)_0 = K^2,$$
$$\mathrm{Hom}_C(T_6, E^{(\lambda)}) \cong D(T_6)_0 = K^2,$$
$$\mathrm{Hom}_C(T_7, E^{(\lambda)}) \cong D(T_7)_0 = K, \text{ and}$$
$$\mathrm{Hom}_C(T_5, E^{(\lambda)}) \cong E^{(\lambda)}_\omega = K.$$

In particular, we get $\mathbf{dim}\,\mathrm{Hom}_C(T, E^{(\lambda)}) = \begin{smallmatrix} & & 1 & & \\ & & 2 & & \\ 1 & 2 & 3 & 2 & 1 \end{smallmatrix}$. Moreover, by applying the above formulae, one shows that there is an A-module isomorphism

$$\mathrm{Hom}_C(T, E^{(\lambda)}) \cong F^{(\lambda)},$$

and hence $\mathrm{Hom}_C(T, \mathcal{T}^C_\lambda) = \mathcal{T}^{\Delta(\widetilde{\mathbb{E}}_6)}_\lambda$, for any $\lambda \in K \setminus \{0, 1\}$. Because, by (XII.2.12), the C-module $E^{(\lambda)}$ is simple regular, then the A-module $\mathrm{Hom}_C(T, E^{(\lambda)}) \cong F^{(\lambda)}$ is simple regular. Furthermore, in view of the isomorphism $\tau_C E^{(\lambda)} \cong E^{(\lambda)}$ in (XII.2.12) and the commutativity of the diagram (2.11), we get the isomorphism $\tau_A F^{(\lambda)} \cong F^{(\lambda)}$. This finishes the proof of (d).

Similarly, by applying the above formulae to any module E in the family

$$E^{(\infty)}_1, \; E^{(\infty)}_2, \; E^{(0)}_1, \; E^{(0)}_2, \; E^{(0)}_3 \; E^{(1)}_1, \; E^{(1)}_2, \; E^{(1)}_3,$$

we get

$$\mathbf{dim}\,\mathrm{Hom}_C(T, E^{(\infty)}_1) = \begin{smallmatrix} & & 0 & & \\ & & 1 & & \\ 0 & 1 & 1 & 1 & 0 \end{smallmatrix}, \qquad \mathbf{dim}\,\mathrm{Hom}_C(T, E^{(\infty)}_2) = \begin{smallmatrix} & & 1 & & \\ & & 1 & & \\ 1 & 1 & 2 & 1 & 1 \end{smallmatrix},$$

$$\mathbf{dim}\,\mathrm{Hom}_C(T, E^{(0)}_1) = \begin{smallmatrix} & & 0 & & \\ & & 0 & & \\ 1 & 1 & 1 & 1 & 0 \end{smallmatrix}, \qquad \mathbf{dim}\,\mathrm{Hom}_C(T, E^{(0)}_2) = \begin{smallmatrix} & & 0 & & \\ & & 1 & & \\ 0 & 0 & 1 & 1 & 1 \end{smallmatrix},$$

$$\mathbf{dim}\,\mathrm{Hom}_C(T, E^{(0)}_3) = \begin{smallmatrix} & & 1 & & \\ & & 1 & & \\ 0 & 1 & 1 & 0 & 0 \end{smallmatrix}, \qquad \mathbf{dim}\,\mathrm{Hom}_C(T, E^{(1)}_1) = \begin{smallmatrix} & & 0 & & \\ & & 0 & & \\ 0 & 1 & 1 & 1 & 1 \end{smallmatrix},$$

$$\mathbf{dim}\,\mathrm{Hom}_C(T, E^{(1)}_2) = \begin{smallmatrix} & & 0 & & \\ & & 1 & & \\ 1 & 1 & 1 & 0 & 0 \end{smallmatrix}, \qquad \mathbf{dim}\,\mathrm{Hom}_C(T, E^{(1)}_3) = \begin{smallmatrix} & & 1 & & \\ & & 1 & & \\ 0 & 0 & 1 & 1 & 0 \end{smallmatrix}.$$

In particular, $\mathbf{dim}\,\mathrm{Hom}_C(T, E^{(s)}_j) = \mathbf{dim}\,F^{(s)}_j$, for all j and $s \in \{0, 1, \infty\}$.

Because, by (XII.2.12), E is a simple regular C-module, then obviously the A-module $\mathrm{Hom}_C(T, E)$ is simple regular.

We recall from (XI.3.3) and (XII.4.2) that the simple regular A-modules lying in the non-homogeneous tubes of $\mathcal{R}(A)$ are uniquely determined by their dimension vectors. Because, by (2.14), the modules presented in Table 2.12 are simple regular, then there is an isomorphism $\mathrm{Hom}_C(T, E^{(s)}_j) \cong F^{(s)}_j$, for all j, and $s \in \{0, 1, \infty\}$.

Therefore, by applying (XII.2.12) and the commutative diagram (2.11), we prove the statements (a)–(c), and we conclude that the tubes

$$\mathcal{T}_\infty^{\Delta(\widetilde{\mathbb{E}}_6)} = \operatorname{Hom}_C(T, \mathcal{T}_\infty^C), \ \mathcal{T}_0^{\Delta(\widetilde{\mathbb{E}}_6)} = \operatorname{Hom}_C(T, \mathcal{T}_0^C), \ \mathcal{T}_1^{\Delta(\widetilde{\mathbb{E}}_6)} = \operatorname{Hom}_C(T, \mathcal{T}_1^C)$$

exhaust all the non-homogeneous tubes of the regular part $\mathcal{R}(A)$ of $\Gamma(\operatorname{mod} A)$. Consequently, the tubes $\mathcal{T}_\lambda^{\Delta(\widetilde{\mathbb{E}}_6)} = \operatorname{Hom}_C(T, \mathcal{T}_\lambda^C)$, with $\lambda \in K \setminus \{0, 1\}$, exhaust all the homogeneous tubes of $\Gamma(\operatorname{mod} A)$. Now the statement (e) follows from (2.1). This completes the proof. □

2.16. Table. Simple regular representations of $\Delta(\widetilde{\mathbb{E}}_7)$

Assume that $\Delta = \Delta(\widetilde{\mathbb{E}}_7)$ is the canonically oriented Euclidean quiver

$$\Delta(\widetilde{\mathbb{E}}_7): \qquad \begin{array}{c} 5 \\ \downarrow \\ 4 \longrightarrow 3 \longrightarrow 2 \longrightarrow 1 \longleftarrow 6 \longleftarrow 7 \longleftarrow 8 \end{array}$$

shown in Table 1.1, and let $A = K\Delta$. Consider the following family of indecomposable regular A-modules, viewed as K-linear representations of $\Delta(\widetilde{\mathbb{E}}_7)$. We show below that they are simple regular.

(a) **Simple regular representations in the tube $\mathcal{T}_\infty^{\Delta(\widetilde{\mathbb{E}}_7)}$**

$$F_1^{(\infty)}: \qquad \begin{array}{c} K \\ \downarrow {\scriptstyle \begin{bmatrix} 1 \\ 1 \end{bmatrix}} \\ 0 \longrightarrow K \underset{1}{\longrightarrow} K \underset{\begin{bmatrix} 1 \\ 0 \end{bmatrix}}{\longrightarrow} K^2 \underset{\begin{bmatrix} 0 \\ 1 \end{bmatrix}}{\longleftarrow} K \underset{1}{\longleftarrow} K \underset{1}{\longleftarrow} K \end{array}$$

$$F_2^{(\infty)}: \qquad \begin{array}{c} K \\ \downarrow {\scriptstyle \begin{bmatrix} 1 \\ 1 \end{bmatrix}} \\ K \underset{1}{\longrightarrow} K \underset{1}{\longrightarrow} K \underset{\begin{bmatrix} 1 \\ 0 \end{bmatrix}}{\longrightarrow} K^2 \underset{\begin{bmatrix} 0 \\ 1 \end{bmatrix}}{\longleftarrow} K \underset{1}{\longleftarrow} K \longleftarrow 0 \end{array}$$

(b) **Simple regular representations in the tube $\mathcal{T}_0^{\Delta(\widetilde{\mathbb{E}}_7)}$**

$$F_1^{(0)}: \qquad \begin{array}{c} K \\ \downarrow {\scriptstyle 1} \\ 0 \longrightarrow 0 \longrightarrow K \underset{1}{\longrightarrow} K \underset{1}{\longleftarrow} 0 \longleftarrow 0 \longleftarrow 0 \end{array}$$

$$F_2^{(0)}: \qquad \begin{array}{c} 0 \\ \downarrow \\ 0 \longrightarrow K \xrightarrow{1} K \xrightarrow{1} K \xleftarrow{1} K \xleftarrow{1} K \longleftarrow 0 \end{array}$$

$$F_3^{(0)}: \qquad \begin{array}{c} K \\ \downarrow \left[\begin{smallmatrix} 1 \\ 1 \end{smallmatrix}\right] \\ K \xrightarrow{1} K \xrightarrow{1} K \xrightarrow{\left[\begin{smallmatrix} 1 \\ 0 \end{smallmatrix}\right]} K^2 \xleftarrow{\left[\begin{smallmatrix} 0 \\ 1 \end{smallmatrix}\right]} K \xleftarrow{1} K \xleftarrow{1} K \end{array}$$

(c) Simple regular representations in the tube $\mathcal{T}_1^{\Delta(\widetilde{\mathbb{E}}_7)}$

$$F_1^{(1)}: \qquad \begin{array}{c} K \\ \downarrow 1 \\ 0 \longrightarrow 0 \longrightarrow 0 \longrightarrow K \xleftarrow{1} K \xleftarrow{1} K \longleftarrow 0 \end{array}$$

$$F_2^{(1)}: \qquad \begin{array}{c} 0 \\ \downarrow \\ 0 \longrightarrow 0 \longrightarrow K \xrightarrow{1} K \xleftarrow{1} K \xleftarrow{1} K \xleftarrow{1} K \end{array}$$

$$F_3^{(1)}: \qquad \begin{array}{c} K \\ \downarrow 1 \\ 0 \longrightarrow K \xrightarrow{1} K \xrightarrow{1} K \longleftarrow 0 \longleftarrow 0 \longleftarrow 0 \end{array}$$

$$F_4^{(1)}: \qquad \begin{array}{c} 0 \\ \downarrow 1 \\ K \xrightarrow{1} K \xrightarrow{1} K \xrightarrow{1} K \xleftarrow{1} K \longleftarrow 0 \longleftarrow 0 \end{array}$$

(d) Simple regular representations in the tube $\mathcal{T}_\lambda^{\Delta(\widetilde{\mathbb{E}}_7)}$,

with $\lambda \in K \setminus \{0, 1\}$

$$F^{(\lambda)}: \qquad \begin{array}{c} K^2 \\ \downarrow \left[\begin{smallmatrix} 1 & \lambda \\ 1 & 0 \\ 1 & 1 \\ 0 & 1 \end{smallmatrix}\right] \\ K \xrightarrow{\left[\begin{smallmatrix} 1 \\ 0 \end{smallmatrix}\right]} K^2 \xrightarrow{\left[\begin{smallmatrix} 1 & 0 \\ 0 & 1 \\ 0 & 0 \end{smallmatrix}\right]} K^3 \xrightarrow{\left[\begin{smallmatrix} 1 & 0 & 0 \\ 0 & 1 & 0 \\ 0 & 0 & 1 \\ 0 & 0 & 0 \end{smallmatrix}\right]} K^4 \xleftarrow{\left[\begin{smallmatrix} 0 & 0 & 0 \\ 1 & 0 & 0 \\ 0 & 1 & 0 \\ 0 & 0 & 1 \end{smallmatrix}\right]} K^3 \xleftarrow{\left[\begin{smallmatrix} 0 & 0 \\ 1 & 0 \\ 0 & 1 \end{smallmatrix}\right]} K^2 \xleftarrow{\left[\begin{smallmatrix} 0 \\ 1 \end{smallmatrix}\right]} K. \end{array}$$

2.17. Definition. Assume that $A = K\Delta$, where Δ is the canonically oriented Euclidean quiver $\Delta(\widetilde{\mathbb{E}}_7)$ shown in Table 1.1. We define the $\mathbb{P}_1(K)$-family (see (2.3))

$$\mathcal{T}^{\Delta(\widetilde{\mathbb{E}}_7)} = \{\mathcal{T}_\lambda^{\Delta(\widetilde{\mathbb{E}}_7)}\}_{\lambda \in \mathbb{P}_1(K)}$$

of connected components $\mathcal{T}_\lambda^{\Delta(\widetilde{\mathbb{E}}_7)}$ of the Auslander–Reiten quiver $\Gamma(\operatorname{mod} A)$ of A as follows:

- $\mathcal{T}_\infty^{\Delta(\widetilde{\mathbb{E}}_7)}$ is the component containing the A-module $F_1^{(\infty)}$,
- $\mathcal{T}_0^{\Delta(\widetilde{\mathbb{E}}_7)}$ is the component containing the A-module $F_1^{(0)}$,
- $\mathcal{T}_1^{\Delta(\widetilde{\mathbb{E}}_7)}$ is the component containing the A-module $F_1^{(1)}$, and
- for each $\lambda \in K \setminus \{0, 1\}$, $\mathcal{T}_\lambda^{\Delta(\widetilde{\mathbb{E}}_7)}$ is the component containing the A-module $F^{(\lambda)}$, where $F_1^{(\infty)}$, $F_1^{(0)}$, $F_1^{(1)}$, and $F^{(\lambda)}$ are the modules shown in Table 2.16.

It follows from (2.19) below that the components $\mathcal{T}_\lambda^{\Delta(\widetilde{\mathbb{E}}_7)}$ of the $\mathbb{P}_1(K)$-family $\mathcal{T}^{\Delta}(\widetilde{\mathbb{E}}_7)$ are stable tubes in $\mathcal{R}(A)$.

We now collect the main properties of the modules listed in (2.16).

2.18. Lemma. *Let Δ be the the canonically oriented Euclidean quiver $\Delta(\widetilde{\mathbb{E}}_7)$ shown in Table 1.1. Let $A = K\Delta$ be the path algebra of Δ. The A-modules*

$$F_1^{(\infty)}, F_2^{(\infty)}, F_1^{(0)}, F_2^{(0)}, F_3^{(0)}, F_1^{(1)}, F_2^{(1)}, F_3^{(1)}, F_4^{(1)}, \text{ and } F^{(\lambda)},$$

with $\lambda \in K \setminus \{0, 1\}$, presented in Table 2.16 are pairwise orthogonal bricks, and each of them is simple regular.

Proof. It is obvious that the A-modules $F_1^{(0)}$, $F_2^{(0)}$, $F_1^{(1)}$, $F_2^{(1)}$, $F_3^{(1)}$, and $F_4^{(1)}$ are bricks. Now we show that there are algebra isomorphisms

$$\operatorname{End}_A F_1^{(\infty)} \cong K, \ \operatorname{End} F_2^{(\infty)} \cong K, \ \operatorname{End} F_3^{(0)} \cong K, \text{ and } \operatorname{End} F^{(\lambda)} \cong K,$$

for each $\lambda \in K \setminus \{0, 1\}$.

Let F be one of the modules $F_1^{(\infty)}$, $F_2^{(\infty)}$, and $F_3^{(0)}$, and let

$$f = (f_1, \ldots, f_8) : F \longrightarrow F$$

be an endomorphism of the A-module F, viewed as a K-linear representation of $\Delta = \Delta(\widetilde{\mathbb{E}}_7)$. Assume that the K-linear map $f_1 : K^2 \longrightarrow K^2$ is defined by the matrix

$$f_1 = \begin{bmatrix} a & b \\ c & d \end{bmatrix}$$

in the standard basis of K^2, where $a, b, c, d \in K$. It follows that

$$\begin{bmatrix} 0 \\ 1 \end{bmatrix} \cdot f_6 = f_1 \cdot \begin{bmatrix} 0 \\ 1 \end{bmatrix} \quad \text{and} \quad f_1 \cdot \begin{bmatrix} 1 \\ 0 \end{bmatrix} = \begin{bmatrix} 1 \\ 0 \end{bmatrix} \cdot f_2.$$

Hence we get $b = 0$ and $c = 0$. Similarly, the equality

$$\begin{bmatrix} 1 \\ 1 \end{bmatrix} \cdot f_5 = f_1 \cdot \begin{bmatrix} 1 \\ 1 \end{bmatrix}$$

yields $a = d$. Hence we get

$$f_1 = \begin{bmatrix} a & 0 \\ 0 & a \end{bmatrix},$$

and obviously the map $\operatorname{End}_A F \longrightarrow K$, assigning to the endomorphism $f : F \to F$ of F the scalar a on the diagonal of the matrix f_1, is a K-algebra isomorphism.

Let

$$g = (g_1, \ldots, g_8) : F^{(\lambda)} \longrightarrow F^{(\lambda)}$$

be an endomorphism of the A-module $F^{(\lambda)}$, viewed as a K-linear representation of $\Delta = \Delta(\widetilde{\mathbb{E}}_7)$, and assume that the K-linear map $g_1 : K^4 \longrightarrow K^4$ is defined by the matrix

$$g_1 = \begin{bmatrix} a_{11} & a_{12} & a_{13} & a_{14} \\ a_{21} & a_{22} & a_{23} & a_{24} \\ a_{31} & a_{32} & a_{33} & a_{34} \\ a_{41} & a_{42} & a_{43} & a_{44} \end{bmatrix}$$

in the standard basis of K^4, where $a_{ij} \in K$. It follows that

$$\begin{bmatrix} 1 \\ 0 \\ 0 \\ 0 \end{bmatrix} \cdot g_4 = g_1 \cdot \begin{bmatrix} 1 \\ 0 \\ 0 \\ 0 \end{bmatrix} \quad \text{and} \quad g_1 \cdot \begin{bmatrix} 0 \\ 0 \\ 0 \\ 1 \end{bmatrix} = \begin{bmatrix} 0 \\ 0 \\ 0 \\ 1 \end{bmatrix} \cdot g_8.$$

Hence we get $a_{21} = 0$, $a_{31} = 0$, $a_{41} = 0$, $a_{14} = 0$, $a_{24} = 0$, and $a_{34} = 0$. Similarly, the equalities

$$\begin{bmatrix} 0 & 0 \\ 0 & 0 \\ 1 & 0 \\ 0 & 1 \end{bmatrix} \cdot g_7 = g_1 \cdot \begin{bmatrix} 0 & 0 \\ 0 & 0 \\ 1 & 0 \\ 0 & 1 \end{bmatrix} \quad \text{and} \quad \begin{bmatrix} 1 & 0 \\ 0 & 1 \\ 0 & 0 \\ 0 & 0 \end{bmatrix} \cdot g_3 = g_1 \cdot \begin{bmatrix} 1 & 0 \\ 0 & 1 \\ 0 & 0 \\ 0 & 0 \end{bmatrix}$$

yield $a_{13} = 0$, $a_{23} = 0$, $a_{32} = 0$, and $a_{42} = 0$. The equalities

$$\begin{bmatrix} 1 & 0 & 0 \\ 0 & 1 & 0 \\ 0 & 0 & 1 \\ 0 & 0 & 0 \end{bmatrix} \cdot g_2 = g_1 \cdot \begin{bmatrix} 1 & 0 & 0 \\ 0 & 1 & 0 \\ 0 & 0 & 1 \\ 0 & 0 & 0 \end{bmatrix} \quad \text{and} \quad \begin{bmatrix} 0 & 0 & 0 \\ 1 & 0 & 0 \\ 0 & 1 & 0 \\ 0 & 0 & 1 \end{bmatrix} \cdot g_6 = g_1 \cdot \begin{bmatrix} 0 & 0 & 0 \\ 1 & 0 & 0 \\ 0 & 1 & 0 \\ 0 & 0 & 1 \end{bmatrix}$$

yield $a_{34} = 0$ and $a_{12} = 0$. Finally, from the equality

$$\begin{bmatrix} 1 & \lambda \\ 1 & 0 \\ 1 & 1 \\ 0 & 1 \end{bmatrix} \cdot g_5 = g_1 \cdot \begin{bmatrix} 1 & \lambda \\ 1 & 0 \\ 1 & 1 \\ 0 & 1 \end{bmatrix}$$

we conclude that $a_{44} = a_{33} = a_{22} = a_{11}$. Consequently, g_1 has the diagonal form

$$g_1 = \begin{bmatrix} a_{11} & 0 & 0 & 0 \\ 0 & a_{11} & 0 & 0 \\ 0 & 0 & a_{11} & 0 \\ 0 & 0 & 0 & a_{11} \end{bmatrix},$$

and obviously the map $\operatorname{End}_A F^{(\lambda)} \longrightarrow K$, assigning to the endomorphism $g : F^{(\lambda)} \longrightarrow F^{(\lambda)}$ of $F^{(\lambda)}$ the scalar a_{11} on the diagonal of the matrix g_1, is a K-algebra isomorphism.

It follows that the modules of Table 2.16 are bricks. Now we show that they are pairwise orthogonal. It is easy to check that $\operatorname{Hom}_A(F_i^{(s)}, F_j^{(t)}) = 0$, for all $i \neq j$ and $s, t \in \{0, 1, \infty\}$. Then, it remains to show that

$$\operatorname{Hom}_A(F^{(\lambda)}, F_j^{(t)}) = 0 \text{ and } \operatorname{Hom}_A(F_j^{(t)}, F^{(\lambda)}) = 0,$$

for all $t \in \{0, 1, \infty\}$, all j, and $\lambda \in \mathbb{P}_1(K) \setminus \{0, 1, \infty\}$.

We only prove that $\operatorname{Hom}_A(F_3^{(1)}, F^{(\lambda)}) = 0$ and $\operatorname{Hom}_A(F^{(\lambda)}, F_3^{(1)}) = 0$, because the remaining equalities follow in a similar way.

Let

$$f = (f_1, \ldots, f_8) : F_3^{(1)} \longrightarrow F^{(\lambda)}$$

be a morphism of representations of the quiver $\Delta = \Delta(\widetilde{\mathbb{E}}_7)$ and let

$$f_1 = \begin{bmatrix} f_{11} \\ f_{21} \\ f_{31} \\ f_{41} \end{bmatrix} : K \longrightarrow K^4 \quad \text{and} \quad f_5 = \begin{bmatrix} f_{51} \\ f_{52} \end{bmatrix} : K \longrightarrow K^2.$$

It follows that

$$f_1 \cdot 1 = \begin{bmatrix} 1 & 0 \\ 0 & 1 \\ 0 & 0 \\ 0 & 0 \end{bmatrix} \cdot f_3 \quad \text{and} \quad f_1 \cdot 1 = \begin{bmatrix} 1 & \lambda \\ 1 & 0 \\ 1 & 1 \\ 0 & 1 \end{bmatrix} \cdot f_5 = \begin{bmatrix} f_{51} + \lambda f_{52} \\ f_{51} \\ f_{51} + f_{52} \\ f_{52} \end{bmatrix}.$$

Hence we get $f_{31} = f_{41} = 0$ and, consequently, $f_{51} = f_{52} = 0$. It follows that $f_5 = f_1 = 0$ and, hence, $f = 0$.

Let

$$g = (g_1, \ldots, g_8) : F^{(\lambda)} \longrightarrow F_3^{(1)}$$

be a morphism of representations of the quiver $\Delta(\widetilde{\mathbb{E}}_7)$ and let

$$g_1 = [g_{11} \ g_{12} \ g_{13} \ g_{14}] : K^4 \longrightarrow K.$$

The equalities

$$0 = 0 \cdot g_6 = g_1 \cdot \begin{bmatrix} 0 & 0 & 0 \\ 1 & 0 & 0 \\ 0 & 1 & 0 \\ 0 & 0 & 1 \end{bmatrix}$$

yield $g_{12} = g_{13} = g_{14} = 0$. On the other hand, the equalities

$$0 = 0 \cdot g_4 = g_1 \cdot \begin{bmatrix} 1 \\ 0 \\ 0 \\ 0 \end{bmatrix}$$

yield $g_{11} = 0$. Hence, $g_1 = 0$ and, consequently, $g = 0$.

We recall from (1.3)(b) that the defect $\partial_A : \mathbb{Z}^8 \longrightarrow \mathbb{Z}$ of A is given by the formula

$$\partial_A(\mathbf{x}) = -4x_1 + x_2 + x_3 + x_4 + 2x_5 + x_6 + x_7 + x_8.$$

It follows that $\partial_A(\dim F) = 0$, for any module F of Table 2.16. Hence, by (XI.2.8), the indecomposable modules of Table 2.16 are regular.

Now we show that they are simple regular. For this purpose, we suppose that F is one of the modules

$$F_1^{(\infty)}, F_2^{(\infty)}, F_1^{(0)}, F_2^{(0)}, F_3^{(0)}, F_1^{(1)}, F_2^{(1)}, F_3^{(1)}, F_4^{(1)}, \text{ or } F^{(\lambda)},$$

with $\lambda \in K \setminus \{0, 1\}$, listed in (2.16) and let X be an indecomposable regular submodule of F. By (XI.2.8), we have $\partial_A(\dim F) = 0$, and a case by case inspection shows that $X = F$ (apply the arguments given in the proof of (2.8)). This is easily seen if $F \neq F^{(\lambda)}$. For the module $F = F^{(\lambda)}$ it is not immediate. However, it follows from the proof of the following theorem that $F^{(\lambda)} \cong \operatorname{Hom}_C(T, E^{(\lambda)})$ and, hence, $F^{(\lambda)}$ is simple regular. This completes the proof.   □

Next, we collect the main properties of the family $\boldsymbol{\mathcal{T}}^{\Delta(\widetilde{\mathbb{E}}_7)}$ of components defined in (2.17).

2.19. Theorem. *Let Δ be the canonically Euclidean quiver $\Delta(\widetilde{\mathbb{E}}_7)$ of Table 1.1. Let $A = K\Delta$ be the path algebra of Δ and let*

$$F_1^{(\infty)}, F_2^{(\infty)}, F_1^{(0)}, F_2^{(0)}, F_3^{(0)}, F_1^{(1)}, F_2^{(1)}, F_3^{(1)}, F_4^{(1)}, \text{ and } F^{(\lambda)},$$

with $\lambda \in K \setminus \{0, 1\}$, be the A-modules presented in Table 2.16. The family

$$\boldsymbol{\mathcal{T}}^{\Delta(\widetilde{\mathbb{E}}_7)} = \{\mathcal{T}_\lambda^{\Delta(\widetilde{\mathbb{E}}_7)}\}_{\lambda \in \mathbb{P}_1(K)}$$

defined in (2.18) has the following properties.

(a) *The component $\mathcal{T}_\infty^{\Delta(\widetilde{\mathbb{E}}_7)}$ is a stable tube of rank 2 containing the modules $F_1^{(\infty)}$ and $F_2^{(\infty)}$, and there are isomorphisms*
$$\tau_A F_2^{(\infty)} \cong F_1^{(\infty)} \text{ and } \tau_A F_1^{(\infty)} \cong F_2^{(\infty)}.$$

(b) *The component $\mathcal{T}_0^{\Delta(\widetilde{\mathbb{E}}_7)}$ is a stable tube of rank 3 containing the modules $F_1^{(0)}$, $F_2^{(0)}$, and $F_3^{(0)}$, and there are isomorphisms*

$$\tau_A F_3^{(0)} \cong F_2^{(0)}, \ \tau_A F_2^{(0)} \cong F_1^{(0)}, \ \tau_A F_1^{(0)} \cong F_3^{(0)}.$$

(c) *The component $\mathcal{T}_1^{\Delta(\widetilde{\mathbb{E}}_7)}$ is a stable tube of rank 4 containing the modules $F_1^{(1)}$, $F_2^{(1)}$, $F_3^{(1)}$, and $F_4^{(1)}$, and there are isomorphisms*
$$\tau_A F_4^{(1)} \cong F_3^{(1)}, \ \tau_A F_3^{(1)} \cong F_2^{(1)}, \ \tau_A F_2^{(1)} \cong F_1^{(1)}, \ \tau_A F_1^{(1)} \cong F_4^{(1)}.$$

(d) *For each $\lambda \in K \setminus \{0,1\}$, the component $\mathcal{T}_\lambda^{\Delta(\widetilde{\mathbb{E}}_7)}$ is a stable tube of rank 1 containing the module $F^{(\lambda)}$, and there is an isomorphism $\tau_A F^{(\lambda)} \cong F^{(\lambda)}$.*

(e) *The Auslander–Reiten quiver $\Gamma(\mathrm{mod}\,A)$ of A consists of a postprojective component $\mathcal{P}(A)$, a preinjective component $\mathcal{Q}(A)$ and the $\mathbb{P}_1(K)$-family $\boldsymbol{\mathcal{T}}^{\Delta(\widetilde{\mathbb{E}}_7)} = \{\mathcal{T}_\lambda^{\Delta(\widetilde{\mathbb{E}}_7)}\}_{\lambda \in \mathbb{P}_1(K)}$ of stable tubes separating the component $\mathcal{P}(A)$ from $\mathcal{Q}(A)$.*

Proof. Let $\Delta = \Delta(\widetilde{\mathbb{E}}_7)$ and let $A = K\Delta$. We consider the canonical algebra $C = C(2,3,4)$ of type $\Delta(\widetilde{\mathbb{E}}_7)$ defined in (XII.1.3), and the family of C-modules

$$E_1^{(\infty)}, E_2^{(\infty)}, E_1^{(0)}, E_2^{(0)}, E_3^{(0)}, E_1^{(1)}, E_2^{(1)}, E_3^{(1)}, E_4^{(1)}, \text{ and } E^{(\lambda)},$$

with $\lambda \in K \setminus \{0,1\}$, defined in Table XII.2.9. It follows from (XII.2.12) that they are simple regular.

By (XII.1.13), the postprojective component $\mathcal{P}(C)$ of $\Gamma(\mathrm{mod}\,C)$ contains the section

$$T_5$$
$$\uparrow$$
$$T_4 \longleftarrow T_3 \longleftarrow T_2 \longleftarrow T_1 \longrightarrow T_6 \longrightarrow T_7 \longrightarrow T_8,$$

where

$$T_1 = \tau^{-3}P(0), \quad T_2 = \tau^{-3}P(b_1), \quad T_3 = \tau^{-3}P(b_2), \quad T_4 = P(\omega),$$
$$T_5 = \tau^{-3}P(a_1), \quad T_6 = \tau^{-3}P(c_1), \quad T_7 = \tau^{-3}P(c_2), \quad T_8 = \tau^{-3}P(c_3).$$

Here we use the notation of (XII.1.2). Moreover,

$$T = T_1 \oplus T_2 \oplus T_3 \oplus T_4 \oplus T_5 \oplus T_6 \oplus T_7 \oplus T_8$$

is a postprojective tilting C-module and there is a K-algebra isomorphism

$$A \cong \mathrm{End}\,T_C.$$

Then, by (VIII.4.5), the functor $\mathrm{Hom}_C(T, -) : \mathrm{mod}\,C \longrightarrow \mathrm{mod}\,A$ restricts to the equivalence

$$\mathrm{Hom}_C(T, -) : \mathrm{add}\,\mathcal{R}(C) \xrightarrow{\ \cong\ } \mathrm{add}\,\mathcal{R}(A)$$

of abelian and serial categories add $\mathcal{R}(C)$ and add $\mathcal{R}(A)$ of regular modules such that the diagram (2.11) is commutative

It follows that the functor $\mathrm{Hom}_C(T, -) : \mathrm{add}\,\mathcal{R}(C) \xrightarrow{\cong} \mathrm{add}\,\mathcal{R}(A)$ carries irreducible morphisms to irreducible ones, and the image $\mathrm{Hom}_C(T, \mathcal{T}_\lambda^C)$ of any stable tube $\mathcal{T}_\lambda^C$ in $\mathcal{R}(C)$ of rank $r_\lambda \geq 1$ is a stable tube in $\mathcal{R}(A)$ of the same rank r_λ. Moreover, any stable tube in $\mathcal{R}(A)$ is of the form $\mathrm{Hom}_C(T, \mathcal{T}_\lambda^C)$.

We claim that, for each $\lambda \in \mathbb{P}_1(K)$, the tubes $\mathrm{Hom}_C(T, \mathcal{T}_\lambda^C)$ and $\mathcal{T}_\lambda^{\Delta(\tilde{\mathbb{E}}_7)}$ coincide. For this purpose, it suffices to show that

$$\mathrm{Hom}_C(T, E_j^{(s)}) \cong F_j^{(s)} \quad \text{and} \quad \mathrm{Hom}_C(T, E^{(\lambda)}) \cong F^{(\lambda)},$$

for all $s \in \{0, 1, \infty\}$, j, and $\lambda \in \mathbb{P}_1(K) \setminus \{0, 1, \infty\}$.

To prove this, we recall that $A \cong \mathrm{End}\,T_C$ and, as in the proof of (XII.1.11), the left hand part of the postprojective component $\mathcal{P}(C)$ of $\Gamma(\mathrm{mod}\,C)$ containing all the indecomposable projective C-modules $P(0), \ldots, P(\omega)$ looks as follows

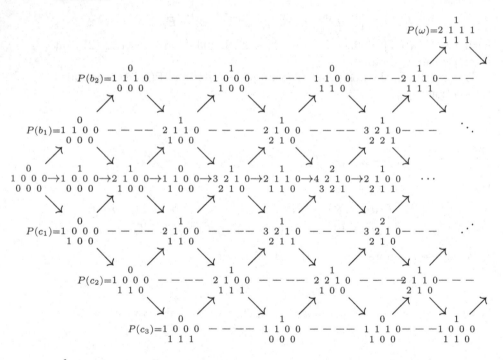

where

$$P(0) = 1\,\underset{0\ 0\ 0}{\overset{0}{0}}\,0\,0, \quad \text{and} \quad P(a_1) = 1\,\underset{0\ 0\ 0}{\overset{1}{0}}\,0\,0,$$

and the indecomposable modules are represented by their dimension vectors.

It follows from (VI.3.10) that $e_i A \cong \operatorname{Hom}_C(T, T_i)$ and, for each regular C-module X and for each $i \in \{1, \dots, 8\}$, we have

$$\begin{aligned}
\operatorname{Hom}_C(T, X)_i &= \operatorname{Hom}_C(T, X)e_i \\
&\cong \operatorname{Hom}_A(e_i A, \operatorname{Hom}_C(T, X)) \\
&\cong \operatorname{Hom}_A(\operatorname{Hom}_C(T, T_i), \operatorname{Hom}_C(T, X)) \\
&\cong \operatorname{Hom}_C(T_i, X),
\end{aligned}$$

and therefore $(\mathbf{dim}\,\operatorname{Hom}_C(T, X))_i = \dim_K \operatorname{Hom}_C(T_i, X)$. Note also that we have

$$\operatorname{Hom}_C(T_4, X) = \operatorname{Hom}_C(P(\omega), X) \cong X_\omega.$$

By consulting the beginning part of the component $\mathcal{P}(C)$ shown above we see that the ω-th coordinate of the dimension vector of each of the C-modules $T_1, T_2, T_3, T_5, T_6, T_7, T_8$ equals zero. Let

$$Q: \quad \begin{array}{c} \overset{a_1}{} \\ {}^{\alpha_1}\nearrow \\ 0 \overset{\beta_1}{\longleftarrow} b_1 \overset{\beta_2}{\longleftarrow} b_2 \\ {}_{\gamma_1}\searrow \\ c_1 \overset{\gamma_2}{\longleftarrow} c_2 \overset{\gamma_3}{\longleftarrow} c_3 \end{array}$$

be the full subquiver of $\Delta(2, 3, 4)$, of the Dynkin type $\mathbb{E}_7$, defined by the points $0, a_1, b_1, b_2, c_1, c_2, c_3$, in the notation of (XII.1.2). Note that there is an isomorphism $KQ \cong e_Q C e_Q$ of algebras and the equivalence of categories $\operatorname{mod} KQ \cong \operatorname{rep}_K(Q)$, where e_Q is the idempotent

$$e_Q = e_0 + e_{a_1} + e_{b_1} + e_{b_2} + e_{c_1} + e_{c_2} + e_{c_3}$$

of the algebra C. It follows that the restriction functor

$$\operatorname{res}_{e_Q} : \operatorname{mod} C \longrightarrow \operatorname{mod} KQ \cong \operatorname{rep}_K(Q),$$

see (I.6.6), defines an isomorphism

$$\operatorname{Hom}_C(T_j, E) \cong \operatorname{Hom}_{KQ}(\operatorname{res}_{e_Q} T_j, \operatorname{res}_{e_Q} E),$$

for $T_j \neq T_4 = P(\omega)$ and for any module E in the family

$$E_1^{(\infty)}, E_2^{(\infty)}, E_1^{(0)}, E_2^{(0)}, E_3^{(0)}, E_1^{(1)}, E_2^{(1)}, E_3^{(1)}, E_4^{(1)}, \text{ and } E^{(\lambda)},$$

with $\lambda \in K \setminus \{0, 1\}$. Because $\operatorname{res}_{e_Q} E^{(\lambda)} = I(0)$ is the injective envelope of the simple representation $S(0)$ in $\operatorname{mod} KQ \cong \operatorname{rep}_K(Q)$ then, for each $j \neq 4$, there is an isomorphism

$$\operatorname{Hom}_C(T_j, E^{(\lambda)}) \cong \operatorname{Hom}_{KQ}(\operatorname{res}_{e_Q} T_j, \operatorname{res}_{e_Q} E^{(\lambda)}) \cong D(T_j)_0,$$

which is functorial at T_j.

By applying the above formulae to the module $E = F^{(\lambda)}$, with $\lambda \in K \setminus \{0,1\}$, we get

$\operatorname{Hom}_C(T_1, E^{(\lambda)}) \cong D(T_1)_0 = K^4,$
$\operatorname{Hom}_C(T_2, E^{(\lambda)}) \cong D(T_2)_0 = K^3,$
$\operatorname{Hom}_C(T_3, E^{(\lambda)}) \cong D(T_3)_0 = K^2,$
$\operatorname{Hom}_C(T_5, E^{(\lambda)}) \cong D(T_5)_0 = K^2,$
$\operatorname{Hom}_C(T_6, E^{(\lambda)}) \cong D(T_6)_0 = K^3,$
$\operatorname{Hom}_C(T_7, E^{(\lambda)}) \cong D(T_7)_0 = K^2,$
$\operatorname{Hom}_C(T_8, E^{(\lambda)}) \cong D(T_8)_0 = K,$ and
$\operatorname{Hom}_C(T_4, E^{(\lambda)}) \cong E_\omega^{(\lambda)} = K.$

In particular, we get

$$\dim \operatorname{Hom}_C(T, E^{(\lambda)}) = {\begin{smallmatrix} & & & 2 & & & \\ 1 & 2 & 3 & 4 & 3 & 2 & 1 \end{smallmatrix}}.$$

Moreover, by applying the above formulae, one shows that there is a module isomorphism $\operatorname{Hom}_C(T, E^{(\lambda)}) \cong F^{(\lambda)}$, and hence

$$\operatorname{Hom}_C(T, \mathcal{T}_\lambda^C) = \mathcal{T}_\lambda^{\Delta(\widetilde{\mathbb{E}}_7)},$$

for any $\lambda \in K \setminus \{0,1\}$. Because, by (XII.2.12), the C-module $E^{(\lambda)}$ is simple regular, then the A-module $\operatorname{Hom}_C(T, E^{(\lambda)}) \cong F^{(\lambda)}$ is simple regular. Furthermore, in view of the isomorphism $\tau_C E^{(\lambda)} \cong E^{(\lambda)}$ in (XII.2.12) and the commutativity of the diagram (2.11), we get the isomorphism $\tau_A F^{(\lambda)} \cong F^{(\lambda)}$. This finishes the proof of (d).

Similarly, by applying the above formulae to any module E in the family $E_1^{(\infty)}$, $E_2^{(\infty)}$, $E_1^{(0)}$, $E_2^{(0)}$, $E_3^{(0)}$, $E_1^{(1)}$, $E_2^{(1)}$, $E_3^{(1)}$, and $E_4^{(1)}$ we get

$$\dim \operatorname{Hom}_C(T, E_1^{(\infty)}) = {\begin{smallmatrix} & & & 1 & & & \\ 0 & 1 & 1 & 2 & 1 & 1 & 1 \end{smallmatrix}}, \qquad \dim \operatorname{Hom}_C(T, E_2^{(\infty)}) = {\begin{smallmatrix} & & & 1 & & & \\ 1 & 1 & 1 & 2 & 1 & 1 & 0 \end{smallmatrix}},$$

$$\dim \operatorname{Hom}_C(T, E_1^{(0)}) = {\begin{smallmatrix} & & & 1 & & & \\ 0 & 0 & 1 & 1 & 1 & 0 & 0 \end{smallmatrix}}, \qquad \dim \operatorname{Hom}_C(T, E_2^{(0)}) = {\begin{smallmatrix} & & & 0 & & & \\ 0 & 1 & 1 & 1 & 1 & 1 & 0 \end{smallmatrix}},$$

$$\dim \operatorname{Hom}_C(T, E_3^{(0)}) = {\begin{smallmatrix} & & & 1 & & & \\ 1 & 1 & 1 & 2 & 1 & 1 & 1 \end{smallmatrix}}, \qquad \dim \operatorname{Hom}_C(T, E_1^{(1)}) = {\begin{smallmatrix} & & & 1 & & & \\ 0 & 0 & 0 & 1 & 1 & 1 & 0 \end{smallmatrix}},$$

$$\dim \operatorname{Hom}_C(T, E_2^{(1)}) = {\begin{smallmatrix} & & & 0 & & & \\ 0 & 0 & 1 & 1 & 1 & 1 & 1 \end{smallmatrix}}, \qquad \dim \operatorname{Hom}_C(T, E_3^{(1)}) = {\begin{smallmatrix} & & & 1 & & & \\ 0 & 1 & 1 & 1 & 0 & 0 & 0 \end{smallmatrix}},$$

$$\dim \operatorname{Hom}_C(T, E_4^{(1)}) = {\begin{smallmatrix} & & & 0 & & & \\ 1 & 1 & 1 & 1 & 1 & 0 & 0 \end{smallmatrix}}.$$

In particular, $\dim \operatorname{Hom}_C(T, E_j^{(s)}) = \dim F_j^{(s)}$, for all j and all $s \in \{0, 1, \infty\}$.

Because, by (XII.2.12), E is a simple regular C-module, then obviously the A-module $\operatorname{Hom}_C(T, E)$ is simple regular.

We recall from (XI.3.3) and (XII.4.2) that the simple regular A-modules lying in the non-homogeneous tubes of $\mathcal{R}(A)$ are uniquely determined by their dimension vectors. Because, by (2.18), the modules presented in Table 2.17 are simple regular, then there is an isomorphism

$$\mathrm{Hom}_C(T, E_j^{(s)}) \cong F_j^{(s)},$$

for all j and $s \in \{0, 1, \infty\}$.

Therefore, by applying (XII.2.12) and the commutative diagram (2.11), we prove the statements (a)–(c), and we conclude that the tubes

$$\mathcal{T}_\infty^{\Delta(\widetilde{\mathbb{E}}_7)} = \mathrm{Hom}_C(T, \mathcal{T}_\infty^C), \quad \mathcal{T}_0^{\Delta(\widetilde{\mathbb{E}}_7)} = \mathrm{Hom}_C(T, \mathcal{T}_0^C), \quad \mathcal{T}_1^{\Delta(\widetilde{\mathbb{E}}_7)} = \mathrm{Hom}_C(T, \mathcal{T}_1^C)$$

exhaust all the non-homogeneous tubes of the regular part $\mathcal{R}(A)$ of the Auslander–Reiten quiver $\Gamma(\mathrm{mod}\, A)$. Consequently, the tubes

$$\mathcal{T}_\lambda^{\Delta(\widetilde{\mathbb{E}}_7)} = \mathrm{Hom}_C(T, \mathcal{T}_\lambda^C), \quad \text{with} \quad \lambda \in K \setminus \{0, 1\},$$

exhaust all the homogeneous tubes of $\Gamma(\mathrm{mod}\, A)$. Now the statement (e) follows from (2.1). This completes the proof.   □

2.20. Table. Simple regular representations of $\Delta(\widetilde{\mathbb{E}}_8)$

Assume that $\Delta = \Delta(\widetilde{\mathbb{E}}_8)$ is the canonically oriented Euclidean quiver

$$\Delta(\widetilde{\mathbb{E}}_8): \qquad \begin{array}{c} 4 \\ \downarrow \\ 3 \longrightarrow 2 \longrightarrow 1 \longleftarrow 5 \longleftarrow 6 \longleftarrow 7 \longleftarrow 8 \longleftarrow 9. \end{array}$$

shown in Table 1.1, and let $A = K\Delta$. Consider the following family of indecomposable regular A-modules, viewed as K-linear representations of $\Delta(\widetilde{\mathbb{E}}_8)$. We show below that they are simple regular.

(a) Simple regular representations in the tube $\mathcal{T}_\infty^{\Delta(\widetilde{\mathbb{E}}_8)}$

$$F_1^{(\infty)}: \qquad \begin{array}{c} K^2 \\ \downarrow {\scriptsize\begin{bmatrix} 1 & 0 \\ 1 & 1 \\ 0 & 1 \end{bmatrix}} \\ \end{array}$$

$$K \xrightarrow{\ \ } K^2 \xrightarrow{\ \ } K^3 \xleftarrow{\ \ } K^2 \xleftarrow{\ \ } K^2 \xleftarrow{\ \ } K \xleftarrow{\ \ } K \xleftarrow{\ \ } 0$$

$$\begin{bmatrix} 0 \\ 1 \end{bmatrix} \quad \begin{bmatrix} 0 & 0 \\ 1 & 0 \\ 0 & 1 \end{bmatrix} \quad \begin{bmatrix} 1 & 0 \\ 0 & 1 \\ 0 & 0 \end{bmatrix} \quad \begin{bmatrix} 1 & 0 \\ 0 & 1 \end{bmatrix} \quad \begin{bmatrix} 1 \\ 0 \end{bmatrix} \quad {\scriptstyle 1}$$

$$F_2^{(\infty)}: \qquad \begin{array}{c} K \\ \downarrow {\scriptsize\begin{bmatrix} 1 \\ 1 \\ 1 \end{bmatrix}} \\ \end{array}$$

$$K \xrightarrow{\ \ } K^2 \xrightarrow{\ \ } K^3 \xleftarrow{\ \ } K^3 \xleftarrow{\ \ } K^2 \xleftarrow{\ \ } K^2 \xleftarrow{\ \ } K \xleftarrow{\ \ } K$$

$$\begin{bmatrix} 0 \\ 1 \end{bmatrix} \quad \begin{bmatrix} 0 & 0 \\ 1 & 0 \\ 0 & 1 \end{bmatrix} \quad \begin{bmatrix} 1 & 0 & 0 \\ 0 & 1 & 0 \\ 0 & 0 & 1 \end{bmatrix} \quad \begin{bmatrix} 1 & 0 \\ 0 & 1 \\ 0 & 0 \end{bmatrix} \quad \begin{bmatrix} 1 & 0 \\ 0 & 1 \end{bmatrix} \quad \begin{bmatrix} 1 \\ 0 \end{bmatrix} \quad {\scriptstyle 1}$$

(b) Simple regular representations in the tube $\mathcal{T}_0^{\Delta(\widetilde{\mathbb{E}}_8)}$

$F_1^{(0)}$:

$$
\begin{array}{c}
K \\
\downarrow {\scriptstyle\begin{bmatrix}1\\1\end{bmatrix}} \\
K \xrightarrow{} K^2 \xrightarrow{} K^2 \xleftarrow{} K \xleftarrow{1} K \xleftarrow{1} K \xleftarrow{} 0 \xleftarrow{} 0 \\
{\scriptstyle\begin{bmatrix}0\\1\end{bmatrix}} \quad {\scriptstyle\begin{bmatrix}1&0\\0&1\end{bmatrix}} \quad {\scriptstyle\begin{bmatrix}1\\0\end{bmatrix}}
\end{array}
$$

$F_2^{(0)}$:

$$
\begin{array}{c}
K \\
\downarrow {\scriptstyle\begin{bmatrix}1\\1\end{bmatrix}} \\
K \xrightarrow{1} K \xrightarrow{} K^2 \xleftarrow{} K^2 \xleftarrow{} K \xleftarrow{1} K \xleftarrow{1} K \xleftarrow{} 0 \\
\quad {\scriptstyle\begin{bmatrix}0\\1\end{bmatrix}} \quad {\scriptstyle\begin{bmatrix}1&0\\0&1\end{bmatrix}} \quad {\scriptstyle\begin{bmatrix}1\\0\end{bmatrix}}
\end{array}
$$

$F_3^{(0)}$:

$$
\begin{array}{c}
K \\
\downarrow {\scriptstyle\begin{bmatrix}1\\1\end{bmatrix}} \\
0 \longrightarrow K \xrightarrow{} K^2 \xleftarrow{} K^2 \xleftarrow{} K^2 \xleftarrow{} K \xleftarrow{1} K \xleftarrow{1} K \\
\quad {\scriptstyle\begin{bmatrix}0\\1\end{bmatrix}} \quad {\scriptstyle\begin{bmatrix}1&0\\0&1\end{bmatrix}} \quad {\scriptstyle\begin{bmatrix}1&0\\0&1\end{bmatrix}} \quad {\scriptstyle\begin{bmatrix}1\\0\end{bmatrix}}
\end{array}
$$

(c) Simple regular representations in the tube $\mathcal{T}_1^{\Delta(\widetilde{\mathbb{E}}_8)}$

$F_1^{(1)}$:

$$
\begin{array}{c}
K \\
\downarrow {\scriptstyle 1} \\
0 \longrightarrow K \xrightarrow{1} K \xleftarrow{1} K \xleftarrow{} 0 \xleftarrow{} 0 \xleftarrow{} 0 \xleftarrow{} 0
\end{array}
$$

$F_2^{(1)}$:

$$
\begin{array}{c}
0 \\
\downarrow \\
K \xrightarrow{1} K \xrightarrow{1} K \xleftarrow{1} K \xleftarrow{1} K \xleftarrow{} 0 \xleftarrow{} 0 \xleftarrow{} 0
\end{array}
$$

$F_3^{(1)}$:

$$
\begin{array}{c}
K \\
\downarrow {\scriptstyle 1} \\
0 \longrightarrow 0 \longrightarrow K \xleftarrow{1} K \xleftarrow{1} K \xleftarrow{1} K \xleftarrow{} 0 \xleftarrow{} 0
\end{array}
$$

$F_4^{(1)}$:

$$
\begin{array}{c}
0 \\
\downarrow \\
0 \longrightarrow K \xrightarrow{1} K \xleftarrow{1} K \xleftarrow{1} K \xleftarrow{1} K \xleftarrow{1} K \xleftarrow{} 0
\end{array}
$$

$F_5^{(1)}$:

$$
\begin{array}{c}
K \\
\downarrow {\scriptstyle\begin{bmatrix}1\\1\end{bmatrix}} \\
K \xrightarrow{1} K \xrightarrow{} K^2 \xleftarrow{} K \xleftarrow{1} K \xleftarrow{1} K \xleftarrow{1} K \xleftarrow{1} K \\
\quad\quad {\scriptstyle\begin{bmatrix}0\\1\end{bmatrix}} \quad {\scriptstyle\begin{bmatrix}1\\0\end{bmatrix}}
\end{array}
$$

(d) Simple regular representations in the tube $\mathcal{T}_\lambda^{\Delta(\widetilde{\mathbb{E}}_8)}$, with $\lambda \in K \setminus \{0,1\}$

$$
F^{(\lambda)}: \quad
\begin{array}{c}
K^3 \\
\Bigg\downarrow
{\begin{bmatrix} \lambda & 1 & 0 \\ 0 & 0 & 1 \\ 1 & 1 & 0 \\ 1 & 0 & 1 \\ 1 & 1 & 0 \\ 0 & 1 & 0 \end{bmatrix}}
\end{array}
$$

$$
K^2 \xrightarrow{\begin{bmatrix}0&0\\0&0\\1&0\\0&1\end{bmatrix}} K^4 \xrightarrow{\begin{bmatrix}0&0&0&0\\0&0&0&0\\1&0&0&0\\0&1&0&0\\0&0&1&0\\0&0&0&1\end{bmatrix}} K^6 \xleftarrow{\begin{bmatrix}1&0&0&0&0\\0&1&0&0&0\\0&0&1&0&0\\0&0&0&1&0\\0&0&0&0&1\\0&0&0&0&0\end{bmatrix}} K^5 \xleftarrow{\begin{bmatrix}1&0&0&0\\0&1&0&0\\0&0&1&0\\0&0&0&1\\0&0&0&0\end{bmatrix}} K^4 \xleftarrow{\begin{bmatrix}1&0&0\\0&1&0\\0&0&1\\0&0&0\end{bmatrix}} K^3 \xleftarrow{\begin{bmatrix}1&0\\0&1\\0&0\end{bmatrix}} K^2 \xleftarrow{\begin{bmatrix}1\\0\end{bmatrix}} K
$$

2.21. Definition. Assume that $A = K\Delta$, where Δ is the canonically oriented Euclidean quiver $\Delta(\widetilde{\mathbb{E}}_8)$ shown in (2.20). We define the $\mathbb{P}_1(K)$-family (see (2.3))

$$\mathcal{T}^{\Delta(\widetilde{\mathbb{E}}_8)} = \{\mathcal{T}_\lambda^{\Delta(\widetilde{\mathbb{E}}_8)}\}_{\lambda \in \mathbb{P}_1(K)}$$

of connected components $\mathcal{T}_\lambda^{\Delta(\widetilde{\mathbb{E}}_8)}$ of the Auslander–Reiten quiver $\Gamma(\mathrm{mod}\,A)$ of A as follows:

- $\mathcal{T}_\infty^{\Delta(\widetilde{\mathbb{E}}_8)}$ is the component containing the A-module $F_1^{(\infty)}$,
- $\mathcal{T}_0^{\Delta(\widetilde{\mathbb{E}}_8)}$ is the component containing the A-module $F_1^{(0)}$,
- $\mathcal{T}_1^{\Delta(\widetilde{\mathbb{E}}_8)}$ is the component containing the A-module $F_1^{(1)}$, and
- for each $\lambda \in K \setminus \{0,1\}$, $\mathcal{T}_\lambda^{\Delta(\widetilde{\mathbb{E}}_8)}$ is the component containing the A-module $F^{(\lambda)}$, where $F_1^{(\infty)}$, $F_1^{(0)}$, $F_1^{(1)}$, and $F^{(\lambda)}$ are the modules shown in Table 2.20.

It follows from (2.23) below that the components $\mathcal{T}_\lambda^{\Delta(\widetilde{\mathbb{E}}_8)}$ of the $\mathbb{P}_1(K)$-family $\mathcal{T}^\Delta(\widetilde{\mathbb{E}}_8)$ are stable tubes in $\mathcal{R}(A)$.

We now collect the main properties of the modules listed in (2.20).

2.22. Lemma. *Let Δ be the the canonically oriented Euclidean quiver $\Delta(\widetilde{\mathbb{E}}_8)$ shown in Table 1.1. Let $A = K\Delta$ be the path algebra of Δ. The A-modules*

$$F_1^{(\infty)},\ F_2^{(\infty)},\ F_1^{(0)},\ F_2^{(0)},\ F_3^{(0)},\ F_1^{(1)},\ F_2^{(1)},\ F_3^{(1)},\ F_4^{(1)},\ F_5^{(1)},\ F^{(\lambda)},$$

with $\lambda \in K \setminus \{0,1\}$, presented in Table 2.20 are pairwise orthogonal bricks, and each of them is simple regular.

Proof. It is obvious that the A-modules $F_1^{(1)}$, $F_2^{(1)}$, $F_3^{(1)}$, $F_4^{(1)}$ are bricks. Now we show that there is a K-algebra isomorphism $\operatorname{End} F_5^{(1)} \cong K$. Let

$$f = (f_1, \dots, f_9) : F_5^{(1)} \longrightarrow F_5^{(1)}$$

be an endomorphism of the A-module $F_5^{(1)}$, viewed as a K-linear representation of $\Delta = \Delta(\widetilde{\mathbb{E}}_8)$, and assume that the K-linear map $f_1 : K^2 \longrightarrow K^2$ is defined by the matrix

$$f_1 = \begin{bmatrix} a & b \\ c & d \end{bmatrix}$$

in the standard basis of K^2, where $a, b, c, d \in K$. It follows that

$$\begin{bmatrix} 0 \\ 1 \end{bmatrix} \cdot f_2 = f_1 \cdot \begin{bmatrix} 0 \\ 1 \end{bmatrix} \quad \text{and} \quad f_1 \cdot \begin{bmatrix} 1 \\ 0 \end{bmatrix} = \begin{bmatrix} 1 \\ 0 \end{bmatrix} \cdot f_5.$$

Hence we get $b = 0$ and $c = 0$. Similarly, the equality

$$\begin{bmatrix} 1 \\ 1 \end{bmatrix} \cdot f_4 = f_1 \cdot \begin{bmatrix} 1 \\ 1 \end{bmatrix}$$

yields $a = d$. Hence we get

$$f_1 = \begin{bmatrix} a & 0 \\ 0 & a \end{bmatrix},$$

and obviously the K-linear map $\operatorname{End}_A F_5^{(1)} \longrightarrow K$, assigning to the endomorphism f of $F_5^{(1)}$ the scalar a, is a K-algebra isomorphism.

By applying the above arguments we also show that there are K-algebra isomorphisms $\operatorname{End} F_1^{(0)} \cong K$, $\operatorname{End} F_2^{(0)} \cong K$, $\operatorname{End} F_3^{(0)} \cong K$.

Let F be one of the modules $F_1^{(\infty)}$ and $F_2^{(\infty)}$. Let

$$g = (g_1, \dots, g_9) : F \longrightarrow F$$

be an endomorphism of the A-module F, viewed as a K-linear representation of $\Delta = \Delta(\widetilde{\mathbb{E}}_8)$, and assume that the K-linear map $g_1 : K^3 \longrightarrow K^3$ is defined by the matrix

$$g_1 = \begin{bmatrix} a_{11} & a_{12} & a_{13} \\ a_{21} & a_{22} & a_{23} \\ a_{31} & a_{32} & a_{33} \end{bmatrix}$$

in the standard basis of K^3, where $a_{ij} \in K$. It follows, as in the proof of (2.18), that $a_{12} = 0$, $a_{13} = 0$, $a_{23} = 0$, $a_{21} = 0$, $a_{31} = 0$, $a_{32} = 0$, and $a_{33} = a_{22} = a_{11}$. Consequently, g_1 has the diagonal form

$$g_1 = \begin{bmatrix} a_{11} & 0 & 0 \\ 0 & a_{11} & 0 \\ 0 & 0 & a_{11} \end{bmatrix},$$

and obviously the K-linear map $\mathrm{End}_A F \longrightarrow K$, assigning to the endomorphism $g : F \to F$ of F the scalar a_{11} on the diagonal of the matrix g_1, is a K-algebra isomorphism.

Finally, by applying the arguments used in the proof of (2.18), we also show that there is a K-algebra isomorphism $\mathrm{End}_A F^{(\lambda)} \cong K$, for each $\lambda \in K \setminus \{0, 1\}$. This shows that the modules of Table 2.20 are bricks. The proof that they are pairwise orthogonal is similar to those in (2.8), (2.14), and (2.18); we leave it to the reader.

We recall from (1.3)(b) that the defect $\partial_A : \mathbb{Z}^9 \longrightarrow \mathbb{Z}$ of A is given by the formula

$$\partial_A(\mathbf{x}) = -6x_1 + 2x_2 + 2x_3 + 3x_4 + x_5 + x_6 + x_7 + x_8 + x_9.$$

It follows that $\partial_A(\mathbf{dim}\, F) = 0$ for any module F of Table 2.20. Hence, by (XI.2.8), the indecomposable modules of Table 2.20 are regular.

To show that they are simple regular, we suppose that F is one of the modules

$$F_1^{(\infty)}, F_2^{(\infty)}, F_1^{(0)}, F_2^{(0)}, F_3^{(0)}, F_1^{(1)}, F_2^{(1)}, F_3^{(1)}, F_4^{(1)}, F_5^{(1)}, F^{(\lambda)},$$

with $\lambda \in K \setminus \{0, 1\}$, listed in (2.20) and let X be an indecomposable regular submodule of F. By (XI.2.8), we have $\partial_A(\mathbf{dim}\, F) = 0$, and a case by case inspection shows that $X = F$ (apply the arguments given in the proof of (2.8)). This is easily seen if F is neither $F^{(\lambda)}$, nor any of the modules $F_1^{(\infty)}$ and $F_2^{(\infty)}$.

The proof is not immediate in case F is one of the modules $F_1^{(\infty)}$, $F_2^{(\infty)}$, and $F^{(\lambda)}$. However, it follows from the proof of the following theorem that there are isomorphisms of A-modules

$$F_1^{(\infty)} \cong \mathrm{Hom}_C(T, E_1^{(\infty)}),$$
$$F_2^{(\infty)} \cong \mathrm{Hom}_C(T, E_2^{(\infty)}),$$
$$F^{(\lambda)} \cong \mathrm{Hom}_C(T, E^{(\lambda)})$$

and, hence, the modules $F_1^{(\infty)}$, $F_2^{(\infty)}$, and $F^{(\lambda)}$ are simple regular, because the C-modules $E_1^{(\infty)}$, $E_2^{(\infty)}$, and $E^{(\lambda)}$ are simple regular, by (XII.2.13). This completes the proof. $\qquad \square$

Next, we collect the main properties of the family $\mathcal{T}^{\Delta(\widetilde{\mathbb{E}}_8)}$ of components defined in (2.21).

2.23. Theorem. *Let Δ be the canonically oriented Euclidean quiver $\Delta(\widetilde{\mathbb{E}}_8)$ shown in Table 1.1, $A = K\Delta$ be the path algebra of Δ, and let*

$$F_1^{(\infty)}, F_2^{(\infty)}, F_1^{(0)}, F_2^{(0)}, F_3^{(0)}, F_1^{(1)}, F_2^{(1)}, F_3^{(1)}, F_4^{(1)}, F_5^{(1)}, F^{(\lambda)},$$

with $\lambda \in K \setminus \{0, 1\}$, be the A-modules presented in Table 2.20. The family

$$\boldsymbol{\mathcal{T}}^{\Delta(\widetilde{\mathbb{E}}_8)} = \{\mathcal{T}_\lambda^{\Delta(\widetilde{\mathbb{E}}_8)}\}_{\lambda \in \mathbb{P}_1(K)}$$

defined in (2.21) has the following properties.

(a) *The component $\mathcal{T}_\infty^{\Delta(\widetilde{\mathbb{E}}_8)}$ is a stable tube of rank 2 containing the modules $F_1^{(\infty)}$ and $F_2^{(\infty)}$, and there are isomorphisms*

$$\tau_A F_2^{(\infty)} \cong F_1^{(\infty)} \text{ and } \tau_A F_1^{(\infty)} \cong F_2^{(\infty)}.$$

(b) *The component $\mathcal{T}_0^{\Delta(\widetilde{\mathbb{E}}_8)}$ is a stable tube of rank 3 containing the modules $F_1^{(0)}$, $F_2^{(0)}$, and $F_3^{(0)}$, and there are isomorphisms*

$$\tau_A F_3^{(0)} \cong F_2^{(0)}, \ \tau_A F_2^{(0)} \cong F_1^{(0)}, \ \tau_A F_1^{(0)} \cong F_3^{(0)}.$$

(c) *The component $\mathcal{T}_1^{\Delta(\widetilde{\mathbb{E}}_8)}$ is a stable tube of rank 5 containing the modules $F_1^{(1)}$, $F_2^{(1)}$, $F_3^{(1)}$, $F_4^{(1)}$, and $F_5^{(1)}$, and there are isomorphisms*

$$\tau_A F_5^{(1)} \cong F_4^{(1)}, \ \tau_A F_4^{(1)} \cong F_3^{(1)}, \ \tau_A F_3^{(1)} \cong F_2^{(1)}, \ \tau_A F_2^{(1)} \cong F_1^{(1)}, \ \tau_A F_1^{(1)} \cong F_5^{(1)}.$$

(d) *For each $\lambda \in K \setminus \{0, 1\}$, the component $\mathcal{T}_\lambda^{\Delta(\widetilde{\mathbb{E}}_8)}$ is a stable tube of rank 1 containing the module $F^{(\lambda)}$, and there is an isomorphism $\tau_A F^{(\lambda)} \cong F^{(\lambda)}$.*

(e) *The Auslander–Reiten quiver $\Gamma(\bmod A)$ of A consists of a postprojective component $\mathcal{P}(A)$, a preinjective component $\mathcal{Q}(A)$ and the $\mathbb{P}_1(K)$-family $\boldsymbol{\mathcal{T}}^{\Delta(\widetilde{\mathbb{E}}_8)} = \{\mathcal{T}_\lambda^{\Delta(\widetilde{\mathbb{E}}_8)}\}_{\lambda \in \mathbb{P}_1(K)}$ of stable tubes separating the component $\mathcal{P}(A)$ from $\mathcal{Q}(A)$.*

Proof. Let $\Delta = \Delta(\widetilde{\mathbb{E}}_8)$ and let $A = K\Delta$. We consider the canonical algebra $C = C(2, 3, 5)$ of type $\Delta(\widetilde{\mathbb{E}}_8)$ defined in (XII.1.3), and the family of C-modules

$$E_1^{(\infty)}, E_2^{(\infty)}, E_1^{(0)}, E_2^{(0)}, E_3^{(0)}, E_1^{(1)}, E_2^{(1)}, E_3^{(1)}, E_4^{(1)}, E_5^{(1)}, E^{(\lambda)},$$

with $\lambda \in K \setminus \{0, 1\}$, defined in Table XII.2.9. It follows from (XII.2.12) that they are simple regular.

By (XII.1.16), the postprojective component $\mathcal{P}(C)$ of the Auslander–Reiten quiver $\Gamma(\bmod C)$ of C contains the section

$$T_4$$
$$\uparrow$$
$$T_3 \longleftarrow T_2 \longleftarrow T_1 \longrightarrow T_5 \longrightarrow T_6 \longrightarrow T_7 \longrightarrow T_8 \longrightarrow T_9,$$

where

$$T_1 = \tau^{-5}P(0), \ T_2 = \tau^{-5}P(b_1), \ T_3 = \tau^{-5}P(b_2), \ T_4 = \tau^{-5}P(a_1),$$
$$T_5 = \tau^{-5}P(c_1), \ T_6 = \tau^{-5}P(c_2), \ T_7 = \tau^{-5}P(c_3), \ T_8 = \tau^{-5}P(c_4),$$
$$T_9 = P(\omega).$$

Here we use the notation of (XII.1.2). Moreover,

$$T = T_1 \oplus T_2 \oplus T_3 \oplus T_4 \oplus T_5 \oplus T_6 \oplus T_7 \oplus T_8 \oplus T_9$$

is a postprojective tilting C-module and there is a K-algebra isomorphism

$$A \cong \operatorname{End} T_C.$$

Then, by (VIII.4.5), the functor $\operatorname{Hom}_C(T, -) : \operatorname{mod} C \longrightarrow \operatorname{mod} A$ restricts to the equivalence

$$\operatorname{Hom}_C(T, -) : \operatorname{add} \mathcal{R}(C) \xrightarrow{\ \cong\ } \operatorname{add} \mathcal{R}(A)$$

of abelian and serial categories $\operatorname{add} \mathcal{R}(C)$ and $\operatorname{add} \mathcal{R}(A)$ of regular modules such that the diagram (2.11) is commutative.

It follows that the functor $\operatorname{Hom}_C(T, -) : \operatorname{add} \mathcal{R}(C) \xrightarrow{\cong} \operatorname{add} \mathcal{R}(A)$ carries irreducible morphisms to irreducible ones, and the image $\operatorname{Hom}_C(T, \mathcal{T}_\lambda^C)$ of any stable tube $\mathcal{T}_\lambda^C$ in $\mathcal{R}(C)$ of rank $r_\lambda \geq 1$ is a stable tube in $\mathcal{R}(A)$ of the same rank r_λ. Moreover, any stable tube in $\mathcal{R}(A)$ is of the form $\operatorname{Hom}_C(T, \mathcal{T}_\lambda^C)$.

We claim that, for each $\lambda \in \mathbb{P}_1(K)$, the tubes $\operatorname{Hom}_C(T, \mathcal{T}_\lambda^C)$ and $\mathcal{T}_\lambda^{\Delta(\widetilde{\mathbb{E}}_8)}$ coincide. For this purpose, it suffices to show that there are isomorphisms

$$\operatorname{Hom}_C(T, E_j^{(s)}) \cong F_j^{(s)} \text{ and } \operatorname{Hom}_C(T, E^{(\lambda)}) \cong F^{(\lambda)},$$

for all $s \in \{0, 1, \infty\}$, all j, and $\lambda \in \mathbb{P}_1(K) \setminus \{0, 1, \infty\}$.

To prove this, we recall that $A \cong \operatorname{End} T_C$ and, as in the proof of (XII.1.14), the left hand part of the postprojective component $\mathcal{P}(C)$ of $\Gamma(\operatorname{mod} C)$ containing all the indecomposable projective C-modules $P(0), \ldots, P(\omega)$ looks

as follows

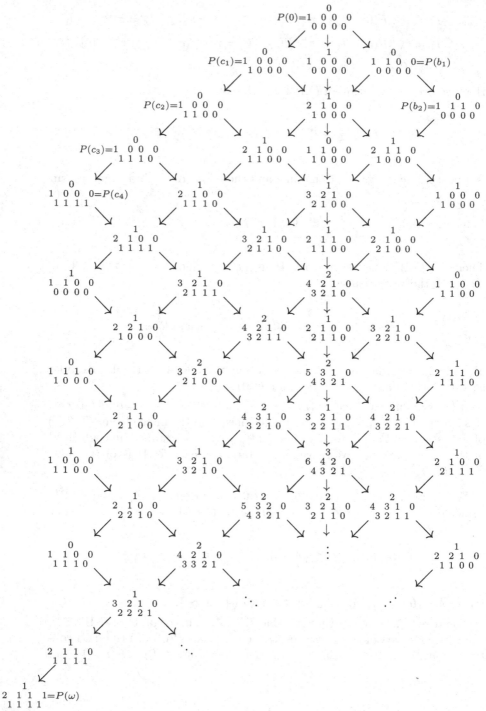

where

$$P(a_1) = \begin{smallmatrix} & & 1 & & & \\ 1 & 0 & 0 & 0 & 0, \\ & 0 & 0 & 0 & 0 & \end{smallmatrix}$$

and the indecomposable modules are represented by their dimension vectors.

It follows from (VI.3.10) that $e_i A \cong \mathrm{Hom}_C(T, T_i)$ and, for each regular C-module X and for each $i \in \{1, \dots, 9\}$, there isomorphisms of A-modules

$$\begin{aligned} \mathrm{Hom}_C(T, X)_i &= \mathrm{Hom}_C(T, X)e_i \\ &\cong \mathrm{Hom}_A(e_i A, \mathrm{Hom}_C(T, X)) \\ &\cong \mathrm{Hom}_A(\mathrm{Hom}_C(T, T_i), \mathrm{Hom}_C(T, X)) \\ &\cong \mathrm{Hom}_C(T_i, X), \end{aligned}$$

and therefore $\mathbf{dim}\,\mathrm{Hom}_C(T, X)_i = \dim_K \mathrm{Hom}_C(T_i, X)$. Note also that we have

$$\mathrm{Hom}_C(T_9, X) = \mathrm{Hom}_C(P(\omega), X) \cong X_\omega.$$

By consulting the beginning part of the component $\mathcal{P}(C)$ shown above we see that the ω-th coordinate of the dimension vector of each of the C-modules T_1, T_2, T_3, T_4, T_5, T_6, T_7, T_8 equals zero. Let

$$Q: \quad \begin{array}{ccccccc} & & a_1 & & & & \\ & \swarrow^{\alpha_1} & & & & & \\ 0 & \xleftarrow{\beta_1} & b_1 & \xleftarrow{\beta_2} & b_2 & & \\ & \nwarrow_{\gamma_1} & & & & & \\ & & c_1 & \xleftarrow{\gamma_2} & c_2 & \xleftarrow{\gamma_3} c_3 & \xleftarrow{\gamma_4} c_4 \end{array}$$

be the full subquiver of $\Delta(2,3,5)$, of the Dynkin type $\mathbb{E}_8$, defined by the points $0, a_1, b_1, b_2, c_1, c_2, c_3, c_4$ (we use the notation of (XII.1.2)).

Note that there is an isomorphism $KQ \cong e_Q C e_Q$ of algebras and the equivalence of categories $\mathrm{mod}\, KQ \cong \mathrm{rep}_K(Q)$, where e_Q is the idempotent

$$e_Q = e_0 + e_{a_1} + e_{b_1} + e_{b_2} + e_{c_1} + e_{c_2} + e_{c_3} + e_{c_4}$$

of the algebra C. It follows that the restriction functor

$$\mathrm{res}_{e_Q} : \mathrm{mod}\, C \longrightarrow \mathrm{mod}\, KQ \cong \mathrm{rep}_K(Q)$$

from the bound quiver $\Delta(2,3,5)$ to its subquiver Q (see (I.6.6)) defines a functorial isomorphism

$$\mathrm{Hom}_C(T_j, E) \cong \mathrm{Hom}_{KQ}(\mathrm{res}_{e_Q} T_j, \mathrm{res}_{e_Q} E),$$

for $T_j \neq T_9 = P(\omega)$ and for any module E in the family

$$E_1^{(\infty)}, \ E_2^{(\infty)}, \ E_1^{(0)}, \ E_2^{(0)}, \ E_3^{(0)}, \ E_1^{(1)}, \ E_2^{(1)}, \ E_3^{(1)}, \ E_4^{(1)}, \ E_5^{(1)}, \ E^{(\lambda)},$$

with $\lambda \in K \setminus \{0, 1\}$. Because $\mathrm{res}_{e_Q} E^{(\lambda)} = I(0)$ is the injective envelope of the simple representation $S(0)$ in $\mathrm{mod}\, KQ \cong \mathrm{rep}_K(Q)$ then, for each $j \in \{1, \dots, 8\}$, there are isomorphisms

$$\mathrm{Hom}_C(T_j, E^{(\lambda)}) \cong \mathrm{Hom}_{KQ}(\mathrm{res}_{e_Q} T_j, \mathrm{res}_{e_Q} E^{(\lambda)}) \cong D(T_j)_0,$$

that are functorial at T_j.

By applying the above formulae to the module $E = F^{(\lambda)}$, with $\lambda \in K \setminus \{0, 1\}$, we get the isomorphisms

$$\mathrm{Hom}_C(T_1, E^{(\lambda)}) \cong D(T_1)_0 = K^6,$$
$$\mathrm{Hom}_C(T_2, E^{(\lambda)}) \cong D(T_2)_0 = K^4,$$
$$\mathrm{Hom}_C(T_3, E^{(\lambda)}) \cong D(T_3)_0 = K^2,$$
$$\mathrm{Hom}_C(T_4, E^{(\lambda)}) \cong D(T_4)_0 = K^3,$$
$$\mathrm{Hom}_C(T_5, E^{(\lambda)}) \cong D(T_5)_0 = K^5,$$
$$\mathrm{Hom}_C(T_6, E^{(\lambda)}) \cong D(T_6)_0 = K^4,$$
$$\mathrm{Hom}_C(T_7, E^{(\lambda)}) \cong D(T_7)_0 = K^3,$$
$$\mathrm{Hom}_C(T_8, E^{(\lambda)}) \cong D(T_8)_0 = K^2, \text{ and}$$
$$\mathrm{Hom}_C(T_9, E^{(\lambda)}) \cong E_\omega^{(\lambda)} = K.$$

In particular, we get

$$\mathbf{dim}\,\mathrm{Hom}_C(T, E^{(\lambda)}) = {}_2\,{}_4\,{}_6\,{\overset{3}{5}}\,{}_4\,{}_3\,{}_2\,{}_1.$$

Moreover, by applying the above formulae, one shows that there is an A-module isomorphism $\mathrm{Hom}_C(T, E^{(\lambda)}) \cong F^{(\lambda)}$, and hence

$$\mathrm{Hom}_C(T, \mathcal{T}_\lambda^C) = \mathcal{T}_\lambda^{\Delta(\widetilde{\mathbb{E}}_8)},$$

for any $\lambda \in K \setminus \{0, 1\}$.

Because, by (XII.2.12), the C-module $E^{(\lambda)}$ is simple regular, then the A-module $\mathrm{Hom}_C(T, E^{(\lambda)}) \cong F^{(\lambda)}$ is simple regular. Furthermore, in view of the isomorphism

$$\tau_C E^{(\lambda)} \cong E^{(\lambda)}$$

in (XII.2.12) and the commutativity of the diagram (2.11), we get the isomorphism

$$\tau_A F^{(\lambda)} \cong F^{(\lambda)}.$$

This finishes the proof of (d).

Similarly, by applying the above formulae to any module E in the family

$$E_1^{(\infty)},\, E_2^{(\infty)},\, E_1^{(0)},\, E_2^{(0)},\, E_3^{(0)},\, E_1^{(1)},\, E_2^{(1)},\, E_3^{(1)},\, E_4^{(1)},\, E_5^{(1)}$$

we get

$$\dim \operatorname{Hom}_C(T, E_1^{(\infty)}) = \begin{smallmatrix}&&2&&&&&\\1&2&3&2&2&1&1&0\end{smallmatrix}, \qquad \dim \operatorname{Hom}_C(T, E_2^{(\infty)}) = \begin{smallmatrix}&&1&&&&&\\1&2&3&3&2&2&1&1\end{smallmatrix},$$

$$\dim \operatorname{Hom}_C(T, E_1^{(0)}) = \begin{smallmatrix}&&1&&&&&\\1&2&2&1&1&1&0&0\end{smallmatrix}, \qquad \dim \operatorname{Hom}_C(T, E_2^{(0)}) = \begin{smallmatrix}&&1&&&&&\\1&1&2&2&1&1&1&0\end{smallmatrix},$$

$$\dim \operatorname{Hom}_C(T, E_3^{(0)}) = \begin{smallmatrix}&&1&&&&&\\0&1&2&2&2&1&1&1\end{smallmatrix}, \qquad \dim \operatorname{Hom}_C(T, E_1^{(1)}) = \begin{smallmatrix}&&1&&&&&\\0&1&1&1&0&0&0&0\end{smallmatrix},$$

$$\dim \operatorname{Hom}_C(T, E_2^{(1)}) = \begin{smallmatrix}&&0&&&&&\\1&1&1&1&1&0&0&0\end{smallmatrix}, \qquad \dim \operatorname{Hom}_C(T, E_3^{(1)}) = \begin{smallmatrix}&&1&&&&&\\0&0&1&1&1&1&0&0\end{smallmatrix},$$

$$\dim \operatorname{Hom}_C(T, E_4^{(1)}) = \begin{smallmatrix}&&0&&&&&\\0&1&1&1&1&1&1&0\end{smallmatrix}, \qquad \dim \operatorname{Hom}_C(T, E_5^{(1)}) = \begin{smallmatrix}&&1&&&&&\\1&1&2&1&1&1&1&1\end{smallmatrix}.$$

In particular, $\dim \operatorname{Hom}_C(T, E_j^{(s)}) = \dim F_j^{(s)}$, for all j, and $s \in \{0, 1, \infty\}$.

Because, by (XII.2.12), E is a simple regular C-module, then obviously the A-module $\operatorname{Hom}_C(T, E)$ is simple regular.

We recall from (XI.3.3) and (XII.4.2) that the simple regular A-modules lying in the non-homogeneous tubes of $\mathcal{R}(A)$ are uniquely determined by their dimension vectors. Because, by (2.22), the modules presented in Table 2.20 are simple regular, then there is an isomorphism

$$\operatorname{Hom}_C(T, E_j^{(s)}) \cong F_j^{(s)},$$

for all j, and $s \in \{0, 1, \infty\}$.

Therefore, by applying (XII.2.12) and the commutatity of the diagram (2.11), we prove the statements (a)–(c), and we conclude that the tubes

$$\mathcal{T}_\infty^{\Delta(\widetilde{\mathbb{E}}_8)} = \operatorname{Hom}_C(T, \mathcal{T}_\infty^C), \;\; \mathcal{T}_0^{\Delta(\widetilde{\mathbb{E}}_8)} = \operatorname{Hom}_C(T, \mathcal{T}_0^C), \;\; \mathcal{T}_1^{\Delta(\widetilde{\mathbb{E}}_8)} = \operatorname{Hom}_C(T, \mathcal{T}_1^C)$$

exhaust all the non-homogeneous tubes of the regular part $\mathcal{R}(A)$ the Auslander–Reiten quiver $\Gamma(\operatorname{mod} A)$ of A. Consequently, the tubes

$$\mathcal{T}_\lambda^{\Delta(\widetilde{\mathbb{E}}_8)} = \operatorname{Hom}_C(T, \mathcal{T}_\lambda^C),$$

with $\lambda \in K \setminus \{0, 1\}$, exhaust all the homogeneous tubes of $\Gamma(\operatorname{mod} A)$. Now the statement (e) follows from (2.1). This completes the proof. $\qquad\square$

Now we summarise the main result of this section.

2.24. Theorem. *Assume that Δ is one of the canonically oriented Euclidean quivers*

$$\Delta(\widetilde{\mathbb{A}}_{p,q}), \text{ with } q \geq p \geq 1, \Delta(\widetilde{\mathbb{D}}_m), \text{ with } m \geq 4, \Delta(\widetilde{\mathbb{E}}_6), \Delta(\widetilde{\mathbb{E}}_7), \text{ and } \Delta(\widetilde{\mathbb{E}}_8),$$

listed in Table 1.1.

(a) *Every simple regular $K\Delta$-module is isomorphic to one of the modules presented in Tables 2.4, 2.6, 2.12, 2.16, and 2.20, if Δ is the quiver $\Delta(\widetilde{\mathbb{A}}_{p,q})$, $\Delta(\widetilde{\mathbb{D}}_m)$, $\Delta(\widetilde{\mathbb{E}}_6)$, $\Delta(\widetilde{\mathbb{E}}_7)$, and $\Delta(\widetilde{\mathbb{E}}_8)$, respectively.*

(b) *The regular part $\mathcal{R}(K\Delta)$ of the Auslander–Reiten quiver $\Gamma(\operatorname{mod} K\Delta)$ of $K\Delta$ is a $\mathbb{P}_1(K)$-family*

$$\mathcal{T}^\Delta = \{\mathcal{T}^\Delta_\lambda\}_{\lambda \in \mathbb{P}_1(K)}$$

of stable tubes, where

- $\mathcal{T}^\Delta_\infty$ *is the stable tube containing the simple regular $K\Delta$-module $E_1^{(\infty)}$,*
- $\mathcal{T}^\Delta_0$ *is the stable tube containing the simple regular $K\Delta$-module $E_1^{(0)}$,*
- $\mathcal{T}^\Delta_1$ *is the stable tube containing the simple regular $K\Delta$-module $E_1^{(1)}$, if $\Delta \neq \Delta(p,q)$,*
- *for each $\lambda \in K \setminus \{0,1\}$, $\mathcal{T}^\Delta_\lambda$ is the homogeneous tube containing the simple regular $K\Delta$-module $E^{(\lambda)}$, if $\Delta \neq \Delta(p,q)$, and*
- *for each $\lambda \in K \setminus \{0\}$, $\mathcal{T}^\Delta_\lambda$ is the homogeneous tube containing the simple regular $K\Delta$-module $E^{(\lambda)}$, if $\Delta = \Delta(p,q)$.*

Here $E_j^{(\infty)}$, $E_i^{(0)}$, $E_s^{(1)}$, $E^{(\lambda)}$ are the simple regular $K\Delta$-modules listed in Tables 2.4, 2.6, 2.12, 2.16, 2.20, if Δ is the quiver $\Delta(\widetilde{\mathbb{A}}_{p,q})$, $\Delta(\widetilde{\mathbb{D}}_m)$, $\Delta(\widetilde{\mathbb{E}}_6)$, $\Delta(\widetilde{\mathbb{E}}_7)$, $\Delta(\widetilde{\mathbb{E}}_8)$, respectively.

(c) *The category $\operatorname{mod} K\Delta$ is controlled by the quadratic form $q_\Delta : \mathbb{Z}^{|\Delta_0|} \longrightarrow \mathbb{Z}$ of the quiver Δ.*

(d) *For any positive vector $\mathbf{x} \in K_0(K\Delta) = \mathbb{Z}^{|\Delta_0|}$, with $q_\Delta(\mathbf{x}) = 0$, there is a $\mathbb{P}_1(K)$-family $\{X_\lambda\}_{\lambda \in \mathbb{P}_1(K)}$ of pairwise non-isomorphic indecomposable $K\Delta$-modules X_λ such that X_λ lies in the tube $\mathcal{T}^\Delta_\lambda$ and $\dim X_\lambda = \mathbf{x}$, for any $\lambda \in \mathbb{P}_1(K)$.*

Proof. The statements (a) and (b) immediately follow by applying (2.5), (2.9), (2.15), (2.19), and (2.23). The statements (c) and (d) are a consequence of (a), (b), and (XII.4.2) applied to the hereditary algebra $K\Delta$, which is obviously concealed of Euclidean type. □

XIII.3. Four subspace problem

The aim of this section is to describe all indecomposable modules over the path algebra

$$A = KQ$$

of the following quiver (of the Euclidean type $\widetilde{\mathbb{D}}_4$), known as a **four subspace quiver**

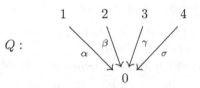

In other words, we consider the problem of classifying the indecomposable K-linear representations

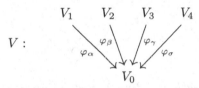

of Q, up to isomorphism. Observe that if V is indecomposable and one of the K-linear maps φ_α, φ_β, φ_γ or φ_σ is not a monomorphism, then V is isomorphic to one of the simple representations $S(1)$, $S(2)$, $S(3)$, $S(4)$. Therefore, up to the known four simple representations, our problem is to classify four subspaces

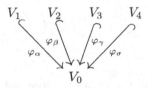

of vector spaces V_0, up to K-linear automorphisms of V_0 preserving the four subspaces.

The problem is known as the **four subspace problem** and is solved in [97] by Gelfand and Ponomarev in case K is an algebraically closed field.

Because any indecomposable K-linear representation of Q is postprojective, preinjective or regular, a solution of the four subspace problem follows from (2.8), (2.11), (3.15), and (3.16) below.

Note that the path algebra $A = KQ$ of the four subspace quiver Q is hereditary and is isomorphic to the matrix subalgebra

$$A = KQ \cong \begin{bmatrix} K & 0 & 0 & 0 & 0 \\ K & K & 0 & 0 & 0 \\ K & 0 & K & 0 & 0 \\ K & 0 & 0 & K & 0 \\ K & 0 & 0 & 0 & K \end{bmatrix} \subseteq \mathbb{M}_5(K) \qquad (3.1)$$

of the full matrix algebra $\mathbb{M}_5(K)$ of 5×5 square matrices with coefficients in the field K.

Throughout, we assume that $A = KQ$ has the lower triangular matrix form (3.1), and that K is an algebraically closed field.

The following fact is a consequence of (III.2.4) and (III.2.6).

3.2. Lemma. *Let Q be the four subspace quiver. Then the path algebra $A = KQ$ of Q has the form (3.1) and*
   (a) *the indecomposable projective A-modules are of the form*

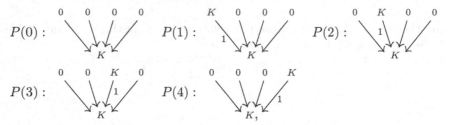

   (b) *the indecomposable injective A-modules are of the form*

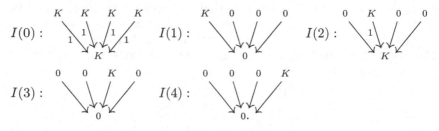

$\square$

By the results in (III.3), the group homomorphism

$$\mathbf{dim} : K_0(A) \overset{\simeq}{\longrightarrow} \mathbb{Z}^5,$$

that associates to the representation V of Q the dimension vector

$$\mathbf{dim}\, V = [\, \dim_K V_0, \dim_K V_1, \dim_K V_2, \dim_K V_3, \dim_K V_4\,]^{\mathrm{t}} = \begin{bmatrix} \dim_K V_0 \\ \dim_K V_1 \\ \dim_K V_2 \\ \dim_K V_3 \\ \dim_K V_4 \end{bmatrix},$$

is an isomorphism of the Grothendieck group $K_0(A)$ of A with the free abelian group $\mathbb{Z}^5$.

The Cartan matrix $\mathbf{C}_A \in \mathbb{M}_5(\mathbb{Z})$ of A, its inverse $\mathbf{C}_A^{-1} \in \mathbb{M}_5(\mathbb{Z})$, the Coxeter matrix $\mathbf{\Phi}_A = -\mathbf{C}_A^t \mathbf{C}_A^{-1} \in \mathbb{M}_5(\mathbb{Z})$ of A and its inverse $\mathbf{\Phi}_A^{-1} \in \mathbb{M}_5(\mathbb{Z})$ are of the forms

$$
\mathbf{C}_A = \begin{bmatrix} 1 & 1 & 1 & 1 & 1 \\ 0 & 1 & 0 & 0 & 0 \\ 0 & 0 & 1 & 0 & 0 \\ 0 & 0 & 0 & 1 & 0 \\ 0 & 0 & 0 & 0 & 1 \end{bmatrix}, \qquad
\mathbf{C}_A^{-1} = \begin{bmatrix} 1 & -1 & -1 & -1 & -1 \\ 0 & 1 & 0 & 0 & 0 \\ 0 & 0 & 1 & 0 & 0 \\ 0 & 0 & 0 & 1 & 0 \\ 0 & 0 & 0 & 0 & 1 \end{bmatrix},
$$

$$
\mathbf{\Phi}_A = \begin{bmatrix} -1 & 1 & 1 & 1 & 1 \\ -1 & 0 & 1 & 1 & 1 \\ -1 & 1 & 0 & 1 & 1 \\ -1 & 1 & 1 & 0 & 1 \\ -1 & 1 & 1 & 1 & 0 \end{bmatrix}, \qquad
\mathbf{\Phi}_A^{-1} = \begin{bmatrix} 3 & -1 & -1 & -1 & -1 \\ 1 & -1 & 0 & 0 & 0 \\ 1 & 0 & -1 & 0 & 0 \\ 1 & 0 & 0 & -1 & 0 \\ 1 & 0 & 0 & 0 & -1 \end{bmatrix}.
$$

We recall that the Euler quadratic form $q_A : \mathbb{Z}^5 \longrightarrow \mathbb{Z}$ of $A = KQ$ is defined by the formula

$$
\begin{aligned}
q_A(\mathbf{x}) = q_Q(\mathbf{x}) &= \mathbf{x}^t (\mathbf{C}_A^{-1})^t \mathbf{x} \\
&= x_0^2 + x_1^2 + x_2^2 + x_3^2 + x_4^2 - x_0 x_1 - x_0 x_2 - x_0 x_3 - x_0 x_4,
\end{aligned}
\tag{3.3}
$$

for any vector

$$
\mathbf{x} = [x_0, x_1, x_2, x_3, x_4]^t = \begin{bmatrix} x_0 \\ x_4 \\ x_3 \\ x_2 \\ x_1 \end{bmatrix} = \begin{smallmatrix} & x_1 & x_2 & & x_3 & x_4 \\ & & & x_0 & & \end{smallmatrix} \in K_0(A) = \mathbb{Z}^5.
$$

Now we determine the defect $\partial_A : \mathbb{Z}^5 \longrightarrow \mathbb{Z}$ of the hereditary algebra $A = KQ$ and the radical $\operatorname{rad} q_A = \{\mathbf{y} \in \mathbb{Z}^5;\ q_A(\mathbf{y}) = 0\}$ of q_A.

3.4. Lemma. *Let $A = KQ$ be the path algebra* (3.1) *of the four subspace quiver Q.*

(i) *The radical $\operatorname{rad} q_A$ of the Euler quadratic form $q_A : \mathbb{Z}^5 \longrightarrow \mathbb{Z}$ of A is the infinite cyclic subgroup $\operatorname{rad} q_A = \mathbb{Z} \cdot \mathbf{h}_A$ of $\mathbb{Z}^5$, where*

$$
\mathbf{h}_A = [2, 1, 1, 1, 1]^t = \begin{smallmatrix} 1 & 1 & & 1 & 1 \\ & & 2 & & \end{smallmatrix}.
$$

(ii) *The defect $\partial_A : \mathbb{Z}^5 \longrightarrow \mathbb{Z}$ of A is given by the formula*

$$
\partial_A(\mathbf{x}) = -2x_0 + x_1 + x_2 + x_3 + x_4,
$$

for all $\mathbf{x} = [x_0, x_1, x_2, x_3, x_4]^t \in K_0(A) = \mathbb{Z}^5$.

Proof. (i) It is easy to check that the quadratic form q_A has the canonical form

$$q_A(\mathbf{x}) = (x_1 - \frac{1}{2}x_0)^2 + (x_2 - \frac{1}{2}x_0)^2 + (x_3 - \frac{1}{2}x_0)^2 + (x_4 - \frac{1}{2}x_0)^2.$$

It follows that $q_A(\mathbf{x}) \geq 0$, for all vectors $\mathbf{x} \in \mathbb{Z}^5$. Moreover, $q_A(\mathbf{x}) = 0$ if and only if $x_0 = 2x_1 = 2x_2 = 2x_3 = 2x_4$, that is, if and only if $x_1 = x_2 = x_3 = x_4$ and $\mathbf{x} = x_1 \cdot \mathbf{h}_A$. Hence (i) follows.

(ii) We have shown already that the Coxeter matrix $\mathbf{\Phi}_A \in \mathbb{M}_5(\mathbb{Z})$ of A has the form

$$\mathbf{\Phi}_A = \begin{bmatrix} -1 & 1 & 1 & 1 & 1 \\ -1 & 0 & 1 & 1 & 1 \\ -1 & 1 & 0 & 1 & 1 \\ -1 & 1 & 1 & 0 & 1 \\ -1 & 1 & 1 & 1 & 0 \end{bmatrix}.$$

An easy matrix calculation shows that, for each vector $\mathbf{x} = [x_0, x_1, x_2, x_3, x_4]^t$ in $K_0(A) = \mathbb{Z}^5$, we have

$$\mathbf{\Phi}_A^2 \cdot \mathbf{x} = \begin{bmatrix} -3 & 2 & 2 & 2 & 2 \\ -2 & 2 & 1 & 1 & 1 \\ -2 & 1 & 2 & 1 & 1 \\ -2 & 1 & 1 & 2 & 1 \\ -2 & 1 & 1 & 1 & 2 \end{bmatrix} \cdot \begin{bmatrix} x_0 \\ x_1 \\ x_2 \\ x_3 \\ x_4 \end{bmatrix} = \begin{bmatrix} -3x_0 + 2x_1 + 2x_2 + 2x_3 + 2x_4 \\ -2x_0 + 2x_1 + x_2 + x_3 + x_4 \\ -2x_0 + x_1 + 2x_2 + x_3 + x_4 \\ -2x_0 + x_1 + x_2 + 2x_3 + x_4 \\ -2x_0 + x_1 + x_2 + x_3 + 2x_4 \end{bmatrix}$$

$$= \begin{bmatrix} x_0 \\ x_1 \\ x_2 \\ x_3 \\ x_4 \end{bmatrix} + (-2x_0 + x_1 + x_2 + x_3 + x_4) \cdot \begin{bmatrix} 2 \\ 1 \\ 1 \\ 1 \\ 1 \end{bmatrix}$$

and

$$\mathbf{\Phi}_A \cdot \mathbf{x} - \mathbf{x} = \begin{bmatrix} -2 & 1 & 1 & 1 & 1 \\ -1 & -1 & 1 & 1 & 1 \\ -1 & 1 & -1 & 1 & 1 \\ -1 & 1 & 1 & -1 & 1 \\ -1 & 1 & 1 & 1 & -1 \end{bmatrix} \cdot \begin{bmatrix} x_0 \\ x_1 \\ x_2 \\ x_3 \\ x_4 \end{bmatrix} = \begin{bmatrix} -2x_0 + x_1 + x_2 + x_3 + x_4 \\ -x_0 - x_1 + x_2 + x_3 + x_4 \\ -x_0 + x_1 - x_2 + x_3 + x_4 \\ -x_0 + x_1 + x_2 - x_3 + x_4 \\ -x_0 + x_1 + x_2 + x_3 - x_4 \end{bmatrix}.$$

It follows that $\mathbf{\Phi}_A \cdot \mathbf{x} - \mathbf{x} \notin \mathbb{Z} \cdot \mathbf{h}_A$ and $\mathbf{\Phi}_A^2 \cdot \mathbf{x} = \mathbf{x} + (-2x_0 + x_1 + x_2 + x_3 + x_4) \cdot \mathbf{h}_A$, for any $\mathbf{x} = [x_0, x_1, x_2, x_3, x_4]^t \in \mathbb{Z}^5$. This shows that

$$\partial_A(\mathbf{x}) = -2x_0 + x_1 + x_2 + x_3 + x_4,$$

for all $\mathbf{x} = [x_0, x_1, x_2, x_3, x_4]^t \in \mathbb{Z}^5$, and finishes the proof. $\square$

Now we describe the postprojective component $\mathcal{P}(A)$ of the Auslander–Reiten quiver $\Gamma(\text{mod } A)$ of A, and the dimension vectors of modules in $\mathcal{P}(A)$.

3.5. Corollary. *Let $A = KQ$ be the path algebra (3.1) of the four subspace quiver Q and let $\partial_A : \mathbb{Z}^5 \longrightarrow \mathbb{Z}$ be the defect of A.*

(a) *An indecomposable A-module V is postprojective if and only if $\partial_A(\dim V) < 0$.*

(b) *The postprojective component $\mathcal{P}(A)$ of $\Gamma(\operatorname{mod} A)$ is of the form*

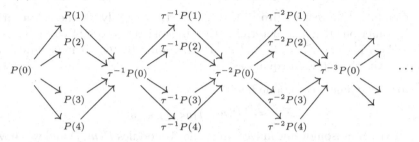

(c) *For each integer $m \geq 0$, we have*

$$\dim \tau_A^{-m} P(0) = [2m+1, m, m, m, m]^t,$$
$$\dim \tau_A^{-2m-1} P(1) = [2m+2, m, m+1, m+1, m+1]^t,$$
$$\dim \tau_A^{-2m-1} P(2) = [2m+2, m+1, m, m+1, m+1]^t,$$
$$\dim \tau_A^{-2m-1} P(3) = [2m+2, m+1, m+1, m, m+1]^t,$$
$$\dim \tau_A^{-2m-1} P(4) = [2m+2, m+1, m+1, m+1, m]^t,$$
$$\dim \tau_A^{-2m} P(1) = [2m+1, m+1, m, m, m]^t,$$
$$\dim \tau_A^{-2m} P(2) = [2m+1, m, m+1, m, m]^t,$$
$$\dim \tau_A^{-2m} P(3) = [2m+1, m, m, m+1, m]^t,$$
$$\dim \tau_A^{-2m} P(4) = [2m+1, m, m, m, m+1]^t$$

Proof. The statement (a) follows from (XI.2.3).

(b) Because the four subspace quiver Q is acyclic then the postprojective component $\mathcal{P}(A)$ of $\Gamma(\operatorname{mod} A)$ presented in (b) can be constructed by applying the technique described in the proof of (VIII.2.1), see also (VIII.2.3) and Example (VIII.2.4)(b).

(c) Because the algebra $A = KQ$ is hereditary then, according to (IV.2.9), we have

$$\dim \tau_A^{-m} P(j) = \Phi_A^{-m} \dim P(j),$$

for all $m \geq 0$ and $j \in \{0, 1, 2, 3, 4\}$. Hence, by a straightforward calculation of $\Phi_A^{-m} \dim P(j)$, we get the equalities listed in (c) by induction on $m \geq 0$, compare (VIII.2.3) and Example (VIII.2.4)(b).

The equalities listed in (c) can be established alternatively by induction as follows (compare with (VIII.2.4)(b)). Because the dimension vectors of the indecomposable projective modules $P(0)$, $P(1)$, $P(2)$, $P(3)$, and $P(4)$ are known then, by applying the additivity of the function (see (III.3.3))

$$\mathbf{dim} : \operatorname{mod} A \longrightarrow \mathbb{Z}^5$$

we compute the vector $\mathbf{dim}\,\tau_A^{-1}P(0)$. Next, by applying the additivity of the function $\mathbf{dim}$, we determine the dimension vectors of the modules $\tau_A^{-1}P(1)$, $\tau_A^{-1}P(2)$, $\tau_A^{-1}P(3)$, and $\tau_A^{-1}P(4)$. Similarly, the obvious inductive step proves the required equalities. $\qquad\square$

Now we define a countable family

$$\mathcal{FP}_A = \{P(m,j)\}_{0 \le j \le 4,\, m \ge 0}$$

of pairwise non-isomorphic indecomposable A-modules $P(m,j)$ and we show that it is just a complete family of the indecomposable postprojective A-modules.

3.6. Table. Indecomposable postprojective representations
of the four subspace quiver

Let $A = KQ$ be the path algebra (3.1) of the four subspace quiver Q. We define three series of indecomposable postprojective A-modules

(a) $P(m,0)$, with $m \ge 0$,
(b) $P(2m+1,1)$, $P(2m+1,2)$, $P(2m+1,3)$, $P(2m+1,4)$, with $m \ge 1$,
(c) $P(2m,1)$, $P(2m,2)$, $P(2m,3)$, $P(2m,4)$, with $m \ge 0$,

viewed as K-linear representations of the four subspace quiver Q.

3.6(a) For each $m \ge 0$, we define

$$P(m,0) :$$

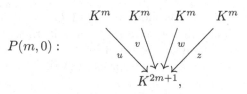

where
$$u([x_1, \ldots, x_m]^t) = [x_1, \ldots, x_m, 0, 0, \ldots, 0]^t,$$
$$v([x_1, \ldots, x_m]^t) = [0, \ldots, 0, 0, x_1, \ldots, x_m]^t,$$
$$w([x_1, \ldots, x_m]^t) = [x_1, \ldots, x_m, x_1, \ldots, x_m, 0]^t, \text{ and}$$
$$z([x_1, \ldots, x_m]^t) = [0, x_1, \ldots, x_m, x_1, \ldots, x_m]^t,$$
for any vector $[x_1, \ldots, x_m]^t \in K^m$.

3.6(b) For each $m \geq 1$, we define

$P(2m+1,1):$

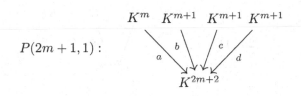

$P(2m+1,2):$

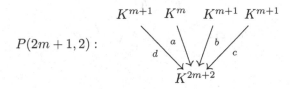

$P(2m+1,3):$

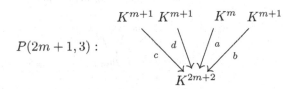

$P(2m+1,4):$

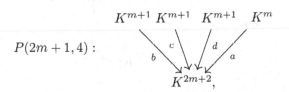

where

$$a([x_1,\ldots,x_m]^t) = [0,x_1,\ldots,x_m,x_1,\ldots,x_m,0]^t, \text{ for } [x_1,\ldots,x_m]^t \in K^m,$$
$$b([y_1,\ldots,y_{m+1}]^t) = [y_1,\ldots,y_{m+1},0,\ldots,0]^t,$$
$$c([y_1,\ldots,y_{m+1}]^t) = [0,\ldots,0,y_1,\ldots,y_{m+1}]^t, \quad \text{ and}$$
$$d([y_1,\ldots,y_{m+1}]^t) = [y_1,\ldots,y_{m+1},y_1,\ldots,y_{m+1}]^t, \text{ for } [y_1,\ldots,y_{m+1}]^t \in K^{m+1}.$$

3.6(c) For each $m \geq 0$, we define the following four A-modules

$P(2m,1)$:
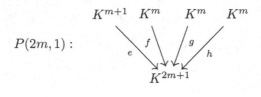

$P(2m,2)$:
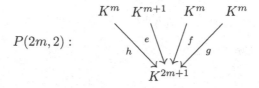

$P(2m,3)$:
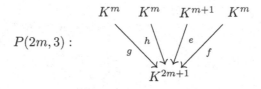

$P(2m,4)$:
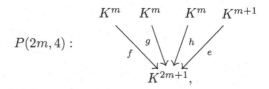

where

$$e([x_1,\ldots,x_{m+1}]^t) = [0,x_1,\ldots,x_{m+1},0,\ldots,0]^t,$$
$$\text{for } [x_1,\ldots,x_{m+1}]^t \in K^{m+1},$$
$$f([y_1,\ldots,y_m]^t) = [0,0,\ldots,0,y_1,\ldots,y_m]^t,$$
$$g([y_1,\ldots,y_m]^t) = [0,y_1,\ldots,y_m,y_1,\ldots,y_m]^t, \text{ and}$$
$$h([y_1,\ldots,y_m]^t) = [y_1,\ldots,y_m,0,y_1,\ldots,y_m]^t, \text{ for } [y_1,\ldots,y_m]^t \in K^m.$$

Now we show that the family of modules of Table 3.6 is a complete set of pairwise non-isomorphic indecomposable postprojective A-modules.

3.7. Proposition. *Let $A = KQ$ be the path algebra* (3.1) *of the four subspace quiver Q and let*

$$\mathcal{FP}_A = \{P(m, j)\}_{0 \leq j \leq 4,\, m \geq 0}$$

be the family of A-modules listed in Table 3.6.

(a) *For each pair of integers $m \geq 0$ and $j \in \{0, 1, 2, 3, 4\}$, there is an isomorphism*

$$\tau^{-m} P(j) \cong P(m, j)$$

of A-modules, where $P(j)$ is the projective A-module of (3.2) *corresponding to the vertex j of Q.*

(b) *For any $m \geq 0$ and $j \in \{0, 1, 2, 3, 4\}$, the module $P(m, j)$ is a postprojective brick and satisfies*

$$\partial_A(\mathbf{dim}\, P(m, j)) < 0 \quad \text{and} \quad q_A(\mathbf{dim}\, P(m, j)) = 1.$$

(c) *The modules $P(m, j)$ of the family $\mathcal{FP}_A$ are pairwise non-isomorphic.*

(d) *Up to isomorphism, every indecomposable postprojective A-module is of the form $P(m, j)$, where $m \geq 0$ and $j \in \{0, 1, 2, 3, 4\}$.*

Proof. It follows from (3.5)(a) that every postprojective indecomposable A-module is directing and has the form $\tau^{-m} P(j)$, where $m \geq 0$ and $j \in \{0, 1, 2, 3, 4\}$. Hence, by (IX.3.1), the module $\tau^{-m} P(j)$ is uniquely determined by its dimension vector. Because, by (3.5)(b),

$$\mathbf{dim}\, P(m, j) = \mathbf{dim}\, \tau^{-m} P(j)$$

then it remains to check that each of the modules $P(m, j)$ is a brick, for $m \geq 0$ and $j \in \{0, 1, 2, 3, 4\}$.

We leave it as an exercise. We only remark that by a simple matrix calculation technique applied in the proof of (2.18) one shows that there is an isomorphism of algebras $\operatorname{End} P(m, j) \cong K$, for each $m \geq 0$ and $j \in \{0, 1, 2, 3, 4\}$. $\qquad\square$

Now we describe the preinjective component $\mathcal{Q}(A)$ of the Auslander–Reiten quiver $\Gamma(\operatorname{mod} A)$ of A, and the dimension vectors of modules in $\mathcal{Q}(A)$.

3.8. Corollary. *Let $A = KQ$ be the path algebra* (3.1) *of the four subspace quiver Q and let $\partial_A : \mathbb{Z}^5 \longrightarrow \mathbb{Z}$ be the defect of A.*

(a) *An indecomposable A-module V is preinjective if and only if $\partial_A(\mathbf{dim}\, V) > 0$.*

(b) *The preinjective component $\mathcal{Q}(A)$ of $\Gamma(\operatorname{mod} A)$ is of the form*

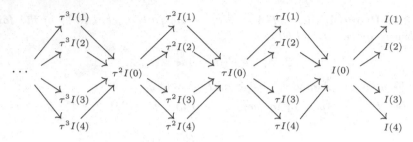

(c) *For each integer $m \geq 0$, we have*

$$\dim \tau_A^m I(0) = [2m+1, m+1, m+1, m+1, m+1]^t,$$
$$\dim \tau_A^{2m+1} I(1) = [2m+1, m, m+1, m+1, m+1]^t,$$
$$\dim \tau_A^{2m+1} I(2) = [2m+1, m+1, m, m+1, m+1]^t,$$
$$\dim \tau_A^{2m+1} I(3) = [2m+1, m+1, m+1, m, m+1]^t,$$
$$\dim \tau_A^{2m+1} I(4) = [2m+1, m+1, m+1, m+1, m]^t,$$
$$\dim \tau_A^{2m} I(1) = [2m, m+1, m, m, m]^t,$$
$$\dim \tau_A^{2m} I(2) = [2m, m, m+1, m, m]^t,$$
$$\dim \tau_A^{2m} I(3) = [2m, m, m, m+1, m]^t,$$
$$\dim \tau_A^{2m} I(4) = [2m, m, m, m, m+1]^t.$$

Proof. The statement (a) follows from (XI.2.3).

(b) Because the four subspace quiver Q is acyclic then the preinjective component $\mathcal{Q}(A)$ of $\Gamma(\mathrm{mod}\, A)$ presented in (b) can be constructed by applying the technique described in the proof of (VIII.2.1), see also (VIII.2.3) and Example (VIII.2.4)(b).

(c) Because the algebra $A = KQ$ is hereditary then, according to (IV.2.9), we have

$$\dim \tau_A^m I(j) = \boldsymbol{\Phi}_A^m \dim I(j),$$

for all $m \geq 0$ and $j \in \{0, 1, 2, 3, 4\}$. Hence, by a straightforward calculation of $\boldsymbol{\Phi}_A^m \dim I(j)$, we get the equalities listed in (c) by induction on $m \geq 0$, compare with (VIII.2.3) and Example (VIII.2.4)(b). $\square$

Now we define a countable family

$$\mathcal{FQ}_A = \{I(m, j)\}_{0 \leq j \leq 4,\, m \geq 0}$$

of pairwise non-isomorphic indecomposable A-modules $I(m, j)$ and we show that it is just a complete family of the indecomposable preinjective A-modules.

3.9. Table. Indecomposable preinjective representations of the four subspace quiver

Let $A = KQ$ be the path algebra (3.1) of the four subspace quiver Q. We define three series of indecomposable preinjective A-modules

(a) $I(m, 0)$, with $m \geq 0$,

(b) $I(2m+1, 1)$, $I(2m+1, 2)$, $I(2m+1, 3)$, $I(2m+1, 4)$, with $m \geq 1$,

(c) $I(2m, 1)$, $I(2m, 2)$, $I(2m, 3)$, $I(2m, 4)$, with $m \geq 0$,

viewed as K-linear representations of the four subspace quiver Q.

<u>**3.9(a)**</u> For each $m \geq 0$, we define

$$I(m, 0):$$

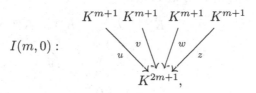

where, for each $[x_1, \ldots, x_{m+1}]^t \in K^{m+1}$, we set

$u([x_1, \ldots, x_{m+1}]^t) = [0, \ldots, 0, x_1, \ldots, x_{m+1}]^t$,

$v([x_1, \ldots, x_{m+1}]^t) = [x_1, \ldots, x_{m+1}, 0, \ldots, 0]^t$,

$w([x_1, \ldots, x_{m+1}]^t) = [x_1, \ldots, x_m, x_1, \ldots, x_m, x_{m+1}]^t$,

$z([x_1, \ldots, x_{m+1}]^t) = [x_1, \ldots, x_{m+1}, x_1, \ldots, x_m]^t$.

<u>**3.9(b)**</u> For each $m \geq 1$, we define

$$I(2m+1, 1):$$

$$I(2m+1, 2):$$

$$I(2m+1, 3):$$

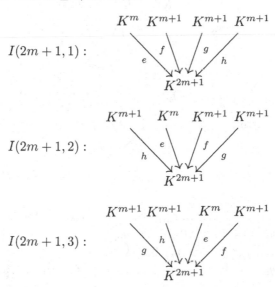

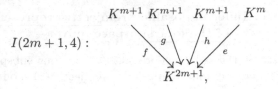

$$I(2m+1,4):$$

where
$$e([x_1,\ldots,x_m]^t) = [0,\ldots,0,x_1,\ldots,x_m]^t, \quad \text{for all } [x_1,\ldots,x_m]^t \in K^m,$$
$$f([y_1,\ldots,y_{m+1}]^t) = [y_1,\ldots,y_{m+1},0,\ldots,0]^t,$$
$$g[y_1,\ldots,y_{m+1}]^t) = [y_1,\ldots,y_{m+1},y_2,\ldots,y_m]^t,$$
$$h([y_1,\ldots,y_{m+1}]^t) = [y_1,\ldots,y_m,y_{m+1},y_1,\ldots,y_m]^t, \text{ for all } [y_1,\ldots,y_{m+1}]^t.$$

3.9(c) For each $m \geq 0$, we define

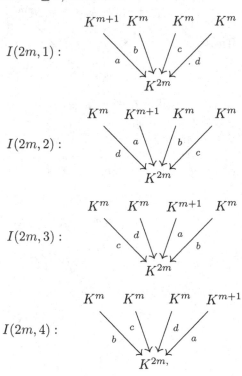

$$I(2m,1):$$

$$I(2m,2):$$

$$I(2m,3):$$

$$I(2m,4):$$

where

$$a([x_1,\ldots,x_{m+1}]^t) = [x_1,\ldots,x_m,x_2,\ldots,x_m,x_{m+1}]^t, \text{ for } [x_1,\ldots,x_{m+1}]^t,$$
$$b([y_1,\ldots,y_m]^t) = [0,\ldots,0,y_1,\ldots,y_m]^t,$$
$$c[y_1,\ldots,y_m]^t) = [y_1,\ldots,y_m,0,\ldots,0]^t,$$
$$d([y_1,\ldots,y_m]^t) = [y_1,\ldots,y_m,y_1,\ldots,y_m]^t, \text{ for } [y_1,\ldots,y_m]^t \in K^m.$$

Now we show that the family $\mathcal{F}\mathcal{Q}_A$ of modules of Table 3.9 is a complete set of pairwise non-isomorphic indecomposable preinjective A-modules.

3.10. Proposition. *Let $A = KQ$ be the path algebra* (3.1) *of the four subspace quiver Q and let*

$$\mathcal{F}\mathcal{Q}_A = \{I(m,j)\}_{0 \leq j \leq 4,\, m \geq 0}$$

be the family of A-modules listed in Table 3.9.

(a) *For each pair of integers $m \geq 0$ and $j \in \{0,1,2,3,4\}$, there is an isomorphism*

$$\tau^m I(j) \cong I(m,j)$$

of A-modules, where $I(j)$ is the injective A-module of (3.2) *corresponding to the vertex j of Q.*

(b) *For any $m \geq 0$ and $j \in \{0,1,2,3,4\}$, the module $I(m,j)$ is a preinjective brick and satisfies*

$$\partial_A(\mathbf{dim}\, I(m,j)) > 0 \quad \text{and} \quad q_A(\mathbf{dim}\, I(m,j)) = 1.$$

(c) *The modules $I(m,j)$ of the family $\mathcal{F}\mathcal{Q}_A$ are pairwise non-isomorphic.*

(d) *Up to isomorphism, every indecomposable preinjective A-module is of the form $I(m,j)$, where $m \geq 0$ and $j \in \{0,1,2,3,4\}$.*

Proof. It follows from (3.8)(a) that every preinjective indecomposable A-module is directing and has the form $\tau^m I(j)$, where $m \geq 0$ and $j \in \{0,1,2,3,4\}$. Hence, by (IX.3.1), the module $\tau^m I(j)$ is uniquely determined by its dimension vector. Because, by (3.8)(b),

$$\mathbf{dim}\, I(m,j) = \mathbf{dim}\, \tau^m I(j)$$

then it remains to check that the modules $I(m,j)$, with $m \geq 0$ and $j \in \{0,1,2,3,4\}$, are bricks.

We leave it as an exercise. We only remark that by a simple matrix calculation technique applied in the proof of (2.18) one shows that there is an isomorphism of algebras $\mathrm{End}\, I(m,j) \cong K$, for each $m \geq 0$ and $j \in \{0,1,2,3,4\}$. □

Our next aim in this section is to describe all indecomposable regular modules over the algebra $A = KQ$, where Q is the four subspace quiver.

We recall from (2.1) that the regular part $\mathcal{R}(A)$ of $\Gamma(\mathrm{mod}\, A)$ consists of the three stable tubes $\mathcal{T}_\infty^Q$, $\mathcal{T}_0^Q$, and $\mathcal{T}_1^Q$ of rank 2, and the family of homogeneous tubes $\mathcal{T}_\lambda^Q$, with $\lambda \in K \setminus \{0,1\}$.

Now we present two lists of indecomposable K-linear representations of the four subspace quiver Q. We show in (3.13) below that they form complete sets of the simple regular representations and regular representations of regular length two of Q, up to isomorphism. We do it by applying the tilting reduction to the canonical algebra $C(2, 2, 2)$, like in the proof of (2.8).

3.11. Table. Simple regular representations of the

four subspace quiver

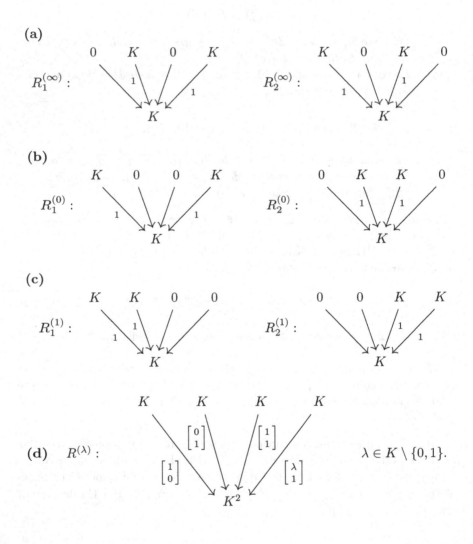

(a)

(b)

(c)

(d) $R^{(\lambda)}$: $\lambda \in K \setminus \{0, 1\}$.

3.12. Table. Regular representations of regular length two

of the four subspace quiver

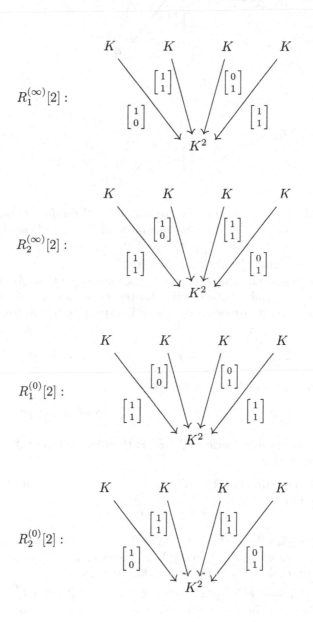

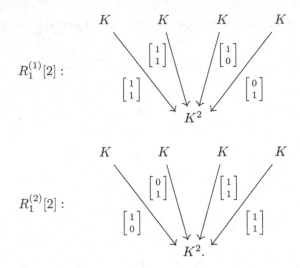

$R_1^{(1)}[2]$:

$R_1^{(2)}[2]$:

We recall that $A = KQ$ is the path algebra of the four subspace quiver Q. It follows from (2.1) that the regular part $\mathcal{R}(A)$ of $\Gamma(\text{mod}\,A)$ consists of the three stable tubes $\mathcal{T}_\infty^Q$, $\mathcal{T}_0^Q$, and $\mathcal{T}_1^Q$ of rank 2, and the family of homogeneous tubes $\mathcal{T}_\lambda^Q$, with $\lambda \in K \setminus \{0,1\}$.

We also recall from Chapter XI that the category add $\mathcal{R}(A)$ of all regular A-modules is a serial (abelian) full subcategory of mod A and, for any indecomposable regular A-module X, there is a path of irreducible monomorphisms

$$R = R[1] \longrightarrow R[2] \longrightarrow \cdots \longrightarrow R[l-1] \longrightarrow R[l] = X$$

and a path of irreducible epimorphisms

$$X = [l]R' \longrightarrow [l-1]R' \longrightarrow \cdots \longrightarrow [2]R' \longrightarrow [1]R' = R',$$

where R is the regular socle of X, R' is the regular top of X, and l is the regular length of X.

3.13. Theorem. *Let $A = KQ$ be the path algebra of the four subspace quiver Q and let*

$$\boldsymbol{\mathcal{T}}^Q = \{\mathcal{T}_\lambda^Q\}_{\lambda \in \mathbb{P}_1(K)}$$

be the $\mathbb{P}_1(K)$-family of standard stable tubes, in the regular part $\mathcal{R}(A)$ of $\Gamma(\text{mod}\,A)$, separating $\mathcal{P}(A)$ from $\mathcal{Q}(A)$, as in (2.1). They are of the form presented in Table 3.17, at the end of this section.

(a) *The family $\boldsymbol{\mathcal{T}}^Q$ contains three tubes $\mathcal{T}_\infty^Q$, $\mathcal{T}_0^Q$, and $\mathcal{T}_1^Q$ of rank two, and the remaining tubes $\mathcal{T}_\lambda^Q$ of $\boldsymbol{\mathcal{T}}^Q$ are homogeneous.*

(b) *For each $\mu \in \{0, 1, \infty\}$, the A-modules $R_1^{(\mu)}$ and $R_2^{(\mu)}$ presented in (3.11) are simple regular and form the mouth of the tube $\mathcal{T}_\mu^Q$ of rank two.*

(c) *For each $\mu \in \{0, 1, \infty\}$, the A-modules $R_1^{(\mu)}[2]$ and $R_2^{(\mu)}[2]$ presented in (3.12) are the unique indecomposable regular modules of regular length 2 in the tube $\mathcal{T}_\mu^Q$ of rank two.*

(d) *The A-modules $R_i^{(\mu)}[2]$, with $\mu \in \{0, 1, \infty\}$, and the A-modules $R^{(\lambda)}$, with $\lambda \in K \setminus \{0, 1\}$, form a complete set of pairwise non-isomorphic indecomposable A-modules of dimension vector*

$$\mathbf{h}_A = {}^{1\,1}_{\;\;2}{}^{1\,1} = [2, 1, 1, 1, 1]^t.$$

(e) *For each $\lambda \in K \setminus \{0, 1\}$, the module $R^{(\lambda)}$ is simple regular and forms the mouth of the homogeneous tube $\mathcal{T}_\lambda^Q$.*

Proof. By applying the tilting technique, like in the proof of (2.9), we reduce the problem to regular modules over the canonical algebra

$$C = C(2, 2, 2) = K\Delta(2, 2, 2)/I(2, 2, 2)$$

of Euclidean type $\Delta(\widetilde{\mathbb{D}}_4)$, defined by the quiver

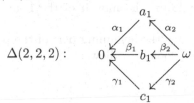

$$\Delta(2, 2, 2):$$

and the ideal $I(2, 2, 2)$ in the path algebra $K\Delta(2, 2, 2)$ of $\Delta(2, 2, 2)$ generated by the zero-relation $\alpha_2\alpha_1 + \beta_2\beta_1 + \gamma_2\gamma_1$, see the notation of (XII.1.2).

We recall from (XII.2.12) that the regular part $\mathcal{R}(C)$ of $\Gamma(\mathrm{mod}\,C)$ consists of the pairwise orthogonal stable tubes $\mathcal{T}_\lambda^C$ of the $\mathbb{P}_1(K)$-family

$$\mathcal{T}^C = \{\mathcal{T}_\lambda^C\}_{\lambda \in \mathbb{P}_1(K)}$$

described in (XII.2.11) and (XII.2.12). The tubes $\mathcal{T}_\infty^C$, $\mathcal{T}_0^C$, and $\mathcal{T}_1^C$ are of rank 2, and the remaining tubes $\mathcal{T}_\lambda^C$ are of rank 1. The following C-modules

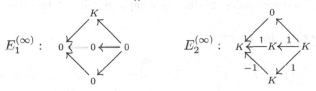

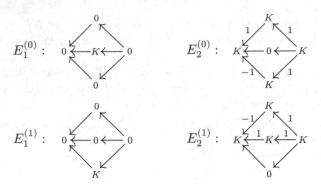

are simple regular and form the mouth of the stable tubes $\mathcal{T}_\infty^C$, $\mathcal{T}_0^C$, and $\mathcal{T}_1^C$, respectively. For each $\lambda \in K \setminus \{0,1\}$, the C-module

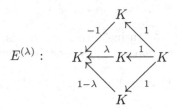

is simple regular and forms the mouth of the homogeneous tube $\mathcal{T}_\lambda^C$ of $\Gamma(\mathrm{mod}\, C)$.

Moreover, by (XII.1.5), the beginning part of the postprojective component $\mathcal{P}(C)$ of $\Gamma(\mathrm{mod}\, C)$ looks as follows:

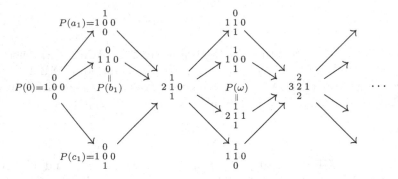

where the indecomposable modules are represented by their dimension vectors.

Observe that the modules

$$T_0 = 2\overset{1}{\underset{1}{1}}0, \quad T_1 = 1\overset{0}{\underset{1}{1}}0, \quad T_2 = 1\overset{1}{\underset{1}{0}}0, \quad T_3 = 2\overset{1}{\underset{1}{1}}1, \quad T_4 = 1\overset{1}{\underset{0}{1}}0$$

form a section in $\mathcal{P}(A)$ of the form

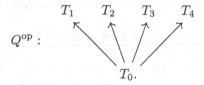

$$Q^{\mathrm{op}}:$$

It follows that $T = T_0 \oplus T_1 \oplus T_2 \oplus T_3 \oplus T_4$ is a postprojective tilting C-module,

$$A = KQ \cong \operatorname{End} T_C$$

and, for each $\lambda \in \mathbb{P}_1(K)$, the image

$$\mathcal{T}_\lambda^Q = \operatorname{Hom}_C(T, \mathcal{T}_\lambda^C)$$

of the tube $\mathcal{T}_\lambda^C$ under the functor $\operatorname{Hom}_C(T, -)$ is a stable tube in $\mathcal{R}(A)$ and the rank of $\mathcal{T}_\lambda^Q$ equals the rank of $\mathcal{T}_\lambda^C$. Moreover, by (2.9), each regular component of $\Gamma(\operatorname{mod} A)$ is one of the tubes $\mathcal{T}_\lambda^Q$, and the $\mathbb{P}_1(K)$-family

$$\mathbf{T}^Q = \{\mathcal{T}_\lambda^Q\}_{\lambda \in \mathbb{P}_1(K)}$$

separates $\mathcal{P}(A)$ from $\mathcal{Q}(A)$.

We have seen in the proof of (2.9) that, for each indecomposable module $X \in \mathcal{R}(C)$, the coordinates of the dimension vector of the A-module $\operatorname{Hom}_C(T, X)$ can be calculated by applying the formula

$$\mathbf{dim} \operatorname{Hom}_C(T, X)_i = \dim_K \operatorname{Hom}_C(T_i, X),$$

for $i \in \{0, 1, 2, 3, 4\}$. Hence, we easily get

$$\mathbf{dim} \operatorname{Hom}_C(T, E_1^{(\infty)}) = {}^{0\ 1}_{\ \ 1}{}^{0\ 1}, \qquad \mathbf{dim} \operatorname{Hom}_C(T, E_2^{(\infty)}) = {}^{1\ 0}_{\ \ 1}{}^{1\ 0},$$

$$\mathbf{dim} \operatorname{Hom}_C(T, E_1^{(0)}) = {}^{1\ 0}_{\ \ 1}{}^{0\ 1}, \qquad \mathbf{dim} \operatorname{Hom}_C(T, E_2^{(0)}) = {}^{0\ 1}_{\ \ 1}{}^{1\ 0},$$

$$\mathbf{dim} \operatorname{Hom}_C(T, E_1^{(1)}) = {}^{1\ 1}_{\ \ 1}{}^{0\ 0}, \qquad \mathbf{dim} \operatorname{Hom}_C(T, E_2^{(1)}) = {}^{0\ 0}_{\ \ 1}{}^{1\ 1},$$

$$\mathbf{dim} \operatorname{Hom}_C(T, E^{(\lambda)}) = {}^{1\ 1}_{\ \ 2}{}^{1\ 1} = \mathbf{h}_A.$$

It follows that

$$\mathbf{dim} \operatorname{Hom}_C(T, E_i^{(\mu)}) = \mathbf{dim} R_i^{(\mu)}, \quad \text{for any } \mu \in \{0, 1, \infty\} \text{ and } i \in \{1, 2\}.$$

Now we are in a position to complete the proof of the theorem.

(a) It follows from (XI.3.3) and (XII.4.2) that the simple regular modules from non-homogeneous tubes of $\Gamma(\operatorname{mod} A)$ are uniquely determined by their dimension vectors. Because $\mathbf{dim} \operatorname{Hom}_C(T, E_i^{(\mu)}) = \mathbf{dim} R_i^{(\mu)}$, for any

$\mu \in \{0, 1, \infty\}$ and $i \in \{1, 2\}$, and the modules of Table 3.11 are bricks, they are indecomposable and, hence, there is an isomorphism

$$R_i^{(\mu)} \cong \operatorname{Hom}_C(T, E_i^{(\mu)})$$

of A-modules, for all $\mu \in \{0, 1, \infty\}$ and $i \in \{1, 2\}$. Then (a) follows from (XII.2.12).

(b) Observe that $\operatorname{Hom}_A(R_i^{(\mu)}, R_i^{(\mu)}[2]) \neq 0$, for any $\mu \in \{0, 1, \infty\}$ and $i \in \{1, 2\}$. Moreover, because

$$\mathbf{dim}\, R_i^{(\mu)}[2] = \mathbf{h}_A$$

then, by (XII.2.13), the module $R_i^{(\mu)}[2]$ is indecomposable regular of regular length 2. Then our claim follows from the fact that different tubes in $\Gamma(\operatorname{mod} A)$ are orthogonal.

(c) Let

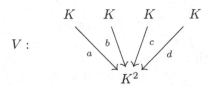

be an indecomposable representation of Q of dimension vector $\mathbf{h}_A = {}^{1\,1}{}_2{}^{1\,1}$. Because the module V is indecomposable, at least 3 of the images $\operatorname{Im} a$, $\operatorname{Im} b$, $\operatorname{Im} c$, $\operatorname{Im} d$ are pairwise different.

If two of them coincide then a simple analysis shows that V is isomorphic to one of the representations $R_i^{(\mu)}[2]$ for some $\mu \in \{0, 1, \infty\}$ and $i \in \{1, 2\}$.

If all four of them are pairwise different then we easily deduce that V is isomorphic to one of the modules $R^{(\lambda)}$, with $\lambda \in K \setminus \{0, 1\}$.

The second claim of (c) follows from the fact that the simple regular modules in homogeneous tubes have dimension vector $\mathbf{h}_A$, see (XII.2.12). This finishes the proof of the theorem.   □

Our final aim in this section is to present a complete description of all indecomposable regular modules over the path algebra $A = KQ$, viewed as K-linear representations of the four subspace quiver Q.

In the tables below we present a list of pairwise non-isomorphic indecomposable regular representations of Q. Next, we prove in (3.15) and (3.16) below that every indecomposable regular representation of Q appears in the list, up to isomorphism.

3.14. Table. Regular indecomposable representations

of the four subspace quiver

Assume that Q is the four subspace quiver and $A = KQ$ is the algebra (3.1). For each integer $m \geq 1$, we consider the following family of indecomposable regular K-linear representations of Q.

(a) Regular representations in the tube $\mathcal{T}_\infty^Q$

$R_1^{(\infty)}[2m-1]:$

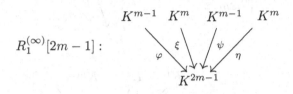

$R_2^{(\infty)}[2m-1]:$

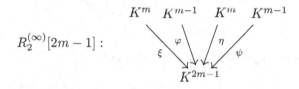

$R_1^{(\infty)}[2m]:$

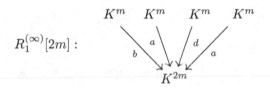

$R_2^{(\infty)}[2m]:$

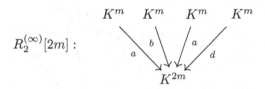

(b) Regular representations in the tube $\mathcal{T}_0^Q$

$R_1^{(0)}[2m-1]:$

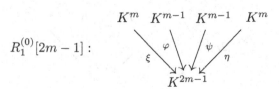

$R_2^{(0)}[2m-1]:$

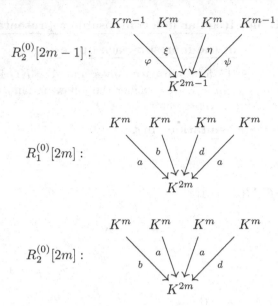

(c) **Regular representations in the tube** $\mathcal{T}_1^Q$

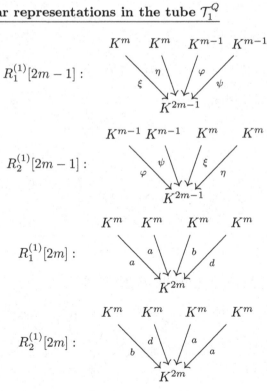

(d) Regular representations in the tube $\mathcal{T}_\lambda^Q$, **with** $\lambda \in K \setminus \{0,1\}$

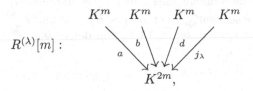

$$R^{(\lambda)}[m] :$$

where

$$\varphi([x_1,\dots,x_{m-1}]^t) = [0,\dots,0,x_1,\dots,x_{m-1}]^t,$$
$$\psi([x_1,\dots,x_{m-1}]^t) = [x_1,\dots,x_{m-1},0,\dots,0]^t, \text{ for } [x_1,\dots,x_{m-1}] \in K^{m-1},$$
$$\xi([y_1,\dots,y_m]^t) = [y_1,\dots,y_m,0,\dots,0]^t,$$
$$\eta([y_1,\dots,y_m]^t) = [y_1,\dots,y_{m-1},y_1,\dots,y_m]^t,$$
$$a([y_1,\dots,y_m]^t) = [y_1,\dots,y_m,0,\dots,0]^t,$$
$$b([y_1,\dots,y_m]^t) = [0,\dots,0,y_1,\dots,y_m]^t,$$
$$d([y_1,\dots,y_m]^t) = [y_1,\dots,y_m,y_1,\dots,y_m]^t,$$
$$j_\lambda([y_1,\dots,y_m]^t) = [\lambda y_1, \lambda y_2 + y_1, \dots, \lambda y_{m-1} + y_m, \lambda y_m, y_1, \dots, y_m]^t,$$
$$\text{for } [y_1,\dots,y_m]^t \in K^m$$

Now we are able to describe the indecomposable regular modules in non-homogeneous tubes of $\Gamma(\text{mod } A)$, up to isomorphism.

3.15. Theorem. *Let* $A = KQ$ *be the path algebra of the four subspace quiver Q, let*

$$\mathcal{T}^Q = \{\mathcal{T}_\lambda^Q\}_{\lambda \in \mathbb{P}_1(K)}$$

be the $\mathbb{P}_1(K)$-family of standard stable tubes as in (2.1), and let $\mu \in \{0,1,\infty\}$.

(a) *The A-modules $R_1^{(\mu)}[m]$ and $R_2^{(\mu)}[m]$, with $m \geq 1$, presented in (3.14) form a complete family of pairwise non-isomorphic indecomposable modules in the tube $\mathcal{T}_\mu^Q$. Moreover, m is the regular length of the modules $R_1^{(\mu)}[m]$ and $R_2^{(\mu)}[m]$.*

(b) *For each $i \in \{1,2\}$, the module $R_i^{(\mu)}[1]$ is the regular socle of all modules $R_i^{(\mu)}[m]$, with $m \geq 1$.*

(c) *For m odd and $i \in \{1,2\}$, the module $R_i^{(\mu)}$ is the regular top of $R_i^{(\mu)}[m]$, that is, $R_i^{(\mu)}[m] \cong [m]R_i^{(\mu)}$.*

(d) *For m even, the module $R_2^{(\mu)}$ is the regular top of the module $R_1^{(\mu)}[m]$ and $R_1^{(\mu)}$ is the regular top of the module $R_2^{(\mu)}[m]$, that is, there are isomorphisms $R_2^{(\mu)}[m] \cong [m]R_1^{(\mu)}$ and $R_1^{(\mu)}[m] \cong [m]R_2^{(\mu)}$ of A-modules.*

Proof. We assume that $\mu \in \{0, 1, \infty\}$. A direct calculation shows that, for each $m \geq 1$, the endomorphism rings of the modules $R_1^{(\mu)}[m]$ and $R_2^{(\mu)}[m]$ are local and, consequently, the modules $R_1^{(\mu)}[m]$ and $R_2^{(\mu)}[m]$ are indecomposable. Moreover, because, by (3.4), the defect $\partial_A : \mathbb{Z}^5 \longrightarrow \mathbb{Z}$ of A is given by the formula

$$\partial_A(\mathbf{x}) = -2x_0 + x_1 + x_2 + x_3 + x_4,$$

for all $\mathbf{x} = [x_0, x_1, x_2, x_3, x_4]^t \in \mathbb{Z}^5$, then

$$\partial_A(\dim R_1^{(\mu)}[m]) = 0 \quad \text{and} \quad \partial_A(\dim R_2^{(\mu)}[m]) = 0,$$

for all $m \geq 1$, and, according to (XI.2.8), the modules $R_1^{(\mu)}[m]$ and $R_2^{(\mu)}[m]$ are regular.

A direct calculation shows also that $\operatorname{Hom}_A(R_i^{(\mu)}, R_i^{(\mu)}[m]) \neq 0$ and, in view of (3.13), the modules $R_1^{(\mu)}[m]$ and $R_2^{(\mu)}[m]$ belong to the tube $\mathcal{T}_\mu^Q$, and $R_i^{(\mu)} = R_i^{(\mu)}[1]$ is the simple regular socle of $R_i^{(\mu)}[m]$, for any $i \in \{1, 2\}$ and all $m \geq 1$. This proves the statements (a) and (b). The statements (c) and (d) follow immediately from (a), (b), and the fact that $\mathcal{T}_\mu^Q$ is a stable tube of rank 2. $\qquad\square$

We finish the section with a description of the indecomposable regular modules lying in any homogeneous tube of $\mathcal{R}(A)$.

3.16. Theorem. *Let $A = KQ$ be the path algebra of the four subspace quiver Q and let*

$$\mathcal{T}^Q = \{\mathcal{T}_\lambda^Q\}_{\lambda \in \mathbb{P}_1(K)}$$

be the $\mathbb{P}_1(K)$-family of standard stable tubes as in (2.1).

(a) *For each $\lambda \in K \setminus \{0, 1\} = \mathbb{P}_1(K) \setminus \{0, 1, \infty\}$, the A-modules $R^{(\lambda)}[m]$, with $m \geq 1$, are regular and form a complete family of pairwise nonisomorphic indecomposable modules in the homogeneous tube $\mathcal{T}_\lambda^Q$.*

(b) *The regular length of the module $R^{(\lambda)}[m]$ equals m.*

Proof. Assume that $\lambda \in K \setminus \{0, 1\}$. A direct calculation shows that, for each $m \geq 1$, the endomorphism ring of the module $R^{(\lambda)}[m]$ is local, and that $\operatorname{Hom}_A(R^{(\lambda)}, R^{(\lambda)}[m]) \neq 0$. It is easy to see that, for any $m \geq 2$, there exists an exact sequence

$$0 \longrightarrow R^{(\lambda)} \longrightarrow R^{(\lambda)}[m] \longrightarrow R^\lambda[m-1] \longrightarrow 0.$$

Because, by (3.13), $R^{(\lambda)}$ is the (unique) simple regular module lying in the homogeneous tube $\mathcal{T}_\lambda^Q$ then, according to (X.2.2) and (X.2.6), the tube $\mathcal{T}_\lambda^Q$ consists of the modules $R^{(\lambda)}[m]$, with $m \geq 1$, and the regular length of $R^{(\lambda)}[m]$ equals m. This completes the proof. $\qquad\square$

3.17. Table. The non-homogeneous tubes of regular representations of the four subspace quiver

Let $A = KQ$ be the path algebra of the four subspace quiver Q and let

$$\mathcal{T}^Q = \{\mathcal{T}_\lambda^Q\}_{\lambda \in \mathbb{P}_1(K)}$$

be the $\mathbb{P}_1(K)$-family of standard stable tubes, in the regular part $\mathcal{R}(A)$ of $\Gamma(\mathrm{mod}\,A)$, separating $\mathcal{P}(A)$ from $\mathcal{Q}(A)$, see (2.1). It follows from (3.13) that the three tubes $\mathcal{T}_\infty^Q$, $\mathcal{T}_0^Q$, and $\mathcal{T}_1^Q$ are of rank two, and the remaining tubes $\mathcal{T}_\lambda^Q$ in the family $\mathcal{T}^Q$ are homogeneous. We recall from (3.13) that

- the modules $R_1^{(\infty)}$ and $R_2^{(\infty)}$ form the mouth of the tube $\mathcal{T}_\infty^Q$ and $\mathbf{dim}\,R_1^{(\infty)} = {}^{0\,1}_{\ \ 1}{}^{0\,1}$ and $\mathbf{dim}\,R_2^{(\infty)} = {}^{1\,0}_{\ \ 1}{}^{1\,0}$,
- the modules $R_1^{(0)}$ and $R_2^{(0)}$ form the mouth of the tube $\mathcal{T}_0^Q$ and $\mathbf{dim}\,R_1^{(0)} = {}^{1\,0}_{\ \ 1}{}^{0\,1}$ and $\mathbf{dim}\,R_2^{(0)} = {}^{0\,1}_{\ \ 1}{}^{1\,0}$, and
- the modules $R_1^{(1)}$ and $R_2^{(1)}$ form the mouth of the tube $\mathcal{T}_1^Q$ and $\mathbf{dim}\,R_1^{(1)} = {}^{1\,1}_{\ \ 1}{}^{0\,0}$ and $\mathbf{dim}\,R_2^{(1)} = {}^{0\,0}_{\ \ 1}{}^{1\,1}$.

If we write $\mathbf{dim}\,X$ instead of the A-module X, then the non-homogeneous tubes $\mathcal{T}_\infty^Q$, $\mathcal{T}_0^Q$, and $\mathcal{T}_1^Q$ have the following forms, compare with [235, p.348].

$$\mathcal{T}_\infty^Q \qquad\qquad\qquad \mathcal{T}_0^Q$$

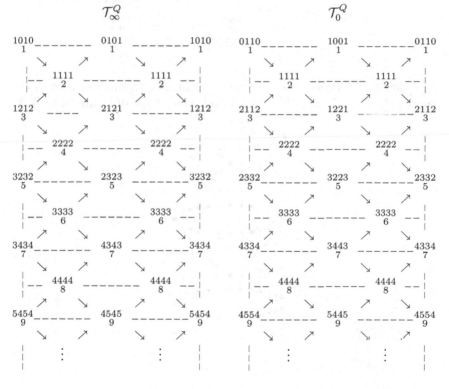

$$\mathcal{T}_1^Q$$

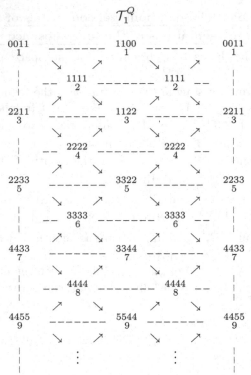

where the vertical broken lines should be identified. In this way each of the components $\mathcal{T}_\infty^Q$, $\mathcal{T}_0^Q$, and $\mathcal{T}_1^Q$ has the following shape of a tube of rank 2:

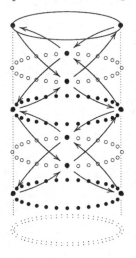

XIII.4. Exercises

1. Assume that $\Delta = \Delta(\widetilde{\mathbb{D}}_m)$, with $m \geq 4$, is the canonically oriented Euclidean quiver shown in Table 1.1. Let

$$\Phi_A : \mathbb{Z}^{m+1} \longrightarrow \mathbb{Z}^{m+1}$$

be the Coxeter transformation of the path algebra $A = K\Delta$ of Δ presented in (1.2), and let $\mathbf{h}_\Delta = {}^1_1 2 \ldots 2^1_1$ be the positive generator of rad q_A.

  (a) Prove that, for $m = 6, 8, 10$, the defect $\partial_A : \mathbb{Z}^{m+1} \longrightarrow \mathbb{Z}$ of A is given by the formula

$$\partial_A(\mathbf{x}) = x_{m+1} + x_m - x_2 - x_1,$$

    for any vector $\mathbf{x} \in \mathbb{Z}^{m+1}$, by showing that

    $\Phi_A^{m-2}(\mathbf{x}) - \mathbf{x} = (x_{m+1} + x_m - x_2 - x_1) \cdot \mathbf{h}_\Delta$, and
    $\Phi_A^j(\mathbf{x}) - \mathbf{x} \neq (x_{m+1} + x_m - x_2 - x_1) \cdot \mathbf{h}_\Delta$,

    for $j = 1, \ldots, m-3$.

  (b) Prove that, for $m = 5, 7, 9$, the defect ∂_A of A is given by the formula $\partial_A(\mathbf{x}) = 2(x_{m+1}+x_m-x_2-x_1)$, for any vector $\mathbf{x} \in \mathbb{Z}^{m+1}$, by showing that

    $\Phi_A^{2(m-2)}(\mathbf{x}) - \mathbf{x} = 2(x_{m+1} + x_m - x_2 - x_1) \cdot \mathbf{h}_\Delta$, and
    $\Phi_A^j(\mathbf{x}) - \mathbf{x} \neq 2(x_{m+1} + x_m - x_2 - x_1) \cdot \mathbf{h}_\Delta$,

    for $j = 1, \ldots, 2m-3$.

2. Assume that $\Delta = \Delta(\widetilde{\mathbb{D}}_m)$, $m \geq 4$, and $A = K\Delta$ is the path algebra of Δ. Let $F_1^{(\infty)}$, $F_2^{(\infty)}$, $F_1^{(0)}$, $F_2^{(0)}$, $F_1^{(1)}$, $F_2^{(1)}$, $\ldots, F_{m-3}^{(1)}, F_{m-2}^{(1)}$ and $F^{(\lambda)}$, with $\lambda \in K \setminus \{0, 1\}$, be the A-modules presented in Table 2.6. Show that

  $\operatorname{Hom}_A(F_i^{(s)}, F_j^{(t)}) = 0$, for all $i \neq j$ and $s, t \in \{0, 1, \infty\}$,

  $\operatorname{Hom}_A(F^{(\lambda)}, F_j^{(t)}) = 0$, and

  $\operatorname{Hom}_A(F_j^{(t)}, F^{(\lambda)}) = 0$, for all $j \leq m-3$, $t \in \{0, 1, \infty\}$ and $\lambda \in \mathbb{P}_1(K) \setminus \{0, 1, \infty\}$.

Hint: Follow the proof of (2.8).

3. Let $A = K\Delta$ be the path algebra of the canonically oriented Euclidean quiver $\Delta = \Delta(\widetilde{\mathbb{E}}_6)$ presented in Table 1.1.

  (a) Show that the indecomposable A-modules $F_\infty^{(2)}$ and $F^{(\lambda)}$, for $\lambda \in K \setminus \{0, 1\}$, shown in Table 2.12 are simple regular.

  (b) Show that the A-modules listed in Table 2.12 are pairwise orthogonal.

Hint: Apply (X.2.12) and the defect ∂_A of A computed in (1.3).

4. Let Δ be one of the canonically oriented Euclidean quivers $\Delta(\widetilde{\mathbb{E}}_7)$ and $\Delta(\widetilde{\mathbb{E}}_8)$, presented in Table 1.1, and let $A = K\Delta$ be the path algebra of Δ. Prove that the indecomposable A-modules shown in Table 2.16 and 2.20, respectively, are pairwise orthogonal bricks and are simple regular.

5. Let $Q = (Q_0, Q_1)$ be an acyclic Euclidean quiver. Let $a \in Q_0$ be a sink of Q and let $Q' = (Q'_0, Q'_1) = \sigma_a Q$ be the reflection of Q at the vertex a, in the sense of Section VII.5. Let $A = KQ$ and $A' = KQ'$ be the path algebras of Q and Q', respectively, and make the identifications $\text{mod } A = \text{rep}_K(Q)$ and $\text{mod } A' = \text{rep}_K(Q')$.

Finally, let $S_a^+ : \text{mod } A \longrightarrow \text{mod } A'$ be the reflection functor at the sink a of Q, let $S_a^- : \text{mod } A' \longrightarrow \text{mod } A$ be the reflection functor at the source vertex a of Q', and let $s_a : \mathbb{Z}^n \longrightarrow \mathbb{Z}^n$ be the reflection homomorphism defined in Section VII.4, page 270, where $n = |Q_0| = |Q'_0|$.

(a) Show that the pair of reflection functors $\text{mod } A \underset{S_a^-}{\overset{S_a^+}{\rightleftarrows}} \text{mod } A'$ restricts to the equivalences

$$\text{add } \mathcal{R}(A) \underset{S_a^-}{\overset{S_a^+}{\rightleftarrows}} \text{add } \mathcal{R}(A')$$

between the categories of regular modules, and the functors are inverse to each other.

(b) Show that the equalities

$$\mathbf{dim}\, S_a^+(M) = s_a(\mathbf{dim}\, M) \quad \text{and} \quad \mathbf{dim}\, S_a^-(M') = s_a(\mathbf{dim}\, M')$$

hold, for any regular indecomposable A-module M and for any regular indecomposable A'-module M'.

6. Let A be the path algebra of the quiver

of the Euclidean type $\widetilde{\mathbb{D}}_4$. Describe the dimension vectors $\mathbf{dim}\, M \in \mathbb{Z}^5$ of all indecomposable regular A-modules M.

Hint: See Section 3 and Exercise 5.

7. Let A be the path algebra of the quiver

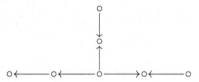

of the Euclidean type $\widetilde{\mathbb{D}}_4$, obtained for the quiver of Exercise 6 by a reflection. Describe the dimension vectors $\mathbf{dim}\, M \in \mathbb{Z}^5$ of all indecomposable regular A-modules M.

Hint: Apply the Exercises 5 and 6.

8. Let A be the path algebra of the quiver

$$
Q: \quad
\begin{array}{c}
\overset{2}{\circ} \longleftarrow \overset{3}{\circ} \\
\nearrow \qquad \searrow \\
1\circ \qquad\qquad \circ 4 \\
\nwarrow \qquad \nearrow \\
\underset{6}{\circ} \longleftarrow \underset{5}{\circ}
\end{array}
$$

of the Euclidean type $\widetilde{\mathbb{A}}_5$. Describe the dimension vectors $\mathbf{dim}\, M \in \mathbb{Z}^6$ of all indecomposable regular A-modules M.

9. Let A be the path algebra of the quiver

of the Euclidean type $\widetilde{\mathbb{D}}_8$. Describe the dimension vectors $\mathbf{dim}\, M \in \mathbb{Z}^9$ of all indecomposable regular A-modules M.

10. Let A be the path algebra of the quiver

of the Euclidean type $\widetilde{\mathbb{E}}_6$. Describe the dimension vectors $\mathbf{dim}\, M \in \mathbb{Z}^7$ of all indecomposable regular A-modules M.

11. Let A be the path algebra of the quiver

of the Euclidean type $\widetilde{\mathbb{E}}_7$. Describe the dimension vectors $\mathbf{dim}\, M \in \mathbb{Z}^8$ of all indecomposable regular A-modules M.

12. Let A be the path algebra of the quiver

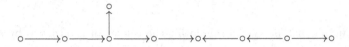

of the Euclidean type $\widetilde{\mathbb{E}}_8$. Describe the dimension vectors $\mathbf{dim}\, M \in \mathbb{Z}^9$ of all indecomposable regular A-modules M.

13. Under the notation of (2.9), show that

(i) the diagram (2.11) is commutative,

(ii) the functor

$$\mathrm{Hom}_C(T, -) : \mathrm{add}\,\mathcal{R}(C) \longrightarrow \mathrm{add}\,\mathcal{R}(C)$$

in the diagram (2.11) carries a tube $\mathcal{T}_\lambda$ of rank r_λ to a tube $\mathcal{T}'_\lambda$ of the same rank r_λ in such a way that the mouth modules of the tube $\mathcal{T}_\lambda$ are carried to the mouth modules of the tube $\mathcal{T}'_\lambda$,

(iii) if $\mathbf{h}_C$ is the positive generator of $\mathrm{rad}\,q_C$, $\mathbf{h}_A$ is the positive generator of $\mathrm{rad}\,q_A$, and M is an indecomposable C-module such that $\mathbf{dim}\, M = m \cdot \mathbf{h}_C$, where $m \geq 1$, then M is regular and $\mathbf{dim}\,\mathrm{Hom}_C(T, M) = m \cdot \mathbf{h}_A$.

Chapter XIV

Minimal representation-infinite algebras

The aim of this chapter is to prove the following criterion for the infinite representation type of algebras, presented in Section 3.

Let A be a basic connected K-algebra (over an algebraically closed field K) such that the Auslander–Reiten quiver $\Gamma(\operatorname{mod} A)$ of A admits a post-projective component. Then A is representation-infinite if and only if there is an idempotent $e \neq 1$ of A such that A/AeA is either the path algebra of the **enlarged Kronecker quiver**

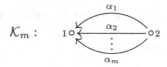

$$\mathcal{K}_m: \quad 1 \circ \mathrel{\mathop{\rightleftarrows}^{\alpha_1}_{}} \circ 2$$

with $m \geq 3$ arrows, or else A/AeA is a concealed algebra of Euclidean type.

In the proof of the criterion we essentially use a characterisation, given in Section 2, of the algebras A that admit a postprojective component in the Auslander–Reiten quiver $\Gamma(\operatorname{mod} A)$ and that are minimal representation-infinite, that is, the algebras A are representation-infinite, and the quotient algebras A/AeA are representation-finite, for any non-zero idempotent $e \neq 1$ of A.

The characterisation is due to Happel–Vossieck [112], and asserts that an algebra A is minimal representation-infinite and admits a postprojective component in $\Gamma(\operatorname{mod} A)$ if and only if A is either the path algebra of an enlarged Kronecker quiver $\mathcal{K}_m$ with at least three arrows, or A is a concealed algebra of Euclidean type.

In the proof, we apply a theorem of Ovsienko [175] on critical integral quadratic forms $q : \mathbb{Z}^n \longrightarrow \mathbb{Z}$, proved in Section 1.

To make the criterion for the infinite representation type of algebras more effective from a practical point of view, we present in Section 4 (with an outline of proof) a complete classification of all basic, connected, concealed algebras

$$B \cong KQ_B/I_B$$

of Euclidean type in terms of their bound quivers (Q_B, I_B). The classification, given independently by Happel–Vossieck [112] and Bongartz [29], is presented in two steps as follows.

(i) We present in Table 4.5 a list of 149 frames $\mathcal{F}r_1, \ldots, \mathcal{F}r_{149}$ of the bound quivers (Q_B, I_B) of a class of concealed algebras B of Euclidean type.

(ii) We define in 4.1 three types of admissible operations $(\mathbf{Op_1})$, $(\mathbf{Op_2})$, and $(\mathbf{Op_3})$ on frames.

Then a bound quiver (Q_B, I_B) of any basic connected concealed algebra B of Euclidean type can be constructed from a frame of Table 4.5 by applying successively a finite number of admissible operations $(\mathbf{Op_1})$, $(\mathbf{Op_2})$, and $(\mathbf{Op_3})$, see (4.3).

We also present a table of the upper bounds of the numbers of the isomorphism classes of the concealed algebras for the Euclidean types $\widetilde{\mathbb{A}}_m$, $\widetilde{\mathbb{D}}_m$, $\widetilde{\mathbb{E}}_6$, $\widetilde{\mathbb{E}}_7$, and $\widetilde{\mathbb{E}}_8$, respectively.

XIV.1. Critical integral quadratic forms

The objective of this section is to prove a result due to Ovsienko [175] on critical integral quadratic forms. Throughout this section we assume that

$$q : \mathbb{Z}^n \longrightarrow \mathbb{Z}$$

is an integral quadratic form in n indeterminates $x_1, \ldots, x_n$ defined by the formula

(1.1) $$q(x_1, \ldots, x_n) = x_1^2 + \ldots + x_n^2 + \sum_{i<j} a_{ij} x_i x_j,$$

where $a_{ij} \in \mathbb{Z}$, for all $i, j \in \{1, \ldots, n\}$, such that $i < j$, see (VII.3.1).

We recall that a vector $\mathbf{x} = [x_1 \, x_2 \, \ldots \, x_n]^t \in \mathbb{Z}^n$ is defined to be **positive** (and is denoted by $\mathbf{x} > 0$), if $\mathbf{x} \neq 0$ and $x_1 \geq 0, x_2 \geq 0, \ldots, x_n \geq 0$. The vector $\mathbf{x} \in \mathbb{Z}^n$ is **sincere**, if $x_1 \neq 0$, $x_2 \neq 0$, $\ldots$, $x_n \neq 0$.

An integral quadratic form $q : \mathbb{Z}^n \longrightarrow \mathbb{Z}$ is defined to be

- **weakly positive**, if $q(\mathbf{x}) > 0$, for each positive vector $\mathbf{x} \in \mathbb{Z}^n$,
- **weakly nonnegative**, if $q(\mathbf{x}) \geq 0$, for each positive vector $\mathbf{x} \in \mathbb{Z}^n$, and
- **positive semidefinite**, if $q(\mathbf{x}) \geq 0$, for each non-zero vector $\mathbf{x} \in \mathbb{Z}^n$.

1.2. Definition. Let $q : \mathbb{Z}^n \longrightarrow \mathbb{Z}$ be an integral quadratic form (1.1).

(i) For each $j \in \{1, \ldots, n\}$, we define the j-th restriction of $q = q(x_1, \ldots, x_n)$ to be the integral quadratic form

$$q^j : \mathbb{Z}^{n-1} \longrightarrow \mathbb{Z},$$

in the indeterminates $x_1, \ldots, x_{j-1}, x_{j+1}, \ldots, x_n$, given by the formula

$$q^j = q(x_1, \ldots, x_{j-1}, 0, x_{j+1}, \ldots, x_n).$$

(ii) The quadratic form $q = q(x_1, \ldots, x_n)$ is defined to be **critical** provided

$q : \mathbb{Z}^n \longrightarrow \mathbb{Z}$ is not weakly positive, but the restrictions

$$q^1, q^2, \ldots, q^n : \mathbb{Z}^{n-1} \longrightarrow \mathbb{Z}$$

of q are weakly positive.

The main properties of the critical integral quadratic forms are collected in the following result of Ovsienko [175].

1.3. Theorem. *Let $q : \mathbb{Z}^n \longrightarrow \mathbb{Z}$ be a critical integral quadratic form in n indeterminates $x_1, \ldots, x_n$.*

(a) *Either $n = 2$ or the quadratic form q is positive semidefinite of corank 1.*

(b) *If $n \geq 3$ then the radical*

$$\operatorname{rad} q = \{\mathbf{x} \in \mathbb{Z}^n; \; q(\mathbf{x}) = 0\}$$

of q has the form $\operatorname{rad} q = \mathbb{Z} \cdot \mathbf{h}_q$, where $\mathbf{h}_q \in \mathbb{Z}^n$ is a positive sincere vector.

Proof. Assume that $n \geq 3$ and $q = q(x_1, \ldots, x_n)$ is a critical integral quadratic form. It follows that

(i) there exists a non-zero vector $\mathbf{y} = [y_1 \; y_2 \; \cdots \; y_n]^t \in \mathbb{N}^n$ such that $q(\mathbf{y}) \leq 0$, because q is not weakly positive, and

(ii) we have $y_1 \geq 1, \ldots, y_n \geq 1$, because the restrictions $q^1, \ldots, q^n : \mathbb{Z}^{n-1} \longrightarrow \mathbb{Z}$ of q are weakly positive.

Choose a non-zero vector $\mathbf{y} = [y_1 \; y_2 \; \cdots \; y_n]^t \in \mathbb{N}^n$ such that the conditions (i) and (ii) are satisfied, and the sum

$$||\mathbf{y}|| = \sum_{i=1}^{n} y_i > n$$

is minimal.

First we prove that $q(\mathbf{y}) = 0$. Denote by

$$(-,-) : \mathbb{Z}^n \times \mathbb{Z}^n \longrightarrow \mathbb{Z}$$

the symmetric bilinear form corresponding to q. It follows from our choice of $\mathbf{y}$ that $q(\mathbf{y} - \mathbf{e}_i) > 0$, for any $i \in \{1, \dots, n\}$, and hence we get

$$(\mathbf{y}, \mathbf{y}) + (\mathbf{e}_i, \mathbf{e}_i) - 2(\mathbf{y}, \mathbf{e}_i) = q(\mathbf{y} - \mathbf{e}_i) \geq 1.$$

It follows that $2(\mathbf{y}, \mathbf{e}_i) \leq (\mathbf{y}, \mathbf{y})$, and therefore

$$2(\mathbf{y}, \mathbf{y}) = \sum_{i=1}^{n} 2(\mathbf{y}, \mathbf{e}_i) y_i \leq \left(\sum_{i=1}^{n} y_i \right) (\mathbf{y}, \mathbf{y}).$$

Hence we conclude that $q(\mathbf{y}) = (\mathbf{y}, \mathbf{y}) = 0$, because $\sum_{i=1}^{n} y_i \geq n \geq 3$ and $(\mathbf{y}, \mathbf{y}) \leq 0$. Then our claim follows.

Note also that

- $0 = (\mathbf{y}, \mathbf{y}) = (\mathbf{y}, \mathbf{e}_1) y_1 + \dots + (\mathbf{y}, \mathbf{e}_n) y_n,$
- $2(\mathbf{y}, \mathbf{e}_i) \leq (\mathbf{y}, \mathbf{y}) = 0,$ and
- $y_1 \geq 1, \dots, y_n \geq 1,$

and therefore $(\mathbf{y}, \mathbf{e}_1) = 0, (\mathbf{y}, \mathbf{e}_2) = 0, \dots, (\mathbf{y}, \mathbf{e}_n) = 0$.

Now we show that q is positive semidefinite and the vector $\mathbf{y}$ is a generator of the subgroup $\operatorname{rad} q$ of $\mathbb{Z}^n$.

Assume that $\mathbf{z} \in \mathbb{Z}^n$ is a vector such that $q(\mathbf{z}) \leq 0$. Then there exists $t \in \{1, \dots, n\}$ such that

$$\frac{z_t}{y_t} \leq \frac{z_i}{y_i},$$

for all $i \in \{1, \dots, n\}$. Consider the vector

$$\mathbf{x} = y_t \mathbf{z} - z_t \mathbf{y}.$$

Clearly, $x_t = 0$ and $x_1 \geq 0, \dots, x_n \geq 0$, for all $i \in \{1, \dots, n\}$. On the other hand, we have

$$q(\mathbf{x}) = (\mathbf{x}, \mathbf{x}) = y_t^2(\mathbf{z}, \mathbf{z}) + z_t^2(\mathbf{y}, \mathbf{y}) - 2 y_t z_t(\mathbf{y}, \mathbf{z}) = y_t^2 q(\mathbf{z}) \leq 0,$$

because $(\mathbf{y}, \mathbf{y}) = 0$ and $(\mathbf{y}, \mathbf{z}) = (\mathbf{y}, \mathbf{e}_1) z_1 + (\mathbf{y}, \mathbf{e}_2) z_2 + \dots + (\mathbf{y}, \mathbf{e}_n) z_n = 0$.

Moreover, we have $q(\mathbf{x}) = q^t(\mathbf{x}) \geq 0$, because the restricted quadratic form $q^t : \mathbb{Z}^{n-1} \longrightarrow \mathbb{Z}$ of q is weakly positive. Consequently, we get $q^t(\mathbf{x}) = 0$, and therefore $\mathbf{x} = 0$, that is,

$$\mathbf{z} = \frac{z_t}{y_t} \cdot \mathbf{y}, \quad \text{and} \quad q(\mathbf{z}) = \left(\frac{z_t}{y_t} \right)^2 \cdot q(\mathbf{y}) = 0.$$

This shows that q is positive semidefinite. Now we show that rational number $\frac{z_t}{y_t}$ is an integer. Let $\bar{z}_t$ and $\bar{y}_t$ be relatively prime integers such that $\frac{z_t}{y_t} = \frac{\bar{z}_t}{\bar{y}_t}$ and $\bar{y}_t \geq 1$. The equality $\mathbf{z} = \frac{z_t}{y_t} \mathbf{y}$ yields $\bar{y}_t \mathbf{z} = \bar{z}_t \mathbf{y}$. It follows that each of the coordinates of the vector $\mathbf{y}$ are divided by $\bar{y}_t$, because $\bar{z}_t$

and $\overline{y}_t$ are relatively prime. Hence, there exists a vector $\mathbf{y}' \in \mathbb{Z}^n$ such that $\mathbf{y} = \overline{y}_t \cdot \mathbf{y}'$. Note that

- the vector $\mathbf{y}'$ is positive, because $\mathbf{y}$ is positive and $\overline{y}_t \geq 1$,
- the equalities $0 = q(\mathbf{y}) = \overline{y}_t^2 \cdot q(\mathbf{y}')$ yield $q(\mathbf{y}') = 0$,

that is, the vector satisfies the conditions (i) and (ii). Hence, the equality $\|\mathbf{y}\| = \overline{y}_t \cdot \|\mathbf{y}'\|$ yields $\overline{y}_t = 1$, because $\mathbf{y}$ was chosen such that $\|\mathbf{y}\|$ is minimal. This shows that $\frac{z_t}{y_t} = \frac{\overline{z}_t}{1} = \overline{z}_t$ is an integer.

It follows that the positive sincere vector $\mathbf{h}_q = \mathbf{y} \in \mathbb{Z}^n$ is a generator of the subgroup $\operatorname{rad} q$ of $\mathbb{Z}^n$. This finishes the proof. $\qquad \square$

XIV.2. Minimal representation-infinite algebras

The main aim of this section is to describe the algebras A with a post-projective component in their Auslander–Reiten quiver $\Gamma(\operatorname{mod} A)$, that are minimal representation-infinite in the following sense.

2.1. Definition. An algebra A is said to be **minimal representation-infinite** if the following two conditions are satisfied:

(i) A is representation-infinite, and
(ii) the quotient algebra A/AeA is representation-finite, for any non-zero idempotent $e \in A$ such that $e \neq 1$.

The following two lemmata are very useful in the study of minimal representation-infinite algebras.

2.2. Lemma. *Let A be a minimal representation-infinite algebra.*

(i) *The algebra A is basic and connected.*
(ii) *The number of pairwise non-isomorphic non-sincere indecomposable A-modules is finite.*
(iii) *If the quotient algebra A/AeA is representation-directed, for any non-zero primitive idempotent $e \neq 1$ of A, then $\operatorname{End}_A M \cong K$ and $\operatorname{Ext}_A^j(M, M) = 0$, for any $j \geq 1$ and any non-sincere indecomposable A-module M.*

Proof. Assume that A is a minimal representation-infinite algebra. It easily follows that the algebra is basic and connected.

Let $e_1, \ldots, e_n$ be a fixed complete set of primitive orthogonal idempotents of A such that $A = e_1 A \oplus \ldots \oplus e_n A$.

Let M be a non-sincere indecomposable A-module. Then there is an idempotent e_j such that $Me_j = 0$ and, therefore, M is a module over the quotient algebra $A/Ae_j A$.

By our hypothesis, A/Ae_jA is a representation-finite algebra. It follows that, for each $j \in \{1, \ldots, n\}$, the number of the non-sincere indecomposable A-modules M satisfying $Me_j = 0$ is finite, up to isomorphism.

Assume, as in (iii), that the algebra A/Ae_jA is representation-directed. Then, according to (IX.3.5), the A/Ae_jA-module M lies in a postprojective component of $\Gamma(\operatorname{mod} A/Ae_jA)$. It follows that the A/Ae_jA-module M is directing and, hence, (IX.1.4) yields

- $\operatorname{End}_A M \cong \operatorname{End}_{A/Ae_A} M = K$, and
- $\operatorname{Ext}_A^j(M, M) \cong \operatorname{Ext}_{A/Ae_A}^j(M, M) = 0$, for all $j \geq 1$.

This completes the proof. $\qquad\square$

The following Ext-vanishing lemma is frequently used.

2.3. Lemma. *Let A be an algebra and X a non-zero A-module satisfying the following two conditions:*

(i) *there is a direct sum decomposition $X = X_1 \oplus \ldots \oplus X_r$, where $r \geq 2$ and the modules $X_1, \ldots, X_r$ are indecomposable, and*

(ii) *$\dim_K \operatorname{End}_A X \leq \dim_K \operatorname{End}_A Y$, for any A-module Y such that $\dim X = \dim Y$.*

Then $\operatorname{Ext}_A^1(X_i, X_j) = 0$, for all $i, j \in \{1, \ldots, r\}$ such that $i \neq j$.

Proof. Assume that X has a decomposition $X = X_1 \oplus \ldots \oplus X_r$, where $r \geq 2$, the modules $X_1, \ldots, X_r$ are indecomposable, and the condition (ii) is satisfied.

We show that, for each $i \in \{1, \ldots, r\}$, we have $\operatorname{Ext}_A^1(X_i, \bigoplus_{j \neq i} X_j) = 0$.

Fix $i \in \{1, \ldots, r\}$ and assume, to the contrary, that $\operatorname{Ext}_A^1(X_i, \bigoplus_{j \neq i} X_j) \neq 0$. Then there exists a non-split short exact sequence

$$0 \longrightarrow \bigoplus_{j \neq i} X_j \longrightarrow Y \longrightarrow X_i \longrightarrow 0$$

in $\operatorname{mod} A$ and obviously,

$$\begin{aligned}
\dim Y &= \dim \bigoplus_{j \neq i} X_j + \dim X_i \\
&= \sum_{j \neq i} \dim X_j + \dim X_i \\
&= \dim X.
\end{aligned}$$

On the other hand, by applying (VIII.2.8) to the preceding non-split exact sequence, we obtain

$$\dim_K \operatorname{End}_A Y < \dim_K \operatorname{End}_A \left(\bigoplus_{j \neq i} X_j \oplus X_i \right) = \dim_K \operatorname{End}_A X,$$

and we get a contradiction with (ii).

It follows that

$$\operatorname{Ext}_A^1\left(X_i, \bigoplus_{j \neq i} X_j\right) = 0,$$

for each $i \in \{1, \dots, r\}$, and hence

$$\operatorname{Ext}_A^1(X_i, X_j) = 0,$$

for each pair $i, j \in \{1, \dots, r\}$ such that $i \neq j$. □

The main result of this section is the following theorem due to Happel and Vossieck [112].

2.4. Theorem. *Given an algebra A, the following two conditions are equivalent.*

    (a) *The algebra A is minimal representation-infinite and has a postprojective component in the Auslander–Reiten quiver $\Gamma(\operatorname{mod} A)$.*

    (b) *A is either the path algebra of the enlarged Kronecker quiver*

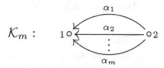

with $m \geq 3$ arrows $\alpha_1, \dots, \alpha_m$, or else, A is a concealed algebra of Euclidean type.

Proof. To prove the implication (b)⇒(a), we consider two cases.

Case 1°. Assume that A is the path algebra of the enlarged Kronecker quiver $\mathcal{K}_m$, with $m \geq 3$ arrows. Then, by (VIII.2.3), the quiver $\Gamma(\operatorname{mod} A)$ has a postprojective component $\mathcal{P}(A)$ and, in view of (VII.5.10), the algebra A is minimal representation-infinite.

Case 2°. Assume that A is a concealed algebra of Euclidean type. Then there is an algebra isomorphism

$$A \cong \operatorname{End} T_H,$$

where T is a postprojective tilting module over the path algebra $H = K\Delta$ of an (acyclic) Euclidean quiver Δ.

We know from (VIII.4.5) that the Auslander–Reiten quiver $\Gamma(\operatorname{mod} A)$ of A contains a postprojective component $\mathcal{P}(A)$ containing all indecomposable projective modules. Further, the torsion part $\mathcal{T}(T)$ of $\operatorname{mod} H$ contains all but finitely many non-isomorphic indecomposable H-modules, and any indecomposable H-module not in $\mathcal{T}(T)$ is postprojective. Because the algebra

H is hereditary then, by (VI.5.7), the tilting H-module T is splitting and, in view of (VI.3.8), we conclude that all but finitely many non-isomorphic indecomposable A-modules are of the form $\mathrm{Hom}_H(T, X)$, where X is an indecomposable H-module in $\mathcal{T}(T)$.

Let P be an indecomposable projective A-module. Then there exists an indecomposable direct summand T_i of T such that $P = \mathrm{Hom}_H(T, T_i)$. Because T_i is postprojective, there exists an isomorphism

$$T_i \cong \tau_H^{-r} P(a),$$

for some $r \geq 0$ and some point a of the quiver Δ. Then, invoking the tilting theorem (VI.3.8) and (IV.2.15), for any indecomposable A-module M of the form $M = \mathrm{Hom}_H(T, X)$, we get the isomorphisms

$$\begin{aligned} \mathrm{Hom}_A(P, M) &= \mathrm{Hom}_A(\mathrm{Hom}_H(T, T_i), \mathrm{Hom}_H(T, X)) \\ &\cong \mathrm{Hom}_H(T_i, X) \\ &\cong \mathrm{Hom}_H(\tau_H^{-r} P(a), X)) \\ &\cong \mathrm{Hom}_H(P(a), \tau_H^r X). \end{aligned}$$

Because Δ is a Euclidean quiver then the quiver $\Delta^{(a)}$, obtained from Δ by removing the point a and the arrows having a as a source or target, is a disjoint union of Dynkin quivers and, consequently, the path algebra $K\Delta^{(a)}$ is representation-finite, by (VII.5.10). It follows that

$$\mathrm{Hom}_H(P(a), Y) \neq 0,$$

for all but finitely many non-isomorphic indecomposable H-modules Y. Clearly, for a fixed $r \geq 0$, all but finitely many indecomposable H-modules Y are of the form $Y = \tau_H^r X$, where X is an indecomposable H-module.

This shows that $\mathrm{Hom}_A(P, M) \neq 0$, for all but finitely many non-isomorphic indecomposable A-modules M. Moreover, A is representation-infinite, as a concealed algebra of Euclidean type. Hence, we conclude that the algebra A is minimal representation-infinite.

(a)$\Rightarrow$(b) Assume that A is a minimal representation-infinite algebra and that $\mathcal{P}$ is a postprojective component of the Auslander–Reiten quiver $\Gamma(\mathrm{mod}\, A)$ of A.

It follows that the algebra A is basic and connected. Moreover, by (IV.5.4), the component $\mathcal{P}$ is infinite, because the algebra A is connected and representation-infinite. Hence we conclude that $\mathcal{P}$ contains all indecomposable projective A-modules, because A is minimal representation-infinite.

Let n be the rank of $K_0(A)$, that is, $K_0(A) \cong \mathbb{Z}^n$.

<u>Case 1°</u>. Suppose that $n = 2$. Because the indecomposable projective modules lie in $\mathcal{P}$ then the quiver Q_A of A has no oriented cycles and, according to (II.3.7), A is the path algebra of the quiver

with $m \geq 1$ arrows. Moreover, $m \geq 2$, because the algebra A is representation-infinite.

<u>Case 2°</u>. Assume that $n \geq 3$. We split the proof into four steps.

<u>Step 2.1°</u>. First we prove that, given a non-zero idempotent $e \neq 1$ of A, we have

$$\operatorname{Ext}_A^1(Z, Z) \cong \operatorname{Ext}_{A/AeA}^1(Z, Z) = 0,$$

for any indecomposable A/AeA-module Z.

Note that the quotient algebra A/AeA of A is representation-finite, because the algebra A is minimal representation-infinite. Moreover, because the postprojective component $\mathcal{P}$ of $\Gamma(\operatorname{mod} A)$ contains all indecomposable projective A-modules, then, by (IX.5.2), the Auslander–Reiten quiver $\Gamma(\operatorname{mod} A/AeA)$ of A/AeA admits postprojective components the union of which contains all indecomposable projective A/AeA-modules. More precisely, if

$$A/AeA = B_1 \times \ldots \times B_m, \quad m \geq 1,$$

is a decomposition of the algebra A/AeA into a direct product of connected subalgebras $B_1, \ldots, B_m$ of A/AeA, the translation quiver $\Gamma(\operatorname{mod} A/AeA)$ is the disjoint union of the translation quivers $\Gamma(\operatorname{mod} B_1), \ldots, \Gamma(\operatorname{mod} B_m)$. Moreover, because the algebra A/AeA is representation-finite, then

- the algebras $B_1, \ldots, B_m$ are representation-finite, and
- for each $j \in \{1, \ldots, m\}$, the translation quiver $\Gamma(\operatorname{mod} B_j)$ is finite, connected, contains all indecomposable projective B_j-modules (which are projective A-modules), and $\Gamma(\operatorname{mod} B_j) = \mathcal{P}(B_j)$ is a postprojective translation quiver.

It follows that $\Gamma(\operatorname{mod} A/AeA)$ is a union of the postprojective components

$$\mathcal{P}(B_1) = \Gamma(\operatorname{mod} B_1), \quad \ldots, \quad \mathcal{P}(B_m) = \Gamma(\operatorname{mod} B_m)$$

and, according to (VIII.2.7), we have

$$\operatorname{Ext}_A^1(Z, Z) \cong \operatorname{Ext}_{A/AeA}^1(Z, Z) = 0,$$

for any indecomposable A/AeA-module Z.

Step 2.2°. Next we prove that the Euler quadratic form

$$q_A : \mathbb{Z}^n \longrightarrow \mathbb{Z}$$

of the algebra A is critical.

We know from (IX.1.5) and (IX.3.1) that the dimension vectors of pairwise non-isomorphic indecomposable modules from $\mathcal{P}$ are pairwise different positive roots of q_A. Because the component $\mathcal{P}$ is infinite then the number of positive roots of q_A is infinite and, according to (VII.3.4), the Euler form q_A is not weakly positive.

Let $\mathbf{x}$ be a positive non-sincere vector from $K_0(A) = \mathbb{Z}^n$. We show that $q_A(\mathbf{x}) \geq 1$. Choose an A-module X with $\dim X = \mathbf{x}$ such that the dimension $\dim_K \operatorname{End}_A X$ is the smallest possible. Decompose the module X into a direct sum

$$X = X_1 \oplus \ldots \oplus X_r$$

of indecomposable A-modules $X_1, \ldots, X_r$. Then, by (2.3), we get

$$\operatorname{Ext}^1_A(X_i, X_j) = 0,$$

for all $i, j \in \{1, \ldots, r\}$ such that $i \neq j$. Moreover, because $\mathbf{x} = \dim X$ is not sincere, then the module X_i is not sincere and, by (2.2), we have

$$\operatorname{Ext}^1_A(X_i, X_i) = 0,$$

for any $i \in \{1, \ldots, r\}$. As a consequence, we get

$$\operatorname{Ext}^1_A(X, X) = \operatorname{Ext}^1_A(X_1 \oplus \ldots \oplus X_r, X_1 \oplus \ldots \oplus X_r) = \bigoplus_{i,j=1}^{r} \operatorname{Ext}^1_A(X_i, X_j) = 0.$$

Hence, in view of (III.3.13), we obtain

$$\begin{aligned}
q_A(\mathbf{x}) &= q_A(\dim X) \\
&= \chi_A(X, X) \\
&= \dim_K \operatorname{End}_A(X) - \dim_K \operatorname{Ext}^1_A(X, X) + \dim_K \operatorname{Ext}^2_A(X, X) \\
&= \dim_K \operatorname{End}_A(X) + \dim_K \operatorname{Ext}^2_A(X, X) > 0,
\end{aligned}$$

because we show in the following step that $\operatorname{gl.dim} A \leq 2$. This shows that the Euler quadratic form $q_A : \mathbb{Z}^n \longrightarrow \mathbb{Z}$ is critical and, by (1.3), the quadratic form q_A is positive semidefinite of corank one.

Step 2.3°. Now we prove that A is a tilted algebra and $\mathcal{P}$ admits a faithful section Δ containing a sincere indecomposable module M.

Because A is representation-infinite and connected then the component $\mathcal{P}$ is infinite. By (2.2), the number of non-isomorphic non-sincere indecomposable A-modules is finite. It follows that $\mathcal{P}$ contains a sincere indecomposable module M and, hence, $\mathrm{Hom}_A(P, M) \neq 0$, for any indecomposable projective A-module P. Then, according to (VIII.2.5), the component $\mathcal{P}$ contains all the indecomposable projective A-modules P. Moreover, $\mathcal{P}$ does not contain a non-zero injective module, because the number of sincere indecomposable A-modules is infinite while, by (VIII.2.5), every module in $\mathcal{P}$ has only finitely many indecomposable predecessors in $\mathrm{mod}\,A$.

Because the module M is obviously directing then, by applying (IX.2.6), we conclude that A is tilted algebra and $\mathcal{P}$ admits a faithful section Δ containing M. In particular, $\mathrm{gl.dim}\,A \leq 2$.

Step 2.4°. We prove that the underlying graph of the section Δ of $\mathcal{P}$ is Euclidean and A is a concealed algebra of type Δ.

Let T be the direct sum of all modules lying on the section Δ of $\mathcal{P}$. We know that the section Δ is faithful and clearly $\mathrm{Hom}_A(T, \tau_A T) = 0$, because $\mathcal{P}$ is a postprojective component of $\Gamma(\mathrm{mod}\,A)$. Applying now the criterion (VIII.5.6), we infer that T is a tilting A-module, the algebra

$$H = \mathrm{End}_A T$$

is hereditary of type Δ^{op}, $T^* = D({}_H T)$ is a tilting H-module, there is an isomorphism

$$\mathrm{End}_H(T^*) \cong A$$

of algebras, and $\mathcal{P}$ is the connecting component of $\Gamma(\mathrm{mod}\,A)$ determined by the tilting H-module T^*.

Moreover, by (VI.4.7), the Euler quadratic form

$$q_H : K_0(H) \longrightarrow \mathbb{Z}$$

of H is $\mathbb{Z}$-congruent to $q_A : \mathbb{Z}^n \longrightarrow \mathbb{Z}$ and, consequently, q_H is positive semidefinite of corank one. Thus, by (VII.4.5), the underlying graph of the quiver Δ is Euclidean. It follows that the algebra H is hereditary of Euclidean type.

Further, because $\mathcal{P}$ does not contain non-zero injective modules then, according to (VIII.4.1), the module T^* has no postprojective direct summands.

We claim that T^* has no non-zero regular direct summands. Suppose, to the contrary, that T^* has an indecomposable direct summand T_i^* which is regular.

Because $\mathrm{Ext}_H^1(T_i^*, T_i^*) = 0$ then, according to (XI.2.8) and (XI.3.5), the module T_i^* belongs to one of the non-homogeneous stable tubes of $\Gamma(\mathrm{mod}\,H)$.

By (XII.3.4), $\Gamma(\mathrm{mod}\,H)$ admits a $\mathbb{P}_1(K)$-family $\boldsymbol{\mathcal{T}}^H = \{\mathcal{T}_\lambda^H\}_{\lambda \in \mathbb{P}_1(K)}$ of pairwise orthogonal standard stable tubes, and at most three of them are non-homogeneous.

Now let X be an indecomposable module from a homogeneous tube of $\Gamma(\mathrm{mod}\,H)$. Because, by (VIII.2.13) and (XI.2.8), we have

$$\mathrm{Hom}_H(\mathcal{Q}(H), \mathcal{R}(H)) = 0$$

and the regular part $\mathcal{R}(H)$ of $\Gamma(\mathrm{mod}\,H)$ is a disjoint union of pairwise orthogonal tubes, we get

$$\mathrm{Hom}_H(T^*, X) = 0 \quad \text{and} \quad \mathrm{Hom}_H(X, \tau T_i^*) = 0.$$

It follows that the H-module X belongs to $\mathcal{F}(T_H^*)$ and, consequently, $\mathrm{Ext}_H^1(T^*, X)$ belongs to $\mathcal{X}(T_H^*)$.

Because the algebra H is hereditary then, by (VI.5.7), T^* is a splitting tilting A-module. Then, by (VI.5.8), $\mathrm{Ext}_H^1(T^*, \tau T_i^*)$ is an indecomposable injective A-module. Moreover, applying again the tilting theorem (VI.3.8), we get

$$\mathrm{Hom}_A(\mathrm{Ext}_H^1(T^*, X), \mathrm{Ext}_H^1(T^*, \tau T_i^*)) \cong \mathrm{Hom}_H(X, \tau T_i^*) = 0.$$

It follows that $\mathrm{Ext}_H^1(T^*, X)$ is a non-sincere A-module. Because every homogeneous tube has infinitely many modules then there are infinitely many pairwise non-isomorphic non-sincere indecomposable modules in $\mathcal{X}(T_H^*)$, and we get a contradiction with the assumption that A is minimal representation-infinite. Therefore, the tilting H-module T^* is preinjective. Hence, by applying (XI.5.2), we conclude that

$$A \cong \mathrm{End}_H(T^*)$$

is a concealed algebra of Euclidean type. This completes the proof. □

We end this section with an example showing that the class of minimal representation-infinite algebras is larger than the class described in (2.4).

2.5. Example. Let Λ be the path K-algebra of the following quiver

bound by the three relations $\alpha^2 = 0$, $\beta^2 = 0$, and $\alpha\beta = \beta\alpha$.

We know from (X.4.8) that Λ is representation-infinite, $\dim_K \Lambda = 4$, and there is a K-algebra isomorphism

$$\Lambda \cong K[t_1, t_2]/(t_1^2, t_2^2).$$

Note that

- Λ is a minimal representation-infinite algebra, because it has only two (trivial) idempotents 0 and 1, and
- the Auslander–Reiten quiver $\Gamma(\mathrm{mod}\, A)$ of A has no postprojective component, because Λ is the unique indecomposable projective Λ-module and Λ is not a brick.

XIV.3. A criterion for the infinite representation type of algebras

We may now prove a useful criterion for the infinite representation type of algebras A whose Auslander–Reiten quiver $\Gamma(\mathrm{mod}\, A)$ admits a postprojective component.

3.1. Theorem. *Let A be a basic connected K-algebra such that the Auslander–Reiten quiver $\Gamma(\mathrm{mod}\, A)$ of A has a postprojective component. Then A is representation-infinite if and only if there is an idempotent $e \neq 1$ of A such that A/AeA is either the path algebra of the quiver*

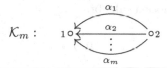

with $m \geq 3$ arrows $\alpha_1, \ldots, \alpha_m$, or else A/AeA is a concealed algebra of Euclidean type.

Proof. The sufficiency. Assume that there is an idempotent $e \neq 1$ of A such that A/AeA is the path algebra of $\mathcal{K}_m$, or A/AeA is a concealed algebra of Euclidean type. It follows that the algebra A/AeA is representation-infinite, and consequently, A is representation-infinite too.

The necessity. Assume that A is a basic, connected, and representation-infinite K-algebra such that $\Gamma(\mathrm{mod}\, A)$ has a postprojective component $\mathcal{P}$. By (IV.5.4), the component $\mathcal{P}$ is infinite.

Let $e_1, \ldots, e_n$ be a fixed complete set of primitive orthogonal idempotents of A such that

$$A = e_1 A \oplus \ldots \oplus e_n A.$$

Let S be the subset of $\{1, \ldots, n\}$ consisting of all $s \in \{1, \ldots, n\}$ such that $Me_s = 0$, for any indecomposable module M in $\mathcal{P}$. Hence, for any

$i \in \{1, \ldots, n\} \setminus S$ there exists an indecomposable A-module N in $\mathcal{P}$ such that $\mathrm{Hom}_A(e_i A, N) \cong N e_i \neq 0$, and hence, the indecomposable projective A-module $e_i A$ belongs to $\mathcal{P}$. If we set

$$f = \sum_{s \in S} e_s \in A,$$

then

- f is an idempotent of A and $Mf = 0$, for any indecomposable module M in $\mathcal{P}$,
- the quotient algebra

$$B = A/AfA$$

of A is representation-infinite,
- the canonical epimorphism $\pi : A \to B$ of algebras induces a fully faithful embedding of module categories $\mathrm{mod}\, B \hookrightarrow \mathrm{mod}\, A$,
- $\mathcal{P}$ is a postprojective component $\Gamma(\mathrm{mod}\, B)$, and
- $\mathcal{P}$ contains all indecomposable projective B-modules.

Hence, by applying (IX.5.2), we conclude that there is an idempotent $f' \in B$ such that the quotient algebra

$$C = B/Bf'B$$

of B is minimal representation-infinite, the Auslander–Reiten quiver $\Gamma(\mathrm{mod}\, C)$ of C admits postprojective components, and their union contains all indecomposable projective C-modules. It follows from (2.4) that C is either the path algebra of the enlarged Kronecker quiver

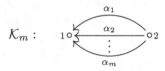

with $m \geq 3$ arrows, or C is a concealed algebra of Euclidean type. It is easy to see that there is a non-zero idempotent e of A such that $C \cong A/AeA$ and $AfA \subseteq AeA$. This finishes the proof. $\square$

Now we illustrate the use of the criterion (3.1) by the following example.

3.2. Example. Let A be the algebra given by the quiver

$$
\begin{array}{c}
9\ \circ \\
\xi \downarrow \\
6\ \circ \quad\ \ \ \ 7\ \longrightarrow 8 \\
\searrow^{\alpha}\ 3\ \nearrow^{\gamma} \\
\circ \\
\swarrow_{\beta}\ \downarrow\ \searrow^{\sigma} \\
4\ \circ\ \ \ 2\ \circ\ \ \ \circ\ 5 \\
\downarrow \\
1\ \circ
\end{array}
$$

$Q_A:$

bound by three zero relations $\xi\alpha = 0$, $\alpha\beta = 0$ and $\gamma\sigma = 0$.

Because the quiver Q_A of A is a tree then, by applying (IX.4.3), we conclude that A satisfies the separation condition. Hence, by (IX.4.5), the Auslander–Reiten quiver $\Gamma(\mathrm{mod}\,A)$ of A admits a postprojective component.

Take the idempotent $e = e_8 + e_9$ of A and consider the quotient algebra

$$B = A/AeA$$

of B. By applying the characterisation given in (XI.5.1), we show that B is a concealed algebra of the Euclidean type $\Delta(\widetilde{\mathbb{E}}_6)$ (and hence, B is representation-infinite).

To see this, we note that the algebra B is given by the quiver

$$
\begin{array}{c}
6\ \circ \quad\quad\quad \circ\ 7 \\
\searrow^{\alpha}\ 3\ \nearrow^{\gamma} \\
\circ \\
\swarrow_{\beta}\ \downarrow\ \searrow^{\sigma} \\
4\ \circ\ \ \ 2\ \circ\ \ \ \circ\ 5 \\
\downarrow \\
1\ \circ
\end{array}
$$

$Q_B:$

bound by two zero relations $\alpha\beta = 0$ and $\gamma\sigma = 0$.

Because the quiver Q_B of the algebra B is a tree then, by applying (IX.4.3) and its dual, we conclude that B satisfies the separation condition and the coseparation condition. Then, by (IX.4.5), the Auslander–Reiten quiver $\Gamma(\mathrm{mod}\,B)$ of B admits a postprojective component and a preinjective component.

Now we show that

- $\Gamma(\mathrm{mod}\,B)$ admits a unique postprojective component $\mathcal{P}$ containing all indecomposable projective B-modules,
- $\Gamma(\mathrm{mod}\,B)$ admits a unique preinjective component $\mathcal{Q}$ containing all indecomposable injective B-modules, and
- B is a concealed algebra of the Euclidean type $\Delta(\widetilde{\mathbb{E}}_6)$.

By the standard mesh calculation technique, we show that the left hand part of the component $\mathcal{P}$ of $\Gamma(\operatorname{mod} B)$, containing the simple projective B-module $P(1)$, looks as follows:

$$
\begin{array}{c}
P(7)=\begin{smallmatrix}0&&1\\&1&\\1&1&0\\&1&\end{smallmatrix}
\;-\;-\;-\;
\begin{smallmatrix}0&&0\\&1&\\0&1&1\\&0&\end{smallmatrix}
\;-\;-\;-\;
\begin{smallmatrix}1&&0\\&2&\\1&1&1\\&1&\end{smallmatrix}
\;-\;-
\end{array}
$$

$$
\begin{array}{c}
P(5)=\begin{smallmatrix}0&&0\\&0&\\0&0&1\\&0&\end{smallmatrix}
\;-\;-\;-\;-
\begin{smallmatrix}0&&0\\&1&\\1&1&0\\&1&\end{smallmatrix}
\quad
\begin{smallmatrix}1&&0\\&1&\\0&1&1\\&1&\end{smallmatrix}
\quad
\begin{smallmatrix}0&&1\\&2&\\1&2&1\\&1&\end{smallmatrix}
\quad
\begin{smallmatrix}0&&0\\&1&\\1&1&0\\&0&\end{smallmatrix}
\quad
\begin{smallmatrix}1&&0\\&3&\\1&2&2\\&1&\end{smallmatrix}
\quad
\begin{smallmatrix}0&&1\\&2&\\1&1&1\\&1&\end{smallmatrix}
\end{array}
$$

$$
\begin{array}{c}
P(4)=\begin{smallmatrix}0&&0\\&0&\\1&0&0\\&0&\end{smallmatrix}\to
\begin{smallmatrix}0&&0\\&1&\\1&1&1\\&1&\end{smallmatrix}\to
\begin{smallmatrix}0&&0\\&1&\\0&1&1\\&1&\end{smallmatrix}\to
\begin{smallmatrix}0&&0\\&2&\\1&2&1\\&1&\end{smallmatrix}\to
\begin{smallmatrix}1&&0\\&2&\\1&2&1\\&1&\end{smallmatrix}\to
\begin{smallmatrix}1&&1\\&4&\\2&3&2\\&2&\end{smallmatrix}\to
\begin{smallmatrix}0&&1\\&3&\\2&2&1\\&1&\end{smallmatrix}\to
\begin{smallmatrix}1&&1\\&5&\\2&4&2\\&2&\end{smallmatrix}\to
\end{array}
$$

$$
\begin{array}{c}
P(2)=\begin{smallmatrix}0&&0\\&0&\\0&1&0\\&1&\end{smallmatrix}
\;-\;-\;-\;-
\begin{smallmatrix}0&&0\\&1&\\1&1&1\\&0&\end{smallmatrix}
\;-\;-\;-\;-
\begin{smallmatrix}0&&0\\&2&\\1&1&1\\&1&\end{smallmatrix}
\;-\;-\;-\;-
\begin{smallmatrix}1&&1\\&3&\\1&3&1\\&2&\end{smallmatrix}
\;-\;-\;-\;-
\end{array}
$$

$$
\begin{array}{c}
P(1)=\begin{smallmatrix}0&&0\\&0&\\0&0&0\\&1&\end{smallmatrix}
\;-\;-\;-
\begin{smallmatrix}0&&0\\&0&\\0&1&0\\&0&\end{smallmatrix}
\;-\;-\;-
\begin{smallmatrix}0&&0\\&1&\\1&0&1\\&0&\end{smallmatrix}
\;-\;-\;-
\begin{smallmatrix}0&&0\\&1&\\0&1&0\\&1&\end{smallmatrix}
\;-\;-\;-
\begin{smallmatrix}1&&1\\&2&\\1&2&1\\&1&\end{smallmatrix}
\;-\;-
\end{array}
$$

where

$$
P(3) = \begin{smallmatrix}0&&0\\&1&\\1&1&1\\&1&\end{smallmatrix} \quad\text{and}\quad P(6) = \begin{smallmatrix}1&&0\\&1&\\0&1&1\\&1&\end{smallmatrix},
$$

and the indecomposable modules are represented by their dimension vectors.

It follows that $\mathcal{P}$ is a postprojective component of $\Gamma(\operatorname{mod} B)$ containing all indecomposable projective B-modules, and hence $\mathcal{P}$ is a unique postprojective component of $\Gamma(\operatorname{mod} B)$. Moreover, $\mathcal{P}$ contains the section

$$
\begin{array}{c}
\qquad\qquad\qquad\nearrow P(7)\\
\qquad\quad \tau^{-1}P(5)\\
\qquad\quad \nearrow\\
P(3) \to\; \tau^{-1}P(4)\; \to\; P(6)\\
\qquad\quad \searrow\\
\qquad\quad \tau^{-1}P(2)\\
\qquad\qquad\qquad \searrow\\
\qquad\qquad\qquad\quad \tau^{-2}P(1)
\end{array}
$$

of the Euclidean type $\Delta(\widetilde{\mathbb{E}}_6)^{\mathrm{op}}$.

Similarly, the standard mesh calculation technique shows that the right hand part of the component $\mathcal{Q}$ of $\Gamma(\operatorname{mod} B)$, containing the simple injective B-module $I(6)$, looks as follows:

$$
\begin{smallmatrix} 1 & & 1 \\ & 2 & \\ 1 & 1 & 0 \\ & 1 & \end{smallmatrix}
\;\; - - - \;\;
\begin{smallmatrix} 0 & & 1 \\ & 1 & \\ 0 & 1 & 0 \\ & 0 & \end{smallmatrix}
\;\; - - - \;\;
\begin{smallmatrix} 1 & & 0 \\ & 1 & \\ 0 & 0 & 1 \\ & 0 & \end{smallmatrix} = I(5)
$$

$$
\begin{smallmatrix} 1 & & 1 \\ & 2 & \\ 0 & 1 & 1 \\ & 1 & \end{smallmatrix}
\;
\begin{smallmatrix} 1 & & 2 \\ & 3 & \\ 1 & 2 & 0 \\ & 1 & \end{smallmatrix}
\;
\begin{smallmatrix} 1 & & 0 \\ & 1 & \\ 0 & 1 & 0 \\ & 0 & \end{smallmatrix}
\;
\begin{smallmatrix} 1 & & 1 \\ & 2 & \\ 0 & 1 & 1 \\ & 0 & \end{smallmatrix}
\;
\begin{smallmatrix} 0 & & 1 \\ & 1 & \\ 1 & 0 & 0 \\ & 0 & \end{smallmatrix}
\;
\begin{smallmatrix} 1 & & 0 \\ & 1 & \\ 0 & 0 & 0 \\ & 0 & \end{smallmatrix}
\;
\begin{smallmatrix} 0 & & 1 \\ & 0 & \\ 0 & 0 & 0 \\ & 0 & \end{smallmatrix} = I(7)
$$

$$
\rightarrow
\begin{smallmatrix} 2 & & 2 \\ & 5 & \\ 2 & 3 & 2 \\ & 1 & \end{smallmatrix}
\rightarrow
\begin{smallmatrix} 2 & & 1 \\ & 3 & \\ 0 & 2 & 1 \\ & 1 & \end{smallmatrix}
\rightarrow
\begin{smallmatrix} 2 & & 2 \\ & 4 & \\ 1 & 2 & 1 \\ & 1 & \end{smallmatrix}
\rightarrow
\begin{smallmatrix} 1 & & 1 \\ & 2 & \\ 1 & 1 & 0 \\ & 0 & \end{smallmatrix}
\rightarrow
\begin{smallmatrix} 1 & & 1 \\ & 2 & \\ 0 & 1 & 0 \\ & 0 & \end{smallmatrix}
\rightarrow
\begin{smallmatrix} 0 & & 1 \\ & 1 & \\ 0 & 0 & 0 \\ & 0 & \end{smallmatrix}
\rightarrow
\begin{smallmatrix} 1 & & 1 \\ & 1 & \\ 0 & 0 & 0 \\ & 0 & \end{smallmatrix}
\rightarrow
\begin{smallmatrix} 1 & & 0 \\ & 0 & \\ 0 & 0 & 0 \\ & 0 & \end{smallmatrix} = I(6)
$$

$$
\begin{smallmatrix} 2 & & 2 \\ & 4 & \\ 2 & 3 & 2 \\ & 1 & \end{smallmatrix}
\; - - - \;
\begin{smallmatrix} 1 & & 1 \\ & 3 & \\ 1 & 1 & 1 \\ & 0 & \end{smallmatrix}
\; - - - \;
\begin{smallmatrix} 1 & & 1 \\ & 2 & \\ 0 & 1 & 0 \\ & 1 & \end{smallmatrix}
\; - - - \;
\begin{smallmatrix} 1 & & 1 \\ & 1 & \\ 0 & 1 & 0 \\ & 0 & \end{smallmatrix} = I(2)
$$

$$
- - - \;
\begin{smallmatrix} 1 & & 1 \\ & 2 & \\ 1 & 1 & 1 \\ & 0 & \end{smallmatrix}
\; - - - \;
\begin{smallmatrix} 0 & & 0 \\ & 1 & \\ 0 & 0 & 0 \\ & 0 & \end{smallmatrix}
\; - - - \;
\begin{smallmatrix} 1 & & 1 \\ & 1 & \\ 0 & 1 & 0 \\ & 1 & \end{smallmatrix} = I(1),
$$

where

$$
I(3) = \begin{smallmatrix} 1 & & 1 \\ & 1 & \\ 0 & 0 & 0 \\ & 0 & \end{smallmatrix}
\quad \text{and} \quad
I(4) = \begin{smallmatrix} 0 & & 1 \\ & 1 & \\ 1 & 0 & 0 \\ & 0 & \end{smallmatrix},
$$

and the indecomposable modules are represented by their dimension vectors.

It follows that $\mathcal{Q}$ is a preinjective component of $\Gamma(\mathrm{mod}\,B)$ containing all indecomposable injective B-modules, and hence $\mathcal{Q}$ is a unique preinjective component of $\Gamma(\mathrm{mod}\,B)$. Moreover, $\mathcal{Q}$ contains the section

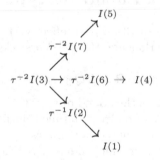

of the Euclidean type $\Delta(\widetilde{\mathbb{E}}_6)^{\mathrm{op}}$.

Because $\mathcal{P} \neq \mathcal{Q}$ then, by applying (XI.5.1), we conclude that B is a concealed algebra of the Euclidean type $\Delta(\widetilde{\mathbb{E}}_6)$.

Next, we note that the algebra B is obtained from the path algebra KQ

of the quiver

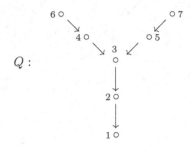

$$Q:$$

by interchanging the subquivers

$$\underset{6}{\circ} \longrightarrow \underset{4}{\circ} \longrightarrow \underset{3}{\circ} \quad \text{and} \quad \underset{3}{\circ} \longleftarrow \underset{5}{\circ} \longleftarrow \underset{7}{\circ}$$

with the subquivers

bound by the zero relations $\alpha\beta = 0$ and $\gamma\sigma = 0$, respectively.

XIV.4. A classification of concealed algebras of Euclidean type

To make the criterion in (3.1) more effective from a practical point of view, we present now the Bongatz–Happel–Vossieck classification of all concealed algebras

$$B \cong KQ_B/I_B$$

of Euclidean type in terms of the bound quivers (Q_B, I_B). We do not present here a complete proof, but we give a detailed outline of the proof, and we illustrate it by several representative examples.

The classification was given independently by Happel–Vossieck [112] and Bongartz [29]. Following [112], to save space and to make the list more accessible, we write down in Table 4.5 below only the possible 149 frames $\mathcal{F}r_1, \dots, \mathcal{F}r_{149}$ of the bound quivers (Q_B, I_B). Then the classification of the bound quivers (Q_B, I_B) is given up to the following three types of admissible operations $(\mathbf{Op}_1)$, $(\mathbf{Op}_2)$, and $(\mathbf{Op}_3)$ on frames.

4.1. Admissible operations on frames

Given a frame in Table 4.5 below, we allow the following three admissible operations.

(Op₁) Replace any non-oriented edge of the frame by an oriented one (except the case the frame is of type $\widetilde{\mathbb{A}}_n$, where we do not allow the cyclic orientation).

(Op₂) Given a frame of the form

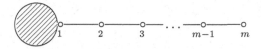

replace its subquiver

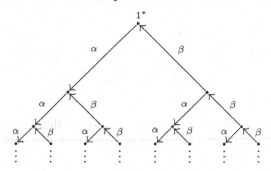

by any full connected subquiver with m vertices of the following tree

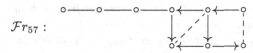

and bound by all possible relations $\beta\alpha = 0$, containing the vertex 1^*, by identifying the vertices 1 and 1^*.

(Op₃) Construct the opposite frame to a given one.   □

Now we illustrate the application of the admissible operations on frames.

4.2. Example. Consider the frame

$$\mathcal{F}r_{57}:$$

of Table 4.5. First, we construct 8 pairwise non-isomorphic quivers by choosing any orientation for the non-oriented edges. Next, by replacing the non-oriented linear graph according to the operation (**Op₂**), we get the following 6 additional frames

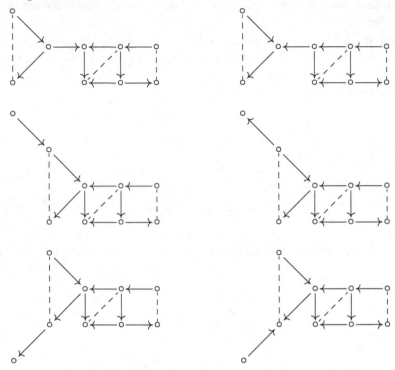

Finally, we construct the opposite quivers to the above ones. Consequently, we construct 28 pairwise non-isomorphic algebras starting from the frame given above.

4.3. Theorem. *Let $A = KQ/I$ be the algebra of a connected bound quiver (Q, I).*

(a) *The algebra A is concealed of Euclidean type if and only if the bound quiver (Q, I) is obtained from any of the frames*

$$\mathcal{F}r_1, \ldots, \mathcal{F}r_{149}$$

*of Table 4.5 below, by applying a sequence of the admissible operations (**Op₁**), (**Op₂**) and (**Op₃**).*

(b) *If the bound quiver (Q, I) is obtained from a frame Δ in Table 4.5, by applying a sequence of the admissible operations then the numbers distributed instead of the vertices of Δ shown in Table 4.5 are the coordinates of the positive generator $\mathbf{h}_A$ of the radical $\operatorname{rad} q_A$ of the Euler quadratic form $q_A : \mathbb{Z}^n \longrightarrow \mathbb{Z}$ of A, where $n = |Q_0|$.*

An outline of proof. The original proof presented in [112] and Bongartz [29] is done by applying a computer program that calculates the endomorphism algebras $\operatorname{End} T_H$ of the multiplicity-free postprojective tilting modules over the path algebras H of the canonically oriented Euclidean quivers

$$\Delta(\widetilde{\mathbb{A}}_{p,q}), \text{ with } 1 \leq p \leq q, \ \Delta(\widetilde{\mathbb{D}}_m), \text{ with } m \geq 4, \ \Delta(\widetilde{\mathbb{E}}_6), \ \Delta(\widetilde{\mathbb{E}}_7), \ \Delta(\widetilde{\mathbb{E}}_8)$$

presented in Table 1.1 of Chapter XIII.

Now we indicate the main steps of the classification of the concealed algebras of Euclidean type.

Step 1°. It follows from (XII.3.1) that any concealed algebra of Euclidean type is a concealed algebra of a Euclidean type Δ, where Δ is one of the canonically oriented Euclidean quivers. Fix such a quiver

$$\Delta = (\Delta_0, \Delta_1) \in \left\{ \Delta(\widetilde{\mathbb{A}}_{p,q}), \Delta(\widetilde{\mathbb{D}}_m), \Delta(\widetilde{\mathbb{E}}_6), \Delta(\widetilde{\mathbb{E}}_7), \Delta(\widetilde{\mathbb{E}}_8) \right\}.$$

Let $n = |\Delta_0|$ be the number of vertices of Δ. We set

$$H = K\Delta,$$

and denote by $\mathcal{P}(H)$ the postprojective component of $\Gamma(\operatorname{mod} H)$.

Step 2°. We show how the multiplicity-free postprojective tilting H-modules T look. First, we note that any such a module T is a direct sum

$$T = T_1 \oplus \ldots \oplus T_n$$

of pairwise non-isomorphic indecomposable H-modules $T_1, \ldots, T_n$ from the component $\mathcal{P}(H)$ such that $\operatorname{Ext}^1_H(T, T) = 0$.

Because there is an isomorphism $\operatorname{Ext}^1_H(T, T) \cong D\operatorname{Hom}_A(T, \tau_H T)$, for any H-module T, and the indecomposable modules in $\mathcal{P}(H)$ are directing then, given an H-module $T = T_1 \oplus \ldots \oplus T_n$, with pairwise non-isomorphic indecomposable H-modules $T_1, \ldots, T_n$ in $\mathcal{P}(H)$, the following two conditions are equivalent.

  (i) T is a tilting H-module,
  (ii) $\operatorname{Hom}_H(T_i, \tau_H T_j) = 0$, for all $i, j \in \{1, \ldots, n\}$ such that $i \neq j$.

Step 3°. We find a subquiver $\mathcal{DP}(H)$ of $\mathcal{P}(H)$ with the following properties:

  (d1) $\mathcal{DP}(H)$ is a finite full translation subquiver of $\mathcal{P}(H)$ and is closed under the predecessors in $\mathcal{P}(H)$,
  (d2) for any multiplicity-free postprojective tilting H-module $T = T_1 \oplus \ldots \oplus T_n$, there exists a postprojective tilting module

$T' = T'_1 \oplus \ldots \oplus T'_n$ such that the H-modules $T'_1, \ldots, T'_n$ are indecomposable, pairwise non-isomorphic, lie in $\mathcal{DP}(H)$, and there is an isomorphism of algebras

$$\text{End}\, T_H \cong \text{End}\, T'_H.$$

We call such a translation subquiver $\mathcal{DP}(H)$ of $\mathcal{P}(H)$ a **concealed domain** of $\mathcal{P}(H)$.

To construct $\mathcal{DP}(H)$, we choose a minimal positive integer r_Δ such that

$$\text{Hom}_H(P(a), \tau_H^{-r} P(b)) \neq 0,$$

for all $r \geq r_\Delta$ and all vertices a and b of Δ, where $P(a) = e_a H$ and $P(b) = e_b H$ are the indecomposable projective H modules corresponding to the vertices a and b, respectively.

We define $\mathcal{DP}(H)$ to be the full translation subquiver of $\mathcal{P}(H)$ whose vertices are the modules

$$\tau_H^{-r} P(a); \quad \text{with } a \in \Delta_0 \text{ and } r \in \{0, 1, \ldots, r_\Delta\}.$$

First we show that the definition of $\mathcal{DP}(H)$ is correct. To see this, we recall that every module in the postprojective component $\mathcal{P}(H)$ is of the form $\tau_H^{-r} P(a)$, for some integer $r \geq 0$ and a vertex $a \in \Delta_0$. Moreover, by (2.2), the number of pairwise non-isomorphic non-sincere indecomposable H-modules in $\mathcal{P}(H)$ is finite, that is, all but a finite number of indecomposable modules in $\mathcal{P}(H)$ are sincere, see also (IX.5.6). It follows that

- given a vertex $a \in \Delta_0$, there exists a minimal positive integer r_a such that
$$\text{Hom}_H(P(a), \tau_H^{-r} P(b)) \neq 0,$$
for all $r \geq r_a$ and all vertices b of Δ, and hence
- there exists a minimal positive integer r_Δ such that
$$\text{Hom}_H(P(a), \tau_H^{-r} P(b)) \neq 0,$$

for all $r \geq r_\Delta$ and all vertices a and b of Δ.

This shows that the definition of $\mathcal{DP}(H)$ is correct.

Now we show that $\mathcal{DP}(H)$ is a concealed domain of $\mathcal{P}(H)$. Note that the indecomposable projective H modules $P(a) = e_a H$, with $a \in \Delta_0$, form a section $\Sigma \cong \Delta^{\text{op}}$ of $\mathcal{DP}(H)$. It follows that the subquiver $\mathcal{DP}(H)$ of $\mathcal{P}(H)$ is closed under the predecessors in $\mathcal{P}(H)$, and therefore the condition (d1) is satisfied.

To prove (d2), assume that $T = T_1 \oplus \ldots \oplus T_n$ is a tilting H-module such that the H-modules $T_1, \ldots, T_n$ are indecomposable, pairwise non-isomorphic, and lie in $\mathcal{P}(H)$. Then, for each $i \in \{1, \ldots, n\}$, there exist a vertex $a_i \in \Delta_0$ and a positive integer m_i such that

$$T_i \cong \tau_H^{-m_i} P(a_i).$$

Let m be the minimal number of the set $\{m_1, \ldots, m_n\}$ and assume, for simplicity, that $m = m_1$. Because, by the equivalence of (i) and (ii) of Step $2°$, we have

- $\mathrm{Hom}_H(T_1, \tau_H T_j) = 0$, for all $j \in \{2, \ldots, n\}$, and
- there is an isomorphism of K-vector spaces

$$0 = \mathrm{Hom}_H(T_1, \tau_H T_j) = \mathrm{Hom}_H(\tau_H^{-m_1} P(a_1),\ \tau_H^{-m_j+1} P(a_j))$$
$$\cong \mathrm{Hom}_H(P(a_1),\ \tau_H^{-(m_j-m_1)+1} P(a_j)),$$

then $m_j - m = m_j - m_1 \leq r_\Delta$, and hence we get $m_j \leq m + r_\Delta$. We set

$$T_1' = \tau_H^m T_1,\ T_2' = \tau_H^m T_2,\ \ldots\ldots,\ \text{and}\ T_n' = \tau_H^m T_n.$$

It is clear that the H-modules $T_1', \ldots, T_n'$ are indecomposable, postprojective, pairwise non-isomorphic, lie in $\mathcal{DP}(H)$ and, under the assumption $m = m_1$, the module T_1' is projective. Moreover, we have $\mathrm{Hom}_H(T_i', \tau_H T_j') \cong \mathrm{Hom}_H(T_i, \tau_H T_j) = 0$, for all $i, j \in \{1, \ldots, n\}$. It follows that

$$T' = T_1' \oplus \ldots \oplus T_n'$$

is a multiplicity-free postprojective tilting H-module such that $\mathrm{End}\, T_H \cong \mathrm{End}\, T_H'$. Consequently, the condition (d2) is satisfied, and therefore $\mathcal{DP}(H)$ is a concealed domain of $\mathcal{P}(H)$, as we claimed.

Step $4°$. Now we describe a procedure for constructing, up to the admissible operations $(\mathbf{Op_1})$, $(\mathbf{Op_2})$ and $(\mathbf{Op_3})$, the frames $\mathcal{F}r_1, \ldots, \mathcal{F}r_{149}$ of concealed algebras B of Euclidean type presented in Table 4.5.

Assume that B is such an algebra. Then, according to Step $1°$ and Step $2°$, the algebra B has the form

$$B \cong \mathrm{End}\, T_H,$$

where

- $H = K\Delta$,
- Δ is one of the canonically oriented Euclidean quivers, and
- $T = T_1 \oplus \ldots \oplus T_n$ is a multiplicity-free postprojective tilting H-module.

It is shown in Step $3°$ that there is a multiplicity-free tilting H-module $T' = T'_1 \oplus \ldots \oplus T'_n$ such that the indecomposable summands $T'_1, \ldots, T'_n$ lie in a finite concealed domain $\mathcal{DP}(H)$ of $\mathcal{P}(H)$ and

$$B \cong \operatorname{End} T_H \cong \operatorname{End} T'_H.$$

Consequently, the problem of finding all concealed algebras B of Euclidean type, reduces to

(a) describing a finite concealed domain $\mathcal{DP}(H)$ of $\mathcal{P}(H)$,
(b) describing all families $T_1, \ldots, T_n$ of pairwise non-isomorphic modules in the domain $\mathcal{DP}(H)$ such that one of the modules $T_1, \ldots, T_n$ is projective and

$$\operatorname{Hom}_H(T_i, \tau_H T_j) = 0,$$

for all $i, j \in \{1, \ldots, n\}$ with $i \neq j$ (see (ii) of Step $2°$), and
(c) describing the frame of the algebra

$$B \cong \operatorname{End}_H(T_1 \oplus \ldots \oplus T_n),$$

for each such a family $T_1, \ldots, T_n$. This means, one should describe a bound quiver (Q, I) such that $B \cong KQ/I$.

Because the concealed domain $\mathcal{DP}(H)$ of $\mathcal{P}(H)$ is finite then, given a canonically oriented Euclidean quiver Δ, the problem of finding all concealed algebras B of the Euclidean type Δ reduces to a finite combinatorial problem that admits an effective computer computation. This is in fact the way the frames of Table 4.5 are computed by Happel–Vossieck [112] in case Δ is of any of the types $\Delta(\widetilde{\mathbb{E}}_6)$, $\Delta(\widetilde{\mathbb{E}}_7)$, and $\Delta(\widetilde{\mathbb{E}}_8)$.

Now we prove the theorem in the cases when Δ is of the type $\Delta(\widetilde{\mathbb{A}}_{p,q})$, with $1 \leq p \leq q$, or Δ is of one of the types $\Delta(\widetilde{\mathbb{D}}_4)$, $\Delta(\widetilde{\mathbb{D}}_5)$, and $\Delta(\widetilde{\mathbb{D}}_6)$.

Case $1°$. Concealed algebras of type $\Delta(\widetilde{\mathbb{A}}_{p,q})$, with $1 \leq p \leq q$. We show that any concealed algebra of type $\Delta(\widetilde{\mathbb{A}}_{p,q})$ is a hereditary algebra of the Euclidean type $\widetilde{\mathbb{A}}_m$, where $m = p + q - 1$.

Assume that p and q are integers such that $1 \leq p \leq q$. Let $H = K\Delta$ be the path algebra of the quiver

$$\Delta = \Delta(\widetilde{\mathbb{A}}_{p,q}):$$

$$
\begin{array}{c}
1 \longleftarrow 2 \longleftarrow \cdots \longleftarrow p-1 \\
\nearrow \qquad\qquad\qquad\qquad\qquad \nwarrow \\
0 \qquad\qquad\qquad\qquad\qquad\qquad p+q-1. \\
\nwarrow \qquad\qquad\qquad\qquad\qquad \swarrow \\
p \longleftarrow p+1 \longleftarrow \cdots \longleftarrow p+q-2
\end{array}
$$

of the Euclidean type $\widetilde{\mathbb{A}}_{p+q-1}$. The standard calculation technique shows that the left hand part of the postprojective component $\mathcal{P}(H)$ of $\Gamma(\operatorname{mod} H)$ looks as follows

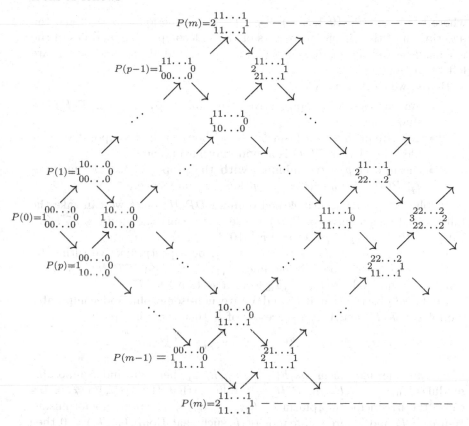

where $m = p + q - 1$, the indecomposable modules are represented by their dimension vectors, and we identify the horizontal dotted lines.

Because $\mathcal{P}(H)$ is a postprojective component of $\Gamma(\operatorname{mod} H)$, the indecomposable modules in $\mathcal{P}(H)$ are of the form $\tau_H^{-r}P(a)$, where $r \geq 0$ and $a \in \{0, 1, \dots, p + q - 1\}$. Moreover, for any indecomposable module M in $\mathcal{P}(H)$ there exists an almost split sequence

$$0 \longrightarrow M \longrightarrow X \oplus Y \longrightarrow N \longrightarrow 0$$

in $\operatorname{mod} H$, where X and Y are indecomposable modules, and we have

$$\dim M < \dim X < \dim N, \text{ and}$$
$$\dim M < \dim Y < \dim N.$$

To prove the inequalities, we apply the fact that M is of the form

$$M \cong \tau_H^{-r} P(a),$$

where $r \geq 0$ and $a \in \{0, 1, \dots, p+q-1\}$, the induction on $r \geq 0$, and the description of the dimension vectors of the indecomposable modules in the left hand part of the component $\mathcal{P}(H)$ are presented above. The details are left to the reader.

Hence, we easily conclude that

- any successor of the projective H-module $P(p+q-1)$ in $\mathcal{P}(H)$ is a sincere module,
- any irreducible morphism $U \longrightarrow V$ between indecomposable modules U and V in $\mathcal{P}(H)$ is a monomorphism, and
- the integer p is the minimal with the property that the modules $\tau_H^{-r} P(a)$, with $a \in \{0, 1, \dots, p+q-1\}$, are sincere.

It follows that for the concealed domain $\mathcal{DP}(H)$ of H we can take the full translation subquiver of $\mathcal{P}(H)$ whose vertices are the modules $\tau_H^{-r} P(a)$, with $a \in \{0, 1, \dots, p+q-1\}$ and $r \in \{0, 1, \dots, p\}$.

Let $T = T_1 \oplus \dots \oplus T_n$, with $n = p+q$, be a postprojective tilting H-module with indecomposable modules $T_1, \dots, T_n$ from $\mathcal{DP}(H)$. Now we prove that the modules $T_1, \dots, T_n$ form a section Σ of $\mathcal{P}(H)$.

First, we observe that if U and V are non-isomorphic indecomposable modules in $\mathcal{P}(H)$ and V is a successor of U then there is a path

$$U = U_0 \xrightarrow{h_0} U_1 \xrightarrow{h_1} \dots \xrightarrow{h_m} U_m \xrightarrow{h_{m+1}} U_{m+1} = V$$

of irreducible morphism $h_0, h_1, \dots, h_m, h_{m+1}$ between indecomposable modules $U_0, U_1, \dots, U_m$ in $\mathcal{P}(H)$. It follows that $\mathrm{Hom}_H(U, V) \neq 0$, because the irreducible morphism $h_0, h_1, \dots, h_m, h_{m+1}$ are monomorphisms. Hence, if T_i and T_j are summands of T such that $\mathrm{Hom}_H(T_i, T_j) \neq 0$ then T_j is a successor of T_i and there is a sectional path from T_i to T_j, because $\mathrm{Hom}_H(T_i, \tau_H Y_j) = 0$. Consequently,

- the modules $T_1, \dots, T_n$ form a section Σ of $\mathcal{P}(H)$, because $n = p+q$ is the number of τ_H-orbits of $\mathcal{P}(H)$, and
- the associated concealed algebra $B = \mathrm{End}\, T_H$ is the path algebra of the quiver $Q = \Sigma^{\mathrm{op}}$.

On the other hand, we have

- the indecomposable projective H-modules

$$P(0), P(1), \dots, P(p+q-1)$$

form a section Δ of the component $\mathcal{P}(H)$,

- the quiver Σ can be obtained from Δ^{op} by applying a finite sequence of reflections, see (VIII.1.8), and
- p is the number of clockwise-oriented arrows in Σ and q is the number of counterclockwise-oriented arrows in Σ.

It follows that q is the number of clockwise-oriented arrows in Q and p is the number of counterclockwise-oriented arrows in Q. Moreover, if Q is an arbitrary quiver such that the underlying graph $\overline{Q}$ of Q coincides with the underlying graph $\overline{\Delta}$ of Δ, and the numbers of clockwise-oriented arrows in Q and counterclockwise-oriented arrows in Q equal q and p, respectively, then the component $\mathcal{P}(H)$ admits a section Σ isomorphic with Q^{op} contained entirely in the concealed domain $\mathcal{DP}(H)$.

Hence we conclude that the concealed algebra $B = \operatorname{End} T_H$ is hereditary of the Euclidean type $\widetilde{\mathbb{A}}_m$, where $m = p + q - 1$. Consequently, by the discussion in Steps 1°-4° of the proof, any concealed algebra of type $\Delta(\widetilde{\mathbb{A}}_{p,q})$ is a hereditary algebra of the Euclidean type $\widetilde{\mathbb{A}}_m$, where $m = p+q-1$. This finishes the proof in Case 1°.

Case 2°. Concealed algebras of type $\Delta(\widetilde{\mathbb{D}}_4)$. We prove that, up to isomorphism, any concealed algebra of type $\Delta(\widetilde{\mathbb{D}}_4)$ is a hereditary algebra of the Euclidean type $\widetilde{\mathbb{D}}_4$ or the canonical algebra $C(2,2,2)$, see (XII.1.2). More precisely, we construct five pairwise non-isomorphic hereditary algebras $A^{(1)}, \ldots, A^{(5)}$ such that any concealed algebra of type $\Delta(\widetilde{\mathbb{D}}_4)$ is isomorphic to one of the algebras $A^{(1)}, \ldots, A^{(5)}$ or to $C(2,2,2)$. Let $H = K\Delta$ be the path algebra of the quiver

$$\Delta = \Delta(\widetilde{\mathbb{D}}_4):$$

of the Euclidean type $\widetilde{\mathbb{D}}_4$.

The standard calculation technique shows that the left hand part of the component $\mathcal{P}(H)$ of $\Gamma(\operatorname{mod} H)$ looks as follows

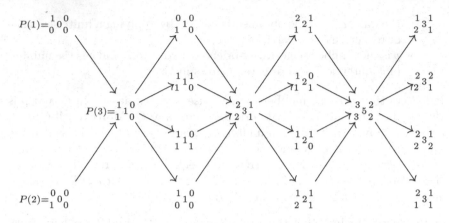

where the indecomposable modules are represented by their dimension vectors and

$$\mathbf{dim}\, P(4) = {}_1^1 1{}_0^1, \quad \mathbf{dim}\, P(5) = {}_0^1 1{}_1^1.$$

By applying the shape of $\mathcal{P}(H)$, we easily check that the H-modules

$$\tau_H^{-2} P(1), \quad \tau_H^{-2} P(2), \quad \tau_H^{-2} P(3), \quad \tau_H^{-2} P(4), \quad \tau_H^{-2} P(5)$$

are sincere and form a section in $\mathcal{P}(H)$ of type Δ^{op}. Moreover, for each $r \geq 2$, the H-modules $\tau_H^{-r} P(1), \tau_H^{-r} P(2), \tau_H^{-r} P(3), \tau_H^{-r} P(4), \tau_H^{-r} P(5)$ are sincere. It follows that the full translation subquiver $\mathcal{DP}(H)$ of $\mathcal{P}(H)$ whose vertices are the modules $\tau_H^{-r} P(1), \tau_H^{-r} P(2), \tau_H^{-r} P(3), \tau_H^{-r} P(4), \tau_H^{-r} P(5)$, with $r \in \{0, 1, 2\}$, is a concealed domain of $\mathcal{P}(H)$. In other words, $\mathcal{DP}(H)$ is the translation subquiver

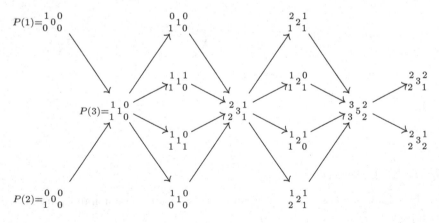

of $\mathcal{P}(H)$.

Let $T = T_1 \oplus T_2 \oplus T_3 \oplus T_4 \oplus T_5$ be an arbitrary postprojective tilting H-module, where the modules T_1, T_2, T_3, T_4, T_5 are indecomposable and lie in $\mathcal{DP}(H)$. By the discussion in Steps 1°-4° of the proof, without loss of generality, we may assume that one of the modules T_1, T_2, T_3, T_4, T_5 is projective.

The problem of describing all such postprojective tilting modules T splits into two cases.

Case 2.1°. The direct summands T_1, T_2, T_3, T_4, T_5 of T form a section Σ of the postprojective component $\mathcal{P}(H)$.

By invoking the condition $\mathrm{Hom}_H(T_i, \tau_H T_j) = 0$, for all $i, j \in \{1, 2, 3, 4, 5\}$ such that $i \neq j$, we easily see that

- each of the following five families
$$\Sigma^{(1)} = \{P(1), P(2), P(3), P(4), P(5)\},$$
$$\Sigma^{(2)} = \{\tau_H^{-1}P(1), P(2), P(3), P(4), P(5)\},$$
$$\Sigma^{(3)} = \{\tau_H^{-1}P(1), \tau_H^{-1}P(2), P(3), P(4), P(5)\},$$
$$\Sigma^{(4)} = \{\tau_H^{-2}P(1), \tau_H^{-1}P(2), \tau_H^{-1}P(3), P(4), P(5)\},$$
$$\Sigma^{(5)} = \{\tau_H^{-1}P(1), \tau_H^{-1}P(2), \tau_H^{-1}P(3), P(4), P(5)\},$$
of modules in $\mathcal{P}(H)$ forms a section of $\mathcal{P}(H)$,
- each of the following five H-modules
$$T^{(1)} = P(1) \oplus P(2) \oplus P(3) \oplus P(4) \oplus P(5) = H,$$
$$T^{(2)} = \tau_H^{-1}P(1) \oplus P(2) \oplus P(3) \oplus P(4) \oplus P(5),$$
$$T^{(3)} = \tau_H^{-1}P(1) \oplus \tau_H^{-1}P(2) \oplus P(3) \oplus P(4) \oplus P(5),$$
$$T^{(4)} = \tau_H^{-2}P(1) \oplus \tau_H^{-1}P(2) \oplus \tau_H^{-1}P(3) \oplus P(4) \oplus P(5),$$
$$T^{(5)} = \tau_H^{-1}P(1) \oplus \tau_H^{-1}P(2) \oplus \tau_H^{-1}P(3) \oplus P(4) \oplus P(5),$$
is a postprojective tilting H-module,
- each of the algebras
$$A^{(1)} = \mathrm{End}\, T_H^{(1)} \cong H \cong H^{\mathrm{op}}, \ A^{(2)} = \mathrm{End}\, T_H^{(2)}, \ A^{(3)} = \mathrm{End}\, T_H^{(3)},$$
$$A^{(4)} = \mathrm{End}\, T_H^{(4)} \cong (A^{(2)})^{\mathrm{op}}, \quad A^{(5)} = \mathrm{End}\, T_H^{(5)} \cong (A^{(3)})^{\mathrm{op}},$$
is concealed and $A^{(j)}$ is the path algebra of the quiver $Q^{(j)} = (\Sigma^{(j)})^{\mathrm{op}}$ of the Euclidean type $\widetilde{\mathbb{D}}_4$, for $j = 1, 2, 3, 4, 5$, where

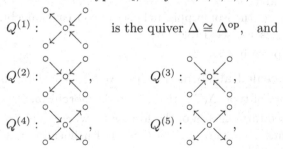

$Q^{(1)} :$ is the quiver $\Delta \cong \Delta^{\mathrm{op}}$, and

- the concealed algebras $A^{(1)}, \ldots, A^{(5)}$ are hereditary and pairwise non-isomorphic, and
- for any quiver Q such that the underlying graph $\overline{Q}$ of Q is the Euclidean graph $\widetilde{\mathbb{D}}_4$, there exists a section Σ of the component $\mathcal{P}(H)$ such that Σ is isomorphic with Q^{op} and is contained entirely in the concealed domain $\mathcal{DP}(H)$.

It follows that the concealed algebra $B = \operatorname{End} T_H$ is hereditary of the Euclidean type $\widetilde{\mathbb{D}}_4$.

Case 2.2°. The direct summands T_1, T_2, T_3, T_4, T_5 of T do not form a section of $\mathcal{P}(H)$. Again, by invoking the condition $\operatorname{Hom}_H(T_i, \tau_H T_j) = 0$, for all $i, j \in \{1, 2, 3, 4, 5\}$ such that $i \neq j$, we conclude that the postprojective tilting H-module $T = T_1 \oplus T_2 \oplus T_3 \oplus T_4 \oplus T_5$ is of one of the forms

- $T' = P(1) \oplus P(4) \oplus P(5) \oplus \tau_H^{-1} P(2) \oplus \tau_H^{-2} P(1)$, or
- $T'' = P(2) \oplus P(4) \oplus P(5) \oplus \tau_H^{-1} P(1) \oplus \tau_H^{-2} P(2)$.

It is not difficult to check that

- the associated concealed algebras $B' = \operatorname{End} T_H'$ and $B'' = \operatorname{End} T_H''$ are isomorphic, and
- the algebra $B' \cong B''$ is isomorphic to the canonical algebra $C(2, 2, 2)$ of the Euclidean type $\Delta(\widetilde{\mathbb{D}}_4)$, defined by the quiver

$$\Delta(2, 2, 2): \qquad$$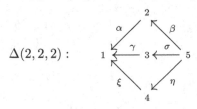

and bound by the relation $\beta\alpha + \sigma\gamma + \eta\xi = 0$.

It follows from the preceding discussion and the discussion in Steps 1°-4° of the proof, that

- any concealed algebra B of type $\Delta(\widetilde{\mathbb{D}}_4)$ is isomorphic either to one of the hereditary algebras $A^{(1)}, \ldots, A^{(5)}$, or to the canonical algebra $C(2, 2, 2)$, and
- the number of the isomorphism classes of concealed algebras of type $\Delta(\widetilde{\mathbb{D}}_4)$ equals six, see Table 4.6.

This finishes the proof in Case 2°.

Case 3°. Concealed algebras of type $\Delta(\widetilde{\mathbb{D}}_5)$. We show that if A is a concealed algebra of type $\Delta(\widetilde{\mathbb{D}}_5)$ then, up to isomorphism,

- A is a hereditary algebra of the form KQ, where Q is a quiver such that the underlying graph $\overline{Q}$ of Q is the Euclidean graph $\widetilde{\mathbb{D}}_5$, or

- A is not hereditary and A is isomorphic to the canonical algebra $C(2,2,3)$ or to one of the three algebras $B^{(1)}, B^{(2)}, B^{(3)}$, constructed in **(B1)-(B3)** below.
- the number of the isomorphism classes of concealed algebras of type $\Delta(\widetilde{\mathbb{D}}_5)$ equals $13 = 2^{5-1} - \frac{5+1}{2}$, see Table 4.6; nine of them are hereditary, and four of them are the non-hereditary algebras $C(2,2,3)$, $B^{(1)}, B^{(2)}, B^{(3)}$.

To prove the statements, we assume that $H = K\Delta$ is the path algebra of the quiver

$$\Delta = \Delta(\widetilde{\mathbb{D}}_5):\qquad \begin{array}{ccc} 1\!\circ & & 5\!\circ \\ \nwarrow & & \swarrow \\ \circ \leftarrow\!\!\!-\!\!\!-\!\!\!-\!\!\!-\!\!\!- \circ \\ \nearrow 3 & & 4 \searrow \\ 2\!\circ & & \circ 6 \end{array}$$

of the Euclidean type $\widetilde{\mathbb{D}}_5$.

The standard calculation technique shows that the left hand part of the postprojective component $\mathcal{P}(H)$ of $\Gamma(\mathrm{mod}\,H)$ looks as follows

Top row:
$P(5)={}_1^{1}11{}_1^{0}$ $-\,-\,-$ ${}_1^{1}21{}_1^{0}$ $-\,-\,-$ ${}_1^{1}22{}_1^{0}$ $-\,-\,-$ ${}_2^{2}33{}_2^{1}$

$P(6)$

Second row:
$P(4)={}_1^{1}11{}_1^{0} \to {}_1^{1}11{}_1^{1} \to {}_2^{2}32{}_1^{1} \to {}_1^{2}21{}_1^{1} \to {}_2^{2}43{}_1^{1} \to {}_1^{1}22{}_1^{0} \to {}_3^{3}55{}_2^{2} \to {}_2^{2}33{}_1^{2}$

Third row:
$P(1)={}_0^{1}00{}_0^{0} \to {}_1^{1}10{}_0^{0} \to {}_1^{0}10{}_0^{0} \to {}_1^{1}21{}_0^{0} \to {}_0^{1}11{}_0^{0} \to {}_2^{2}33{}_1^{0} \to {}_2^{1}22{}_1^{0} \to {}_3^{3}54{}_2^{0} \to {}_1^{2}32{}_1^{0} \to {}_3^{3}65{}_2^{0}$

$P(3)$ $\parallel$

Bottom row:
$P(2)={}_1^{0}00{}_0^{0}$ $-\,-\,-$ ${}_0^{0}10{}_0^{0}$ $-\,-\,-$ ${}_0^{1}11{}_0^{0}$ $-\,-\,-$ ${}_1^{2}22{}_1^{1}$ $-\,-\,-$ ${}_2^{1}32{}_1^{1}$ $-\,-\,-$

where the indecomposable modules are represented by their dimension vectors.

Observe that the modules

$$\tau_H^{-3}P(1),\quad \tau_H^{-3}P(2),\quad \tau_H^{-3}P(3),\quad \tau_H^{-3}P(4),\quad \tau_H^{-3}P(5),\quad \tau_H^{-3}P(6)$$

are sincere and form a section in $\mathcal{P}(H)$ of type Δ^{op}. Moreover, for each integer $r \geq 3$, the H-modules $\tau_H^{-r}P(1), \tau_H^{-r}P(2), \tau_H^{-r}P(3), \tau_H^{-r}P(4), \tau_H^{-r}P(5)$, and $\tau_H^{-r}P(6)$ are sincere. It follows that the full translation subquiver $\mathcal{DP}(H)$ of $\mathcal{P}(H)$ whose vertices are the modules

$$\tau_H^{-r}P(1), \tau_H^{-r}P(2), \tau_H^{-r}P(3), \tau_H^{-r}P(4), \tau_H^{-r}P(5), \tau_H^{-r}P(6),$$

with $r \in \{0,1,2,3\}$, is a concealed domain of $\mathcal{P}(H)$. In other words, $\mathcal{D}\mathcal{P}(H)$ is the translation subquiver

$$
\begin{array}{c}
P(5)={}_1^1 11_0^1 \quad ---\quad {}_1^1 21_1^0 \quad ---\quad {}_1^1 22_0^1 \quad ---\quad {}_2^2 33_2^1 \\
\nearrow \; P(6) \searrow \quad \nearrow \quad \searrow \quad \nearrow \quad \searrow \quad \nearrow \\
P(4)={}_1^1 11_0^0 \to {}_1^1 11_1^0 \to {}_2^2 32_1^1 \to {}_1^1 21_1^0 \to {}_2^2 43_1^1 \to {}_1^1 22_1^0 \to {}_3^3 55_2^2 \to {}_2^2 33_1^2 \\
\nearrow \quad \searrow \quad \nearrow \quad \searrow \quad \nearrow \quad \searrow \quad \nearrow \\
P(1)={}_0^1 00_0^0 \to {}_1^1 10_0^0 \to {}_1^0 10_0^0 \to {}_1^1 21_0^0 \to {}_0^1 11_0^0 \to {}_2^2 33_1^1 \to {}_2^1 22_1^0 \to {}_3^3 54_2^2 \\
\nearrow \; P(3) \searrow \quad \nearrow \quad \searrow \quad \nearrow \quad \searrow \quad \nearrow \\
P(2)={}_1^0 00_0^0 \quad ---\quad {}_0^1 10_0^0 \quad ---\quad {}_1^0 11_0^0 \quad ---\quad {}_1^2 22_1^1
\end{array}
$$

of $\mathcal{P}(H)$.

Following the idea applied in the proof of Case 2.1°, one shows that

- for any quiver Q such that the underlying graph $\overline{Q}$ of Q is the Euclidean graph $\widetilde{\mathbb{D}}_5$, there exists a section Σ of the component $\mathcal{P}(H)$ such that Σ is isomorphic with Q^{op} and is contained entirely in the concealed domain $\mathcal{D}\mathcal{P}(H)$,
- the path algebra KQ is isomorphic to the endomorphism algebra of the direct sum

$$T_\Sigma = T_1 \oplus T_2 \oplus T_3 \oplus T_4 \oplus T_5 \oplus T_6$$

of all indecomposable modules $T_1, T_2, T_3, T_4, T_5, T_6$ lying on Σ,
- there are only nine pairwise non-isomorphic algebras of the form $A_\Sigma = \mathrm{End}_H\, T_\Sigma$, where Σ is a section of the component $\mathcal{P}(H)$ and Σ is isomorphic with Q^{op}, for a quiver Q with $\overline{Q} = \widetilde{\mathbb{D}}_5$.

We recall that the modules on such a section satisfy the vanishing condition by $\mathrm{Hom}_H(T_i, \tau_H T_j) = 0$, for all $i, j \in \{1,2,3,4,5,6\}$ such that $i \neq j$.

To construct the nine algebras, we note that one of such a section is formed by the projective modules $P(1), P(2), P(3), P(4), P(5), P(6)$ and the corresponding concealed algebra is isomorphic with H. The remaining eight algebras are defined by the following eight sections in the concealed domain $\mathcal{D}\mathcal{P}(H)$ that are given by the modules

- $\tau_H^{-1} P(1), P(2), P(3), P(4), P(5), P(6)$,
- $\tau_H^{-1} P(1), \tau_H^{-1} P(2), P(3), P(4), P(5), P(6)$,
- $\tau_H^{-1} P(1), \tau_H^{-1} P(2), \tau_H^{-1} P(3), P(4), P(5), P(6)$,
- $\tau_H^{-2} P(1), \tau_H^{-1} P(2), \tau_H^{-1} P(3), P(4), P(5), P(6)$,
- $\tau_H^{-1} P(1), \tau_H^{-1} P(2), \tau_H^{-1} P(3), \tau_H^{-1} P(4), P(5), P(6)$,

- $\tau_H^{-1}P(1), \tau_H^{-2}P(2), \tau_H^{-1}P(3), \tau_H^{-1}P(4), P(5), P(6),$
- $\tau_H^{-2}P(1), \tau_H^{-3}P(2), \tau_H^{-1}P(3), \tau_H^{-1}P(4), P(5), P(6),$
- $\tau_H^{-2}P(1), \tau_H^{-1}P(2), \tau_H^{-1}P(3), \tau_H^{-1}P(4), \tau_H^{-1}P(5), P(6)$

Now we construct four pairwise non-isomorphic non-hereditary concealed algebras of the Euclidean type $\Delta(\widetilde{\mathbb{D}}_5)$.

(B1) To construct the first one, we consider the module

$$T^{(1)} = P(1) \oplus \tau_H^{-1}P(2) \oplus P(4) \oplus P(5) \oplus P(6) \oplus \tau_H^{-2}P(1),$$

and note that $\mathrm{Hom}_H(T^{(1)}, \tau_H T^{(1)}) = 0$. Hence, by the discussion in Steps $1°$–$4°$ of the proof, $T^{(1)}$ is a postprojective tilting H-module. Then

$$B^{(1)} = \mathrm{End}\, T_H^{(1)}$$

is a concealed algebra of type $\Delta(\widetilde{\mathbb{D}}_5)$, and one can show $B^{(1)}$ is given by the quiver

$$Q_{B^{(1)}} :$$

and bound by the commutativity relation $\alpha\beta = \gamma\delta$, where the numbering of vertices corresponds to the following order $P(1)$, $\tau_H^{-1}P(2)$, $P(4)$, $P(5)$, $P(6)$, $\tau_H^{-2}P(1)$ of the summands of $T^{(1)}$.

(B2) Similarly, we show that

$$T^{(2)} = \tau_H^{-1}P(1) \oplus \tau_H^{-2}P(2) \oplus \tau_H^{-1}P(4) \oplus P(5) \oplus P(6) \oplus \tau_H^{-3}P(1),$$

is a postprojective tilting H-module and

$$B^{(2)} = \mathrm{End}\, T_H^{(2)}$$

is a concealed algebra of type $\Delta(\widetilde{\mathbb{D}}_5)$ given by the quiver

$$Q_{B^{(2)}} :$$

and bound by the commutativity relation $\alpha\beta = \gamma\delta$.

(B3) One also shows that

$$T^{(3)} = \tau_H^{-1}P(1) \oplus \tau_H^{-2}P(2) \oplus \tau_H^{-1}P(4) \oplus \tau_H^{-1}P(5) \oplus P(6) \oplus \tau_H^{-3}P(1)$$

is a postprojective tilting H-module and

$$B^{(3)} = \operatorname{End} T_H^{(3)}$$

is a concealed algebra of type $\Delta(\widetilde{\mathbb{D}}_5)$ given by the quiver

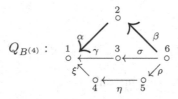

and bound by the commutativity relation $\alpha\beta = \gamma\delta$.

(B4) Finally, one easily shows that

$$T^{(4)} = P(2) \oplus P(5) \oplus P(6) \oplus \tau_H^{-1}P(1) \oplus \tau_H^{-2}P(2) \oplus \tau_H^{-3}P(1),$$

is a postprojective tilting H-module and the concealed algebra

$$B^{(4)} = \operatorname{End} T_H^{(4)}$$

is isomorphic to the canonical algebra $C(2,2,3)$ of type $\Delta(\widetilde{\mathbb{D}}_6)$ given by the quiver

$$Q_{B^{(4)}} :$$

and bound by the relation $\beta\alpha + \sigma\gamma + \rho\eta\xi = 0$.

A simple analysis shows that the four algebras $B^{(1)}$, $B^{(2)}$, $B^{(3)}$, and $B^{(4)} \cong$ $C(2,2,3)$, constructed in **(B1)**–**(B4)** have the following properties:

- the algebras $B^{(1)}$, $B^{(2)}$, $B^{(3)}$, and $B^{(4)}$ are pairwise non-isomorphic,
- the algebras $B^{(1)}$, $B^{(2)}$, $B^{(3)}$, and $B^{(4)}$ are of global dimension two, and
- any non-hereditary concealed algebra of type $\Delta(\widetilde{\mathbb{D}}_5)$ and of the form $\widehat{B} = \operatorname{End} \widehat{T}_H$, where $\widehat{T}_H$ is a tilting module that is a direct sum of six indecomposable modules lying in the concealed domain $\mathcal{DP}(H)$, is isomorphic to one of the algebras $B^{(1)}$, $B^{(2)}$, $B^{(3)}$, and $B^{(4)}$.

Hence, by the discussion in Steps 1°–4° of the proof, any non-hereditary concealed algebra of type $\Delta(\widetilde{\mathbb{D}}_5)$ is isomorphic to one of the four algebras $B^{(1)}$, $B^{(2)}$, $B^{(3)}$, and $B^{(4)} \cong C(2,2,3)$ constructed in **(B1)**–**(B4)**. This finishes the proof in Case 3°.

Case 4°. Concealed algebras of type $\Delta(\widetilde{\mathbb{D}}_6)$. We show that if A is a concealed algebra of type $\Delta(\widetilde{\mathbb{D}}_6)$ then, up to isomorphism,

- A is a hereditary algebra of the form KQ, where Q is a quiver such that the underlying graph $\overline{Q}$ of Q is the Euclidean graph $\widetilde{\mathbb{D}}_6$, or
- A is not hereditary and A is isomorphic to the canonical algebra $C(2,2,3)$ or to one of the 7 algebras $C^{(1)}$–$C^{(7)}$, constructed in **(C1)**–**(C7)** below.
- the number of the isomorphism classes of concealed algebras of type $\Delta(\widetilde{\mathbb{D}}_6)$ equals (see Table 4.6)

$$29 = 2^{6-1} - \frac{6}{2};$$

21 of them are hereditary, and 8 of them are the non-hereditary algebras $C(2,2,4)$ and $C^{(1)}$–$C^{(7)}$.

To prove the statements, we assume that $H = K\Delta$ is the path algebra of the quiver

$$\Delta = \Delta(\widetilde{\mathbb{D}}_6):$$

of the Euclidean type $\widetilde{\mathbb{D}}_6$.

The standard calculation technique shows that the left hand part of the postprojective component $\mathcal{P}(H)$ of $\Gamma(\mathrm{mod}\,H)$ looks as follows

where the indecomposable modules are represented by their dimension vectors.

Observe that the modules

$$\tau_H^{-4}P(1),\ \tau_H^{-4}P(2),\ \tau_H^{-4}P(3),\ \tau_H^{-4}P(4),\ \tau_H^{-4}P(5),\ \tau_H^{-4}P(6),\ \tau_H^{-4}P(7)$$

are sincere and form a section in $\mathcal{P}(H)$ of type Δ^{op}. Moreover, for each $r \geq 4$, the H-modules $\tau_H^{-r}P(1),\ \tau_H^{-r}P(2),\ \tau_H^{-r}P(3),\ \tau_H^{-r}P(4),\ \tau_H^{-r}P(5),$ $\tau_H^{-r}P(6)$, and $\tau_H^{-r}P(7)$ are sincere. It follows that, the full translation subquiver $\mathcal{DP}(H)$ of $\mathcal{P}(H)$ whose vertices are the modules

$$\tau_H^{-r}P(1), \tau_H^{-r}P(2), \tau_H^{-r}P(3), \tau_H^{-r}P(4), \tau_H^{-r}P(5), \tau_H^{-r}P(6), \tau_H^{-r}P(7),$$

with $r \in \{0,1,2,3,4\}$, is a concealed domain of $\mathcal{P}(H)$. In other words, $\mathcal{DP}(H)$ is the translation subquiver

of $\mathcal{P}(H)$.

Following the idea applied in the proof of Case 2.1°, one shows that

- for any quiver Q such that the underlying graph $\overline{Q}$ of Q is the Euclidean graph $\widetilde{\mathbb{D}}_6$, there exists a section Σ of the component $\mathcal{P}(H)$ such that Σ is isomorphic with Q^{op} and is contained entirely in the concealed domain $\mathcal{DP}(H)$,

- the path algebra KQ is isomorphic to the endomorphism algebra of the direct sum

$$T_{\Sigma} = T_1 \oplus T_2 \oplus T_3 \oplus T_4 \oplus T_5 \oplus T_6 \oplus T_7$$

of all indecomposable modules $T_1, T_2, T_3, T_4, T_5, T_6, T_7$ lying on Σ,

- there are 21 pairwise non-isomorphic algebras of the form $A_{\Sigma} = \mathrm{End}_H\,T_{\Sigma}$, where Σ is a section of the component $\mathcal{P}(H)$ and Σ is isomorphic with Q^{op}, for a quiver Q with $\overline{Q} = \widetilde{\mathbb{D}}_6$.

We recall that the modules lying on such a section satisfy the vanishing condition $\mathrm{Hom}_H(T_i, \tau_H T_j) = 0$, for all $i, j \in \{1,2,3,4,5,6,7\}$ such that $i \neq j$.

To construct the 21 algebras, we note that one of such a section is

$$\Sigma^{(1)} = \{P(1), P(2), P(3), P(4), P(5), P(6), P(7)\}$$

formed by the indecomposable projective modules, and the corresponding concealed algebra is isomorphic with H. The remaining 20 algebras are defined by the following 20 sections in the concealed domain $\mathcal{DP}(H)$ that are given by the modules

$$\Sigma^{(2)} = \{\tau_H^{-1}P(1), P(2), P(3), P(4), P(5), P(6), P(7)\},$$
$$\Sigma^{(3)} = \{\tau_H^{-1}P(1), \tau_H^{-1}P(2), P(3), P(4), P(5), P(6), P(7)\},$$
$$\Sigma^{(4)} = \{\tau_H^{-1}P(1), \tau_H^{-1}P(2), \tau_H^{-1}P(3), P(4), P(5), P(6), P(7)\},$$
$$\Sigma^{(5)} = \{\tau_H^{-2}P(1), \tau_H^{-1}P(2), \tau_H^{-1}P(3), P(4), P(5), P(6), P(7)\},$$
$$\Sigma^{(6)} = \{\tau_H^{-2}P(1), \tau_H^{-2}P(2), \tau_H^{-1}P(3), P(4), P(5), P(6), P(7)\},$$
$$\Sigma^{(7)} = \{\tau_H^{-1}P(1), \tau_H^{-1}P(2), \tau_H^{-1}P(3), \tau_H^{-1}P(4), P(5), P(6), P(7)\},$$
$$\Sigma^{(8)} = \{\tau_H^{-2}P(1), \tau_H^{-1}P(2), \tau_H^{-1}P(3), \tau_H^{-1}P(4), P(5), P(6), P(7)\},$$
$$\Sigma^{(9)} = \{\tau_H^{-2}P(1), \tau_H^{-2}P(2), \tau_H^{-1}P(3), \tau_H^{-1}P(4), P(5), P(6), P(7)\},$$
$$\Sigma^{(10)} = \{\tau_H^{-2}P(1), \tau_H^{-2}P(2), \tau_H^{-2}P(3), \tau_H^{-1}P(4), P(5), P(6), P(7)\},$$
$$\Sigma^{(11)} = \{\tau_H^{-3}P(1), \tau_H^{-1}P(2), \tau_H^{-2}P(3), \tau_H^{-1}P(4), P(5), P(6), P(7)\},$$
$$\Sigma^{(12)} = \{\tau_H^{-2}P(1), \tau_H^{-2}P(2), \tau_H^{-1}P(3), \tau_H^{-1}P(4), P(5), P(6), P(7)\},$$
$$\Sigma^{(13)} = \{\tau_H^{-1}P(1), \tau_H^{-1}P(2), \tau_H^{-1}P(3), \tau_H^{-1}P(4), \tau_H^{-1}P(5), P(6), P(7)\},$$
$$\Sigma^{(14)} = \{\tau_H^{-2}P(1), \tau_H^{-2}P(2), \tau_H^{-2}P(3), \tau_H^{-1}P(4), \tau_H^{-1}P(5), P(6), P(7)\},$$
$$\Sigma^{(15)} = \{\tau_H^{-3}P(1), \tau_H^{-2}P(2), \tau_H^{-2}P(3), \tau_H^{-1}P(4), \tau_H^{-1}P(5), P(6), P(7)\},$$
$$\Sigma^{(16)} = \{\tau_H^{-2}P(1), \tau_H^{-2}P(2), \tau_H^{-2}P(3), \tau_H^{-2}P(4), \tau_H^{-1}P(5), P(6), P(7)\},$$
$$\Sigma^{(17)} = \{\tau_H^{-3}P(1), \tau_H^{-2}P(2), \tau_H^{-2}P(3), \tau_H^{-2}P(4), \tau_H^{-1}P(5), P(6), P(7)\},$$
$$\Sigma^{(18)} = \{\tau_H^{-1}P(1), \tau_H^{-1}P(2), \tau_H^{-1}P(3), \tau_H^{-1}P(4), \tau_H^{-1}P(5), \tau_H^{-1}P(6), P(7)\},$$
$$\Sigma^{(19)} = \{\tau_H^{-2}P(1), \tau_H^{-1}P(2), \tau_H^{-1}P(3), \tau_H^{-1}P(4), \tau_H^{-1}P(5), \tau_H^{-1}P(6), P(7)\},$$
$$\Sigma^{(20)} = \{\tau_H^{-3}P(1), \tau_H^{-2}P(2), \tau_H^{-2}P(3), \tau_H^{-1}P(4), \tau_H^{-1}P(5), \tau_H^{-1}P(6), P(7)\},$$
$$\Sigma^{(21)} = \{\tau_H^{-3}P(1), \tau_H^{-2}P(2), \tau_H^{-2}P(3), \tau_H^{-2}P(4), \tau_H^{-1}P(5), \tau_H^{-1}P(6), P(7)\}.$$

Now we construct eight pairwise non-isomorphic non-hereditary concealed algebras of the Euclidean type $\Delta(\widetilde{\mathbb{D}}_6)$.

(C1) To construct the first one, we consider the module

$$T^{(1)} = P(1) \oplus \tau_H^{-1}P(2) \oplus \tau_H^{-2}P(1) \oplus P(4) \oplus P(5) \oplus P(6) \oplus P(7),$$

and note that $\operatorname{Hom}_H(T^{(1)}, \tau_H T^{(1)}) = 0$. Hence, by the discussion in Steps 1°–4° of the proof, $T^{(1)}$ is a postprojective tilting H-module and

$$C^{(1)} = \operatorname{End} T_H^{(1)}$$

is a concealed algebra of type $\Delta(\widetilde{\mathbb{D}}_6)$. One shows that $C^{(1)}$ is given by the quiver

$$Q_{C^{(1)}} : \quad$$

and bound by the commutativity relation $\alpha\beta = \gamma\delta$, where the numbering of the vertices corresponds to the following order $P(1)$, $\tau_H^{-1}P(2)$, $\tau_H^{-2}P(1)$, $P(4)$, $P(5)$, $P(6)$, $P(7)$ of the summands of $T^{(1)}$.

(C2) The module

$$T^{(2)} = \tau_H^{-1}P(1)\oplus\tau_H^{-2}P(2)\oplus\tau_H^{-3}P(1)\oplus\tau_H^{-1}P(4)\oplus\tau_H^{-1}P(5)\oplus P(6)\oplus P(7),$$

is a postprojective tilting H-module and

$$C^{(2)} = \operatorname{End} T_H^{(2)}$$

is a concealed algebra of type $\Delta(\widetilde{\mathbb{D}}_6)$ given by the quiver

$$Q_{C^{(2)}} : \quad$$

and bound by the commutativity relation $\alpha\beta = \gamma\delta$.

(C3) The module

$$T^{(3)} = \tau_H^{-1}P(1)\oplus\tau_H^{-2}P(2)\oplus\tau_H^{-3}P(1)\oplus\tau_H^{-1}P(4)\oplus\tau_H^{-1}P(5)\oplus P(6)\oplus\tau_H^{-1}P(7)$$

is a postprojective tilting H-module and

$$C^{(3)} = \operatorname{End} T_H^{(3)}$$

is a concealed algebra of type $\Delta(\widetilde{\mathbb{D}}_6)$ given by the quiver

$$Q_{C^{(3)}} : \quad$$

and bound by the commutativity relation $\alpha\beta = \gamma\delta$.

(C4) The module
$$T^{(4)} = \tau_H^{-1}P(1) \oplus \tau_H^{-2}P(2) \oplus \tau_H^{-3}P(1) \oplus \tau_H^{-1}P(4) \oplus P(5) \oplus P(6) \oplus P(7)$$
is a postprojective tilting H-module and
$$C^{(4)} = \operatorname{End} T_H^{(4)}$$
is a concealed algebra of type $\Delta(\widetilde{\mathbb{D}}_6)$ given by the quiver

and bound by the commutativity relation $\alpha\beta = \gamma\delta$.

(C5) The module
$$T^{(5)} = \tau_H^{-2}P(1) \oplus \tau_H^{-3}P(2) \oplus \tau_H^{-4}P(1) \oplus \tau_H^{-2}P(4) \oplus \tau_H^{-1}P(5) \oplus P(6) \oplus P(7)$$
is a postprojective tilting H-module and
$$C^{(5)} = \operatorname{End} T_H^{(5)}$$
is a concealed algebra of type $\Delta(\widetilde{\mathbb{D}}_6)$ given by the quiver

and bound by the commutativity relation $\alpha\beta = \gamma\delta$.

(C6) Finally, the module
$$T^{(6)} = \tau_H^{-2}P(1) \oplus \tau_H^{-3}P(2) \oplus \tau_H^{-4}P(1) \oplus \tau_H^{-2}P(4) \oplus \tau_H^{-1}P(5) \oplus P(6) \oplus \tau_H^{-1}P(7)$$
is a postprojective tilting H-module and
$$C^{(6)} = \operatorname{End} T_H^{(6)}$$
is a concealed algebra of type $\Delta(\widetilde{\mathbb{D}}_6)$ given by the quiver

and bound by the commutativity relation $\alpha\beta = \gamma\delta$.

(C7) Now, consider the module

$$T^{(7)} = \tau_H^{-2} P(1) \oplus \tau_H^{-3} P(2) \oplus \tau_H^{-4} P(1) \oplus \tau_H^{-2} P(4) \oplus P(7) \oplus \tau_H^{-1} P(6) \oplus \tau_H^{-2} P(7).$$

It is easy to see that $T^{(7)}$ is a postprojective tilting H-module and

$$C^{(7)} = \operatorname{End} T_H^{(7)}$$

is a concealed algebra of type $\Delta(\widetilde{\mathbb{D}}_6)$ given by the quiver

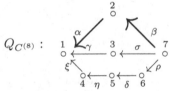

and bound by two commutativity relations $\alpha\beta = \gamma\sigma$ and $\mu\delta = \xi\eta$.

(C8) Finally, we consider the postprojective tilting H-module

$$T^{(8)} = P(2) \oplus P(6) \oplus P(7) \oplus \tau_H^{-1} P(1) \oplus \tau_H^{-2} P(2) \oplus \tau_H^{-3} P(1) \oplus \tau_H^{-4} P(2).$$

One can easily show that the concealed algebra

$$C^{(8)} = \operatorname{End} T_H^{(8)}$$

is isomorphic to the canonical algebra $C(2,2,4)$ of type $\Delta(\widetilde{\mathbb{D}}_6)$. The algebra $C^{(8)}$ is given by the quiver

$$
Q_{C^{(8)}} : \quad
\begin{array}{c}
\overset{2}{\circ} \\
\overset{\alpha}{\swarrow} \quad \overset{\beta}{\nwarrow} \\
\overset{1}{\circ} \xleftarrow{\gamma} \overset{3}{\circ} \xleftarrow{\sigma} \overset{7}{\circ} \\
\overset{\xi}{\nwarrow} \quad \swarrow \rho \\
\underset{4}{\circ} \xleftarrow{\eta} \underset{5}{\circ} \xleftarrow{\delta} \underset{6}{\circ}
\end{array}
$$

and bound by the relation $\beta\alpha + \sigma\gamma + \rho\delta\eta\xi = 0$.

A simple analysis shows that the eight algebras $C^{(1)} - C^{(7)}$, and $C^{(8)} \cong C(2,2,4)$, constructed in **(C1)**–**(C8)** have the following properties:

- the algebras $C^{(1)} - C^{(7)}$, $C^{(8)} \cong C(2,2,4)$, are pairwise non-isomorphic,
- the algebras $C^{(1)} - C^{(7)}$, and $C^{(8)} \cong C(2,2,4)$ are of global dimension two, and
- any non-hereditary concealed algebra of type $\Delta(\widetilde{\mathbb{D}}_6)$ and of the form $\widehat{B} = \operatorname{End} \widehat{T}_H$, where $\widehat{T}_H$ is a tilting module that is a direct sum of seven indecomposable modules lying in the concealed domain $\mathcal{DP}(H)$, is isomorphic to one of the algebras $C^{(1)} - C^{(7)}$, $C^{(8)} \cong C(2,2,4)$.

Hence, by the discussion in Steps 1°–4° of the proof, any non-hereditary concealed algebra of type $\Delta(\widetilde{\mathbb{D}}_6)$ is isomorphic to one of the eight algebras $C^{(1)}\!-\!C^{(7)}$, $C^{(8)} \cong C(2,2,4)$ constructed in **(C1)–(C8)**. This finishes the proof in Case 4°.   □

The following corollary is a consequence of the proof of the preceding theorem.

4.4. Corollary. *Let A be a concealed algebra of the Euclidean type*

$$\Delta \in \left\{ \Delta(\widetilde{\mathbb{A}}_{p,q}), \text{ with } 1 \leq p \leq q, \ \Delta(\widetilde{\mathbb{D}}_4), \Delta(\widetilde{\mathbb{D}}_5), \Delta(\widetilde{\mathbb{D}}_6) \right\}.$$

(a) *If $\Delta = \Delta(\widetilde{\mathbb{A}}_{p,q})$, where $1 \leq p \leq q$, then A is a hereditary algebra of the Euclidean type $\widetilde{\mathbb{A}}_m$, where $m = p + q - 1$.*

(b) *If $\Delta = \Delta(\widetilde{\mathbb{D}}_4)$ then A is isomorphic either to one of the five hereditary algebras $A^{(1)}$, $A^{(2)}$, $A^{(3)}$, $A^{(4)}$, $A^{(5)}$ of the Euclidean type $\widetilde{\mathbb{D}}_4$ corresponding to the five different orientations of the Euclidean graph $\widetilde{\mathbb{D}}_4$, or A is isomorphic to the canonical algebra $C(2,2,2)$. The algebra $C(2,2,2)$ has the frame $\mathcal{F}r_3$.*

(c) *If $\Delta = \Delta(\widetilde{\mathbb{D}}_5)$, then A satisfies one of the following conditions*

• *A is a hereditary algebra of the Euclidean type $\widetilde{\mathbb{D}}_5$ and A is isomorphic either to one of the 9 hereditary algebras of the Euclidean type $\widetilde{\mathbb{D}}_5$ corresponding to the nine different orientations of the Euclidean graph $\widetilde{\mathbb{D}}_5$, or*

• *A is not hereditary, gl.dim $A = 2$, and A is isomorphic to the canonical algebra $C(2,2,3)$, or to one of the algebras $B^{(1)}$, $B^{(2)}$, and $B^{(3)}$ constructed in **(B1)–(B3)** of the proof of (4.3). The algebra $C(2,2,3)$ has the frame $\mathcal{F}r_3$, and each of the algebras $B^{(1)}$, $B^{(2)}$, $B^{(3)}$ has the frame $\mathcal{F}r_4$.*

(d) *If $\Delta = \Delta(\widetilde{\mathbb{D}}_6)$, then A satisfies one of the following conditions*

• *A is a hereditary algebra of the Euclidean type $\widetilde{\mathbb{D}}_6$ and A is isomorphic either to one of the 21 hereditary algebras of the Euclidean type $\widetilde{\mathbb{D}}_6$ corresponding to the 21 different orientations of the Euclidean graph $\widetilde{\mathbb{D}}_6$, or*

• *A is not hereditary, gl.dim $A = 2$, and A is isomorphic to the canonical algebra $C(2,2,4)$, or to one of the seven algebras*

$$C^{(1)}, \ C^{(2)}, \ C^{(3)}, \ C^{(4)}, \ C^{(5)}, \ C^{(6)}, \ and \ C^{(7)}$$

*constructed in **(C1)–(C7)** of the proof of (4.3). The algebra $C(2,2,4)$ has the frame $\mathcal{F}r_3$, and each of the algebras $C^{(1)}\!-\!C^{(7)}$ has the frame $\mathcal{F}r_4$.*

Proof. Apply the proof of (4.3).   □

4.5. Table. A complete list of frames of tame concealed algebras
of Euclidean type

The table contains a list of **frames** $\mathcal{F}r_1, \dots, \mathcal{F}r_{149}$. Each of the frames (equipped with an orientation) is a bound quiver $\mathcal{F}r_j = (Q^{(j)}, I^{(j)})$ such that the bound quiver algebra $B^{(j)} = KQ^{(j)}/I^{(j)}$ is concealed of Euclidean type.

By the dotted lines we indicate the relation ρ in $I^{(j)} \subseteq KQ^{(j)}$ that the sum of all paths from the starting point of ρ to the ending point of ρ is zero.

The positive integers that appear instead of the vertices of the quiver $Q^{(j)}$ are the coordinates of the positive generator $\mathbf{h}_{B^{(j)}} \in \operatorname{rad} q_{B^{(j)}} \subseteq \mathbb{Z}^{n_j}$ of the radical $\operatorname{rad} q_{B^{(j)}}$ of the Euler quadratic form $q_{B^{(j)}} : \mathbb{Z}^{n_j} \longrightarrow \mathbb{Z}$ of $B^{(j)}$, where $n_j = |Q_0^{(j)}|$.

PART 4.5.1. THE FRAME $\mathcal{F}r_1$ OF TYPE $\widetilde{\mathbb{A}}_n$, $n \geq 1$

PART 4.5.2. THE FRAMES $\mathcal{F}r_2, \dots, \mathcal{F}r_5$ OF TYPE $\widetilde{\mathbb{D}}_n$, $n \geq 4$

Here the relation ρ defined earlier is denoted by the long vertical dotted line.

PART 4.5.3. THE FRAMES $\mathcal{F}r_6, \dots, \mathcal{F}r_{10}$ OF TYPE $\widetilde{\mathbb{E}}_6$

PART 4.5.4. THE FRAMES $\mathcal{F}r_{11}, \ldots, \mathcal{F}r_{32}$ OF TYPE $\widetilde{\widetilde{\mathbb{E}}}_7$

PART 4.5.5. THE FRAMES $\mathcal{F}r_{33},\dots,\mathcal{F}r_{149}$ OF TYPE $\widetilde{\mathbb{E}}_8$

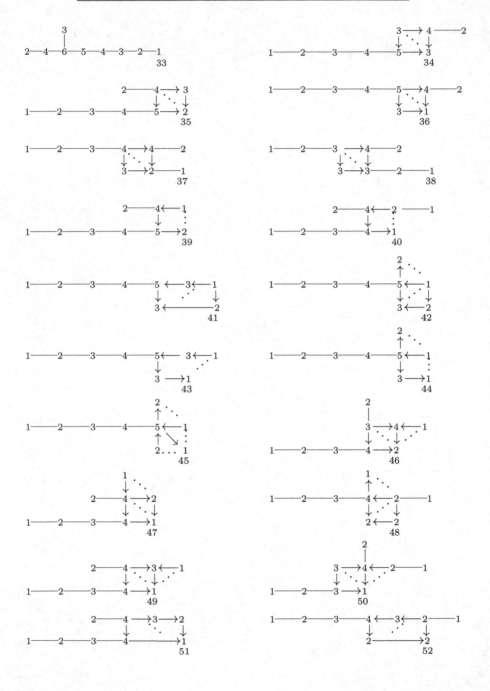

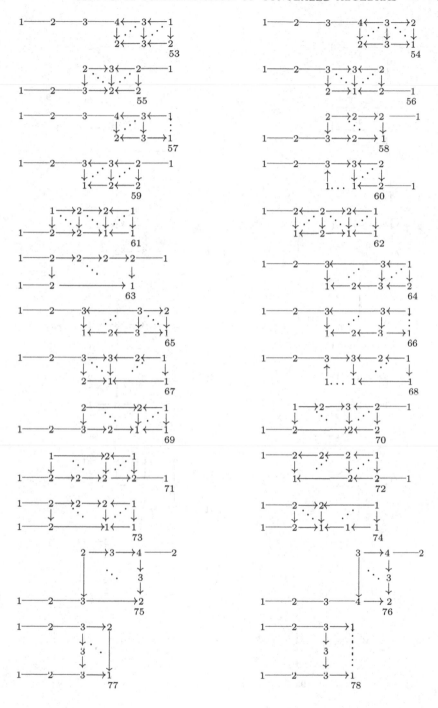

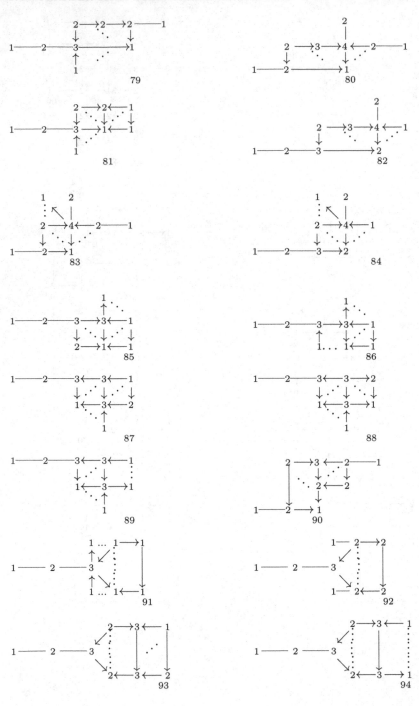

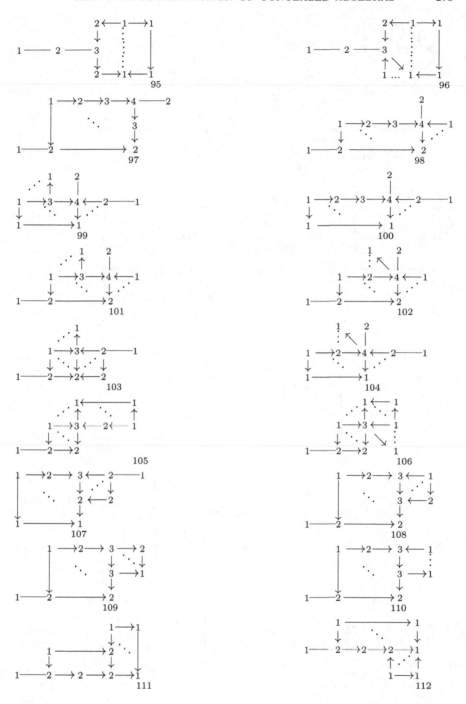

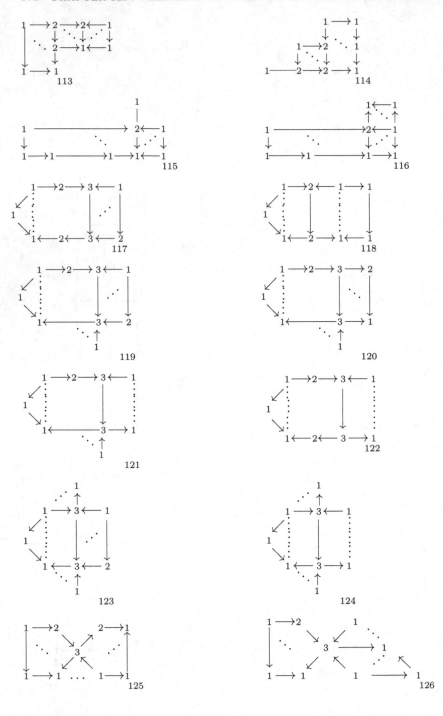

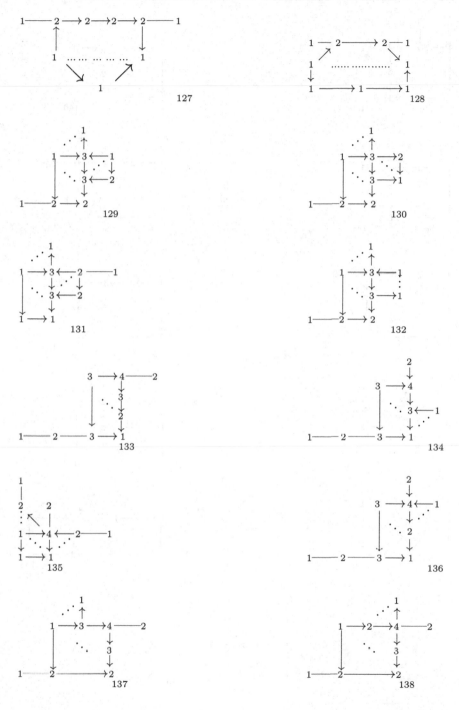

127

128

129

130

131

132

133

134

135

136

137

138

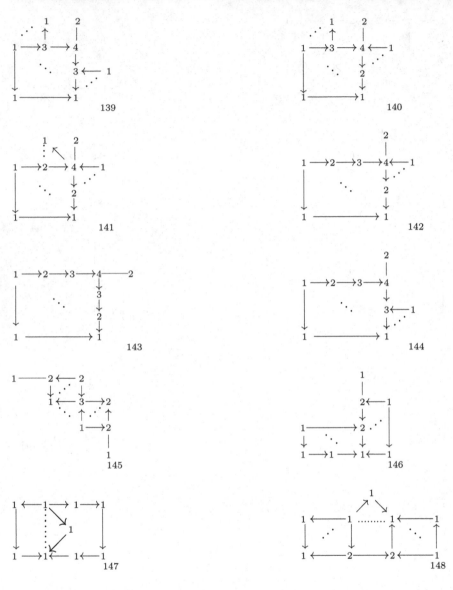

For the convenience of the reader, we complete the list of Table 4.5 with the following one given in [112].

4.6. Table. The number of the isomorphism classes of concealed algebras for the different Euclidean types

TYPE   AN ESTIMATION FOR THE NUMBER OF THE ISOMORPHISM
CLASSES OF CONCEALED ALGEBRAS

$\widetilde{\mathbb{A}}_n$: $= \left(\frac{1}{2(n+1)} \sum_{d \mid n+1} \varphi(d) \cdot 2^{\frac{n+1}{d}} \right) + \begin{cases} -1, & \text{if } n \geq 1 \text{ is even,} \\ 2^{\frac{n-3}{2}} - 1, & \text{if } n \geq 1 \text{ is odd,} \end{cases}$

where $\varphi(d)$ denotes the Euler's φ-function (see [233]);

$\widetilde{\mathbb{D}}_n$: $\begin{cases} = 6, & \text{if } n = 4, \\ \leq 2^{n-1} - \dfrac{n+1}{2}, & \text{if } n \geq 5 \text{ and } n \text{ is odd,} \\ \leq 2^{n-1} - \dfrac{n+2}{2} + 2^{\frac{n-2}{2}}, & \text{if } n \geq 6 \text{ and } n \text{ is even;} \end{cases}$

$\widetilde{\mathbb{E}}_6$: $= 56$;

$\widetilde{\mathbb{E}}_7$: $= 437$;

$\widetilde{\mathbb{E}}_8$: $= 3809$.

We end this section with an illustration of the use of the criterion (3.1) by two additional examples.

4.7. Example. Let A be the algebra given by the quiver

and bound by the four commutativity relations

$$\alpha_1\gamma_1 = \gamma_2\beta_1, \quad \alpha_2\gamma_2 = \gamma_3\beta_2, \quad \alpha_3\gamma_3 = \gamma_4\beta_3, \quad \text{and} \quad \alpha_4\gamma_4 = \gamma_5\beta_4.$$

Observe that the algebra A has separated radical, and consequently, the Auslander–Reiten quiver $\Gamma(\text{mod } A)$ of A has a postprojective component. Let $e = e_a + e_b$ be the sum of two primitive idempotents e_a and e_b corresponding to the unique source a and the unique sink b of the above quiver. It is easy to see that the quotient algebra A/AeA is isomorphic with the algebra B given by the quiver and bound by the two commutativity relations $\alpha_2\gamma_2 = \gamma_3\beta_2$ and $\alpha_3\gamma_3 = \gamma_4\beta_3$. A simple inspection of the Table 4.5, with the frames of concealed algebras of Euclidean type, shows that B is a concealed algebra of the Euclidean type $\widetilde{\mathbb{E}}_7$ given by the frame $\mathcal{F}r_{18}$. In particular, by applying the criterion (3.1), we conclude that the algebra A is representation-infinite.

4.8. Example. Let A be the algebra given by the quiver

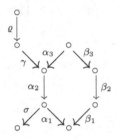

and bound by the following three relations

$$\alpha_3\alpha_2\alpha_1 = \beta_3\beta_2\beta_1 \ , \ \gamma\alpha_2\sigma = 0, \ \text{ and } \ \varrho\gamma = 0.$$

It is easy to see that the algebra A has the separated radical and therefore the Auslander–Reiten quiver $\Gamma(\mathrm{mod}\,A)$ of A contains a postprojective component.

On the other hand, a simple inspection of the frames of concealed algebras of Euclidean types shows that they do not appear as the algebras of the form A/AeA, where e is an idempotent of A. Therefore, the algebra A is representation-finite, the Auslander–Reiten quiver $\Gamma(\mathrm{mod}\,A)$ of A is finite, and coincides with its postprojective component.

XIV.5. Exercises

1. Let $Q = (Q_0, Q_1)$ be an acyclic quiver whose underlying graph is one of the Euclidean diagrams $\widetilde{\mathbb{A}}_m$, $\widetilde{\mathbb{D}}_m$, $\widetilde{\mathbb{E}}_6$, $\widetilde{\mathbb{E}}_7$, and $\widetilde{\mathbb{E}}_8$. Let $n = |Q_0| + 1$. Show that the integral quadratic form $q_Q : \mathbb{Z}^n \longrightarrow \mathbb{Z}$ of the quiver Q is a critical one, see Section VII.4.

2. Show that each of the following integral quadratic forms $q : \mathbb{Z}^n \longrightarrow \mathbb{Z}$ is a critical one.

(a) $n = 5$ and $q(\mathbf{x}) = x_1^2 + x_2^2 + x_3^2 + x_4^2 + x_5^2 - x_1x_5 - x_2x_5 - x_3x_5 - x_4x_5$.

(b) $n = 7$ and $q(\mathbf{x}) = x_1^2 + x_2^2 + x_3^2 + x_4^2 + x_5^2 + x_6^2 + x_7^2$
$$+ x_1x_2 + x_3x_4 + x_5x_6 - (x_1 + x_2 + x_3 + x_4 + x_5 + x_6)x_7.$$

(c) $n = 8$ and $q(\mathbf{x}) = x_1^2 + x_2^2 + x_3^2 + x_4^2 + x_5^2 + x_6^2 + x_7^2 + x_8^2 + x_2(x_3 + x_4)$
$\quad -(x_1 + x_2 + x_3)x_6 - (x_2 + x_3 + x_4)x_7 - (x_2 + x_4 + x_5)x_8.$

(d) $n = 9$ and $q(\mathbf{x}) = x_1^2 + x_2^2 + x_3^2 + x_4^2 + x_5^2 + x_6^2 + x_7^2 + x_8^2 + x_9^2$
$\quad + x_1 x_2 + x_1 x_4 + x_3 x_4 + x_5 x_6 + x_5 x_7$
$\quad + x_5 x_8 + x_6 x_7 + x_6 x_8 + x_7 x_8$
$\quad - (x_1 + x_2 + x_3 + x_4 + x_5 + x_6 + x_7 + x_8)x_9.$

(e) $n = 9$ and $q(\mathbf{x}) = x_1^2 + x_2^2 + x_3^2 + x_4^2 + x_5^2 + x_6^2 + x_7^2 + x_8^2 + x_9^2$
$\quad + x_4 x_5 + x_4 x_6 + x_5 x_6 + x_7 x_9 + x_8 x_9$
$\quad - x_1(x_4 + x_5 + x_6 + x_7 + x_9)$
$\quad - x_2(x_7 + x_8 + x_9) - x_3(x_8 + x_9).$

For each of the forms (a)-(e), find a positive sincere vector $\mathbf{h}_q \in \mathbb{Z}^n$ such that $\operatorname{rad} q = \mathbb{Z} \cdot \mathbf{h}_q$.

3. Show that each of the following integral quadratic forms $q : \mathbb{Z}^n \longrightarrow \mathbb{Z}$ is not critical

(a) $n = 5$ and $q(\mathbf{x}) = x_1^2 + x_2^2 + x_3^2 + x_4^2 + x_5^2$
$\quad + (x_1 + x_2)(x_3 + x_4) - (x_1 + x_2 + x_3 + x_4)x_5.$

(b) $n = 5$ and $q = x_1^2 + x_2^2 + x_3^2 + x_4^2 + x_5^2 + x_2 x_4$
$\quad - (x_1 + x_2)x_3 - (x_1 + x_2 + x_4)x_5.$

(c) $n = 6$ and $q(\mathbf{x}) = x_1^2 + x_2^2 + x_3^2 + x_4^2 + x_5^2 + x_6^2 + x_1(x_2 + x_3 + x_4 + x_5)$
$\quad + (x_2 + x_3)(x_4 + x_5) - (x_1 + x_2 + x_3 + x_4 + x_5)x_6.$

(d) $n = 6$ and $q(\mathbf{x}) = x_1^2 + x_2^2 + x_3^2 + x_4^2 + x_5^2 + x_6^2 + x_4 x_1 + x_5 x_1 + x_6 x_2 + x_6 x_3$
$\quad - x_2 x_1 - x_3 x_1 - x_6 x_1 - x_4 x_2 - x_5 x_2 - x_4 x_3$
$\quad - x_5 x_3 - x_6 x_4 - x_6 x_5.$

(e) $n = 6$ and $q(\mathbf{x}) = x_1^2 + x_2^2 + x_3^2 + x_4^2 + x_5^2 + x_6^2$
$\quad + x_1 x_3 + x_2 x_3 + x_2 x_4$
$\quad - (x_1 + x_2 + x_3)x_5) - (x_1 + x_2 + x_4)x_6.$

Show that the quadratic forms of (b) and (c) are $\mathbb{Z}$-congruent.

4. Let B be a K-algebra given by the quiver

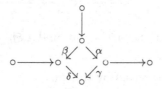

and bound by the commutativity relation $\alpha\gamma = \beta\delta$. Prove that B is a concealed algebra of the Euclidean type $\widetilde{\mathbb{E}}_6$.

5. Let B be a K-algebra given by the quiver

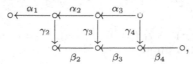

and bound by two commutativity relations $\alpha_2\gamma_2 = \gamma_3\beta_2$ and $\alpha_3\gamma_3 = \gamma_4\beta_3$. Prove that B is a concealed algebra of the Euclidean type $\widetilde{\mathbb{E}}_7$.

6. Let B be a K-algebra given by the quiver

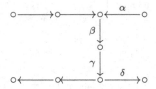

and bound by the zero relation $\alpha\beta\gamma\delta = 0$. Prove that B is a concealed algebra of the Euclidean type $\widetilde{\mathbb{E}}_8$.

7. Let A be an algebra given by the quiver

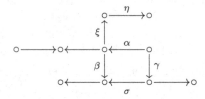

and bound by the zero relations $\alpha\beta = 0$, $\gamma\sigma = 0$, $\xi\eta = 0$, and $\alpha\eta\nu = 0$. Prove that A is representation-infinite.

8. Let A be an algebra given by the quiver

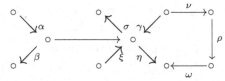

and bound by the relations $\alpha\beta = \gamma\sigma$ and $\alpha\xi\eta = 0$. Prove that A is representation-infinite.

9. Let A be an algebra given by the quiver

and bound by the relations $\alpha\beta = 0$, $\gamma\sigma = 0$, $\xi\eta = 0$, and $\gamma\eta = \nu\rho\omega$. Prove that A is representation-infinite.

10. Let A be the algebra given by the quiver

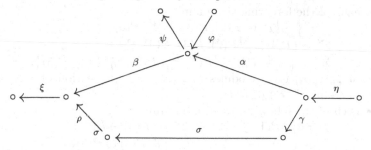

and bound by the relations $\alpha\beta = \gamma\sigma\rho$, $\rho\xi = 0$, $\eta\gamma = 0$ and $\varphi\psi = 0$. Prove that A is representation-finite and $\Gamma(\mathrm{mod}\, A)$ coincides with its postprojective component.

11. Let A be the algebra given by the quiver

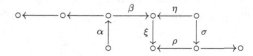

bound by the relations $\alpha\beta = 0$ and $\eta\xi = \sigma\rho$. Prove that A is representation-finite.

12. Let A be the algebra given by the quiver

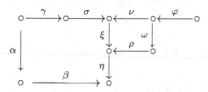

bound by the relations $\alpha\beta = \gamma\sigma\xi\eta$, $\nu\xi = \omega\rho$, and $\varphi\nu = 0$. Prove that A is representation-finite.

13. Prove that any minimal representation-infinite algebra A is basic and connected. **Hint:** To see that A is basic, look at the algebra A/AeA, where $e = 1 - e_A$ and e_A is the idempotent of A defined in (I.6.3).

14. Let $H = K\Delta$ be the path algebra of the quiver

$$\Delta = \Delta(\widetilde{\mathbb{D}}_4) :$$

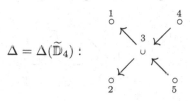

of the Euclidean type $\widetilde{\mathbb{D}}_4$. Prove that

- each of the following three families
$$\Sigma = \{\tau_H^{-1}P(1), P(2), P(3), P(4), P(5)\},$$
$$\Sigma' = \{P(1), \tau_H^{-1}P(2), P(3), P(4), P(5)\},$$
$$\Sigma'' = \{\tau_H^{-2}P(1), \tau_H^{-2}P(2), \tau_H^{-1}P(3), \tau_H^{-1}P(4), P(5)\},$$
of postprojective modules in $\mathcal{P}(H)$ forms a section of a concealed domain $\mathcal{DP}(H)$ of $\mathcal{P}(H)$,

- each of the following three H-modules
$$T = \tau_H^{-1}P(1) \oplus P(2) \oplus P(3) \oplus P(4) \oplus P(5),$$
$$T' = P(1) \oplus \tau_H^{-1}P(2) \oplus P(3) \oplus P(4) \oplus P(5),$$
$$T'' = \tau_H^{-2}P(1) \oplus \tau_H^{-2}P(2) \oplus \tau_H^{-1}P(3) \oplus \tau_H^{-1}P(4) \oplus P(5),$$
is a postprojective tilting H-module,

- the algebras $B = \operatorname{End} T_H$, $B' = \operatorname{End} T'_H$, $B'' = \operatorname{End} T''_H$ are concealed hereditary of the Euclidean type $\widetilde{\mathbb{D}}_4$, and

- there are isomorphisms of algebras $B \cong B' \cong B'' \cong KQ$, where Q is the Euclidean quiver

$$Q: \quad$$

Hint: Consult the concealed domain $\mathcal{DP}(H)$ of the postprojective component $\mathcal{P}(H)$ of H presented in the proof of Theorem 4.3.

15. Let $H = K\Delta$ be the path algebra of the quiver

$$\Delta = \Delta(\widetilde{\mathbb{D}}_5):$$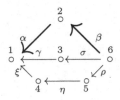

of the Euclidean type $\widetilde{\mathbb{D}}_5$.

(a) Prove that
$$T = P(2) \oplus P(5) \oplus P(6) \oplus \tau_H^{-1}P(1) \oplus \tau_H^{-2}P(2) \oplus \tau_H^{-3}P(1),$$

is a postprojective tilting H-module and the concealed algebra
$$B = \operatorname{End} T_H$$

is isomorphic to the canonical algebra $C(2, 2, 3)$ of type $\Delta(\widetilde{\mathbb{D}}_5)$ given by the quiver and bound by the relation $\beta\alpha + \sigma\gamma + \rho\eta\xi = 0$.

(b) Construct two non-isomorphic postprojective tilting H-modules T' and T'' such that

- the indecomposable direct summands of T' and T'' lie in the concealed domain $\mathcal{DP}(H)$ of the postprojective component $\mathcal{P}(H)$ of H,
- the H-modules T' and T'' are not isomorphic to the H-module T presented in (a),
- each of the concealed algebras $B' = \mathrm{End}\, T'_H$ and $B'' = \mathrm{End}\, T''_H$ is isomorphic to the canonical algebra $C(2, 2, 3)$ of type $\Delta(\widetilde{\mathbb{D}}_5)$.

Hint: Consult the concealed domain $\mathcal{DP}(H)$ of the postprojective component $\mathcal{P}(H)$ of H presented in the proof of Theorem 4.3.

16. Let $H = K\Delta$ be the path algebra of the quiver

$$\Delta = \Delta(\widetilde{\mathbb{D}}_6):$$

of the Euclidean type $\widetilde{\mathbb{D}}_6$.

(a) Prove that

- the modules

$$\tau_H^{-2} P(2), \tau_H^{-3} P(1), \tau_H^{-4} P(2), \tau_H^{-2} P(4), P(6), \tau_H^{-1} P(7), \tau_H^{-2} P(6)$$

form a section of a concealed domain $\mathcal{DP}(H)$ of $\mathcal{P}(H)$,
- the module

$$T = \tau_H^{-2} P(2) \oplus \tau_H^{-3} P(1) \oplus \tau_H^{-4} P(2) \oplus \tau_H^{-2} P(4) \oplus P(6)$$
$$\oplus \tau_H^{-1} P(7) \oplus \tau_H^{-2} P(6)$$

is a postprojective tilting H-module,
- $B = \mathrm{End}\, T_H$ is a concealed algebra of type $\Delta(\widetilde{\mathbb{D}}_6)$, and
- the algebra B is isomorphic to the concealed algebra $C^{(7)} = \mathrm{End}\, T_H^{(7)}$ (constructed in **(C7)** of the proof of Theorem 4.3) given by the quiver

and bound by two commutativity relations $\alpha\beta = \gamma\sigma$ and $\mu\delta = \xi\eta$.

(b) Prove that the H-module

$$T = P(2)\oplus P(6)\oplus P(7)\oplus\tau_H^{-1}P(1)\oplus\tau_H^{-2}P(2)\oplus\tau_H^{-3}P(1)\oplus\tau_H^{-4}P(2)$$

is a postprojective tilting H-module, and the concealed algebra

$$B = \operatorname{End} T_H$$

is isomorphic to the canonical algebra $C(2,2,4)$ of type $\Delta(\widetilde{\mathbb{D}}_6)$ given by the quiver

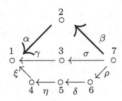

and bound by the relation $\beta\alpha + \sigma\gamma + \rho\delta\eta\xi = 0$.

(c) Construct two non-isomorphic postprojective tilting H-modules T' and T'' such that

- the indecomposable direct summands of T' and T'' lie in the concealed domain $\mathcal{DP}(H)$ of the postprojective component $\mathcal{P}(H)$ of H,
- the H-modules T' and T'' are not isomorphic to the H-module T constructed in (b),
- each of the concealed algebras $B' = \operatorname{End} T'_H$ and $B'' = \operatorname{End} T''_H$ is isomorphic to the canonical algebra $C(2,2,4)$ of type $\Delta(\widetilde{\mathbb{D}}_6)$.

Hint: Consult the concealed domain $\mathcal{DP}(H)$ of the postprojective component $\mathcal{P}(H)$ of H presented in the proof of Theorem 4.3.

Bibliography

[1] J. L. Alperin, *Local Representation Theory*, Cambridge Studies in Advanced Mathematics 11, Cambridge University Press, 1986.

[2] F. W. Anderson and K. R. Fuller, *Rings and Categories of Modules*, Graduate Texts in Mathematics 13, Springer-Verlag, New York, Heidelberg, Berlin, 1973 (new edition 1991).

[3] I. Assem, D. Simson and A. Skowroński, *Elements of the Representation Theory of Associative Algebras, Volume 1: Techniques of Representation Theory*, London Mathematical Society Student Texts 65, Cambridge University Press, 2006.

[4] I. Assem and A. Skowroński, Iterated tilted algebras of type $\widetilde{A}_n$, *Math. Z.*, 195(1987), 269–290.

[5] I. Assem and A. Skowroński, Algebras with cycle-finite derived categories, *Math. Ann.*, 280(1988), 441–463.

[6] I. Assem and A. Skowroński, Quadratic forms and iterated tilted algebras, *J. Algebra*, 128(1990), 55–85.

[7] M. Auslander, R. Bautista, M. Platzeck, I. Reiten and S. Smalø, Almost split sequences whose middle term has at most two indecomposable summands, *Canad. J. Math.*, 31(1979), 942–960.

[8] M. Auslander, M. I. Platzeck and I. Reiten, Coxeter functors without diagrams, *Trans. Amer. Math. Soc.*, 250(1979), 1–46.

[9] M. Auslander and I. Reiten, DTr-periodic modules and functors, In: *Representation Theory of Algebras*, Canad. Math. Soc. Conf. Proc., AMS, Vol. 18, 1996, pp. 39–50.

[10] M. Auslander, I. Reiten and S. Smalø, *Representation Theory of Artin Algebras*, Cambridge Studies in Advanced Mathematics 36, Cambridge University Press, 1995.

[11] M. Barot and J. A. de la Peña, Algebras whose Euler form is non-negative, *Colloq. Math.*, 79(1999), 119–131.

[12] R. Bautista, P. Gabriel, A. V. Roiter and L. Salmeron, Representation-finite algebras and multiplicative basis, *Invent. Math.*, 81(1985), 217–285.

285

[13] R. Bautista and S. Smalø, Nonexistent cycles, *Comm. Algebra*, 11(1983), 1755–1767.

[14] G. Belitskii and V. V. Sergeichuk, Complexity of matrix problems, *Linear Algebra Appl.*, 361(2003), 203–222.

[15] D. J. Benson, *Representations and Cohomology I: Basic Representation Theory of Finite Groups and Associative Algebras*, Cambridge Studies in Advanced Mathematics 30, Cambridge University Press, 1991.

[16] D. J. Benson, *Representations and Cohomology II: Cohomology of Groups and Modules*, Cambridge Studies in Advanced Mathematics 31, Cambridge University Press, 1991.

[17] I. N. Bernstein, I. M. Gelfand and V. A. Ponomarev, Coxeter functors and Gabriel's theorem, *Uspiehi Mat. Nauk*, 28(1973), 19–33 (in Russian); English translation in *Russian Math. Surveys*, 28(1973), 17–32.

[18] J. Białkowski and A. Skowroński, Selfinjective algebras of tubular type, *Colloq. Math.*, 94(2002), 175–194.

[19] J. Białkowski, A. Skowroński and K. Yamagata, Cartan matrices of symmetric algebras having generalized standard stable tubes, *Osaka J. Math.*, 2007, in press.

[20] G. Bobiński, P. Dräxler and A. Skowroński, Domestic algebras with many nonperiodic Auslander-Reiten components, *Comm. Algebra*, 31(2003), 1881–1926.

[21] R. Bocian, T. Holm and A. Skowroński, Derived equivalence classification of weakly symmetric algebras of Euclidean type, *J. Pure Appl. Algebra*, 191(2004), 43–74.

[22] R. Bocian and A. Skowroński, Symmetric special biserial algebras of Euclidean type, *Colloq. Math.*, 96(2003), 121–148.

[23] R. Bocian and A. Skowroński, Weakly symmetric algebras of Euclidean type, *J. reine angew. Math.*, 580(2005), 157–199.

[24] V. M. Bondarenko, Representations of dihedral groups over a field of characteristic 2, *Math. Sbornik*, 96(1975), 63–74 (in Russian).

[25] V. M. Bondarenko, Representations of bundles of semichained sets and their applications, *Algebra i Analiz*, 3(1991), 38–61 (in Russian); English translation: *St. Petersburg Math. J.*, 3(1992), 973–996.

[26] V. M. Bondarenko and Ju. A. Drozd, The representation type of finite groups, In: *Modules and Representations, Zap. Nauchn. Sem. Leningrad. Otdel. Mat. Inst. Steklov. (LOMI)*, 57 (1977), 24–41 (in Russian).

[27] K. Bongartz, Tilted algebras, In: *Representations of Algebras*, Lecture Notes in Math., No. 903, Springer-Verlag, Berlin, Heidelberg, New York, 1981, pp. 26–38.

[28] K. Bongartz, Algebras and quadratic forms, *J. London Math. Soc.*, 28(1983), 461–469.

[29] K. Bongartz, Critical simply connected algebras, *Manuscr. Math.*, 46(1984), 117–136.

[30] K. Bongartz, A criterion for finite representation type, *Math. Ann.*, 269(1984), 1–12.

[31] K. Bongartz, Some geometric aspects of representation theory, In: *Algebras and Modules I*, Canad. Math. Soc. Conf. Proc., AMS, Vol. 23, 1998, pp. 1–27.

[32] K. Bongartz and D. Dudek, Decomposition classes for representations of tame quivers, *J. Algebra*, 240(2001), 268–288.

[33] K. Bongartz and T. Fritzsche, On minimal disjoint degenerations for preprojective representations of quivers, *Math. Comp.*, 72(2003), 2013–2042.

[34] K. Bongartz and P. Gabriel, Covering spaces in representation theory, *Invent. Math.*, 65(1982), 331–378.

[35] K. Bongartz and C. M. Ringel, Representation-finite tree algebras, In: *Representations of Algebras*, Lecture Notes in Math., No. 903, Springer-Verlag, Berlin, Heidelberg, New York, 1981, pp. 39–54.

[36] N. Bourbaki, *Algèbres de Lie*, Chapitre IV, Masson, Paris, 1968.

[37] S. Brenner, On four subspaces of a vector space, *J. Algebra*, 29(1974), 100–114.

[38] S. Brenner, Quivers with commutativity conditions and some phenomenology of forms, In: *Representations of Algebras*, Lecture Notes in Math. No. 488, Springer-Verlag, Berlin, Heidelberg, New York, 1975, pp. 29–53.

[39] S. Brenner and M. C. R. Butler, Generalisations of the Bernstein–Gelfand–Ponomarev reflection functors, In: *Representation Theory II*, Lecture Notes in Math. No. 832, Springer-Verlag, Berlin, Heidelberg, New York, 1980, pp. 103–169.

[40] T. Brüstle, On positive roots of pg-critical algebras, *Linear Alg. Appl.* 365(2003), 107–114.

[41] T. Brüstle, Tame tree algebras, *J. reine angew. Math.*, 567(2004), 51–98.

[42] T. Brüstle and Y. Han, Tame two-point algebras without loops, *Comm. Algebra*, 29(2001), 4683–4692.

[43] T. Brüstle, J. A. de la Peña and A. Skowroński, Tame algebras and Tits quadratic forms, Preprint, Toruń, 2007.

[44] M. C. R. Butler and C. M. Ringel, Auslander-Reiten sequences with few middle terms and applications to string algebras, *Comm. Algebra*, 15(1987), 145–179.

[45] C. Chang and J. Weyman, Representations of quivers with free module of covariants, *J. Pure Appl. Algebra*, 192(2004), 69–94.

[46] F. U. Coelho, D. Happel, and L. Unger, Tilting up algebras of small homological dimensions, *J. Pure Appl. Algebra*, 174(2002), 219–241.

[47] F. U. Coelho and M. Lanzilotta, Algebras with small homological dimensions, *Manuscr. Math.*, 100(1999), 1–11.

[48] F. U. Coelho, E. N. Marcos, H. A. Merklen and A. Skowroński, Module categories with infinite radical square zero are of finite type, *Comm. Algebra*, 22(1994), 4511–4517.

[49] F. U. Coelho, E. N. Marcos, H. Merklen and A. Skowroński, Module categories with infinite radical cube zero, *J. Algebra*, 183(1996), 1–23.

[50] F. U. Coelho, J. A. de la Peña and B. Tomé, Algebras whose Tits form weakly controls the module category, *J. Algebra*, 191(1997), 89–108.

[51] F. U. Coelho and A. Skowroński, On Auslander-Reiten components for quasi-tilted algebras, *Fund. Math.*, 149(1996), 67–82.

[52] P. M. Cohn, *Skew Fields. Theory of General Division Rings,* Encyclopedia of Mathematics and its Applications, 57, Cambridge University Press, 1995.

[53] W. W. Crawley-Boevey, Tameness of biserial algebras, *Archiv Math. (Basel)*, 65(1995), 399–407.

[54] W. W. Crawley-Boevey, Infinite-dimensional modules in the representation theory of finite-dimensional algebras, In: *Algebras and Modules I*, Canad. Math. Soc. Conf. Proc., AMS, Vol. 23, 1998, pp. 29–54.

[55] W. W. Crawley-Boevey and C. M. Ringel, Algebras whose Auslander-Reiten quivers have large regular components, *J. Algebra*, 153(1992), 494–516.

[56] W. W. Crawley-Boevey and M. Van den Bergh, Absolutely indecomposable representations and Kac-Moody Lie algebras. With an appendix by Hiraki Nakajima, *Invent. Math.*, 155(2004), 537–539.

[57] C. W. Curtis and I. Reiner, *Representation Theory of Finite Groups and Associative Algebras*, Wiley (Interscience), New York, 1962.

[58] B. Deng, On a problem of Nazarova and Roiter, *Comment. Math. Helvetici*, 75(2000), 368–409.

[59] H. Derksen and J. Weyman, Semi-invariants of quivers and saturation for Littlewood-Richardson coefficients, *J. Amer. Math. Soc.*, 13(2000), 467–479.

[60] H. Derksen and J. Weyman, On the canonical decomposition of quiver representations, *Compositio Math.*, 133(2002), 245–265.

[61] G. D'Este and C. M. Ringel, Coherent tubes, *J. Algebra*, 87(1984), 150–201.

[62] V. Dlab and C. M. Ringel, On algebras of finite representation type, *J. Algebra*, 33(1975), 306–394.

[63] V. Dlab and C. M. Ringel, Indecomposable representations of graphs and algebras, *Memoirs Amer. Math. Soc.*, Vol. 173, 1976.

[64] V. Dlab and C. M. Ringel, Eigenvalues of Coxeter transformations and the Gelfand-Kirilov dimension of the preprojective algebras, *Proc. Amer. Math. Soc.*, 87(1981), 228–232.

[65] M. Domokos and H. Lenzing, Invariant theory of canonical algebras, *J. Algebra*, 228(2000), 738–762.

[66] P. Donovan and M. R. Freislich, The representation theory of finite graphs and associated algebras, *Carleton Lecture Notes*, 5, Ottawa, 1973.

[67] P. Dowbor, C. M. Ringel and D. Simson, Hereditary artinian rings of finite representation type, In: *Representation Theory II*, Lecture Notes in Math. No. 832, Springer-Verlag, Berlin, Heidelberg, New York, 1980, pp. 232–241.

[68] P. Dowbor and D. Simson, A characterization of hereditary rings of finite representation type, *Bull. Amer. Math. Soc.* 2(1980), 300–302.

[69] P. Dowbor and A. Skowroński, On Galois coverings of tame algebras, *Archiv Math. (Basel)*, 44(1985), 522–529.

[70] P. Dowbor and A. Skowroński, On the representation type of locally bounded categories, *Tsukuba J. Math.*, 10(1986), 63–77.

[71] P. Dowbor and A. Skowroński, Galois coverings of representation-infinite algebras, *Comment. Math. Helvetici*, 62(1987), 311–337.

[72] P. Dräxler, Completely separating algebras, *J. Algebra*, 165(1994), 550–565.

[73] P. Dräxler, Generalized one-point extensions, *Math. Ann.*, 304(1996), 645–667.

[74] P. Dräxler, N. Golovachtchuk, S. Ovsienko and J. A. de la Peña, Coordinates of maximal roots of weakly non-negative unit forms, *Colloq. Math.*, 78(1998), 163–193.

[75] P. Dräxler and J. A. de la Peña, On the existence of preprojective component in the Auslander-Reiten quiver of an algebra, *Tsukuba J. Math.*, 20(1996), 457–469.

[76] Yu. A. Drozd, Coxeter transformations and representations of partially ordered sets, *Funkc. Anal. i Priložen.*, 8(1974), 34–42 (in Russian).

[77] Yu. A. Drozd, On tame and wild matrix problems, In: *Matrix Problems*, Akad. Nauk Ukr. S.S.R., Inst. Mat., Kiev, 1977, pp. 104–114 (in Russian).

[78] Yu. A. Drozd, Tame and wild matrix problems, In: *Representations and Quadratic Forms*, Akad. Nauk USSR, Inst. Matem., Kiev, 1979, pp. 39–74 (in Russian).

[79] Yu. A. Drozd and V. V. Kirichenko, *Finite Dimensional Algebras*, Springer-Verlag, Berlin, Heidelberg, New York, 1994.

[80] S. Eilenberg, Abstract description of some basic functors, *J. Indian Math. Soc.*, 24(1960), 231–234.

[81] K. Erdmann, *Blocks of Tame Representation Type and Related Algebras*, Lecture Notes in Math. No. 1428, Springer-Verlag, Berlin-Heidelberg-New York, 1990.

[82] K. Erdmann, On Auslander-Reiten components for group algebras, *J. Pure Appl. Algebra,* 104(1995), 149–160.

[83] K. Erdmann and D. K. Nakano, Representation type of Hecke algebras of type A, *Trans. Amer. Math. Soc.*, 354(2002), 275–285.

[84] K. Erdmann and A. Skowroński, On Auslander-Reiten components of blocks and selfinjective biserial algebras, *Trans. Amer. Math. Soc.*, 330(1992), 165–189.

[85] R. Farnsteiner, On the Auslander-Reiten quiver of an infinitesimal group, *Nagoya. J. Math.*, 160(2000), 103–121.

[86] P. Gabriel, Unzerlegbare Darstellungen I, *Manuscr. Math.*, 6(1972), 71–103.

[87] P. Gabriel, Indecomposable representations II, *Symposia Mat. Inst. Naz. Alta Mat.*, 11(1973), 81–104.

[88] P. Gabriel, Représentations indécomposables, In: *Séminaire Bourbaki* (1973–74), Lecture Notes in Math., No. 431, Springer-Verlag, Berlin, Heidelberg, New York, 1975, pp. 143–169.

[89] P. Gabriel, Auslander–Reiten sequences and representation-finite algebras, In: *Representation Theory* I, Lecture Notes in Math., No. 831, Springer-Verlag, Berlin, Heidelberg, New York, 1980, pp. 1–71.

[90] P. Gabriel and C. Riedtmann, Group representations without groups, *Coment. Math. Helvetici*, 54(1979), 240–287.

[91] P. Gabriel and A. V. Roiter, *Representations of Finite Dimensional Algebras*, Algebra VIII, Encyclopaedia of Math. Sc., Vol. 73, Springer-Verlag, Berlin, Heidelberg, New York, 1992.

[92] W. Geigle and H. Lenzing, Perpendicular categories with applications to representations and sheaves, *J. Algebra*, 144(1991), 273–343.

[93] C. Geiss, On degenerations of tame and wild algebras, *Archiv Math. (Basel)*, 64(1995), 11–16.

[94] C. Geiss, Introduction to moduli spaces associated to quivers (with an appendix by Lieven Le Bruyn and Markus Reineke), In: *Trends in Representation Theory of Algebras and Related Topics*, Contemp. Math., 406(2006), 31–50.

[95] C. Geiss and J. A. de la Peña, Auslander-Reiten components for clans, *Bol. Soc. Mat. Mexicana*, 5(1999), 307–326.

[96] I. M. Gelfand and V. A. Ponomarev, Indecomposable representations of the Lorentz group, *Uspechi Mat. Nauk*, 2(1968), 1–60 (in Russian).

[97] I. M. Gelfand and V. A. Ponomarev, Problems of linear algebra and classification of quadruples of subspaces in a finite-dimensional vector space. *Coll. Math. Soc. Bolyai*, Tihany (Hungary), 5(1970), 163–237.

[98] E. L. Green, Group-graded algebras and the zero relation problem, In: *Representations of Algebras*, Lecture Notes in Math., No. 903, Springer-Verlag, Berlin, Heidelberg, New York, 1981, pp. 106–115.

[99] E. L. Green and I. Reiner, Integral representations and diagrams, *Michigan Math. J.*, 25(1968), 53–84.

[100] J. A. Green, Locally finite representations, *J. Algebra* 41(1976), 137–171.

[101] D. Happel, On the derived category of a finite dimensional algebra, *Comment. Math. Helvetici*, 62(1987), 339–388.

[102] D. Happel, A characterization of hereditary categories with tilting object, *Invent. Math.*, 144(2001), 381–398.

[103] D. Happel and S. Liu, Module categories without short cycles are of finite type, *Proc. Amer. Math. Soc.*, 120(1994), 371–375.

[104] D. Happel and C. M. Ringel, Construction of tilted algebras, In: *Representations of Algebras*, Lecture Notes in Math., No. 903, Springer-Verlag, Berlin, Heidelberg, New York, 1981, pp. 125–144.

[105] D. Happel and C. M. Ringel, Tilted algebras, *Trans. Amer. Math. Soc.*, 274(1982), 399–443.

[106] D. Happel and I. Reiten, Hereditary abelian categories with tilting object over arbitrary basic fields, *J. Algebra*, 256(2002), 414–432.

[107] D. Happel, I. Reiten and S. Smalø, Tilting in abelian categories and quasitilted algebras, *Memoirs Amer. Math. Soc.*, Vol. 575, 1996.

[108] D. Happel, I. Reiten and S. Smalø, Piecewise hereditary algebras, *Archiv Math. (Basel)*, 66(1996), 182–186.

[109] D. Happel and L. Unger, Almost complete tilting modules, *Proc. Amer. Math. Soc.*, 107(1989), 603–610.

[110] D. Happel and L. Unger, On the quiver of tilting modules, *J. Algebra*, 284(2005), 857–868.

[111] D. Happel and L. Unger, On partial order of tilting modules, *Algebras and Representation Theory*, 8(2005), 147–156.

[112] D. Happel and D. Vossieck, Minimal algebras of infinite representation type with preprojective component, *Manuscr. Math.*, 42(1983), 221–243.

[113] T. Holm and A. Skowroński, Derived equivalence classification of symmetric algebras of domestic type, *J. Math. Soc. Japan*, 58(2006), 1133–1149.

[114] H. J. von Höhne, On weakly positive unit forms, *Comment Math. Helvetici*, 63(1988), 312–336.

[115] H. J. von Höhne, On weakly non-negative unit forms and tame algebras, *Proc. London Math. Soc.*, 73(1996), 47–67.

[116] B. Huisgen-Zimmermann, Purity, algebraic compactness, direct sum decompositions, and representation type In: *Infinite Length Modules*, (Eds: H. Krause and C.M. Ringel), Trends in Mathematics, Birkhäuser Verlag, Basel-Boston-Berlin, 2000, pp. 331–367.

[117] D. Hughes and J. Waschbüsch, Trivial extensions of tilted algebras, *Proc. London Math. Soc.*, 46(1983), 347–364.

[118] O. Iyama, Finiteness of representation dimension, *Proc. Amer. Math. Soc.*, 131(2003), 1011–1014.

[119] O. Iyama, The relationship between homological properties and representation theoretic realization of artin algebras, *Trans. Amer. Math. Soc.*, 357(2004), 709–734.

[120] K. Igusa, M. I. Platzeck, G. Todorov and D. Zacharia, Auslander algebras of finite representation type, *Comm. Algebra*, 15(1987), 377–424.

[121] C. U. Jensen and H. Lenzing, *Model Theoretic Algebra With Particular Emphasis on Fields, Rings, Modules*, Algebra, Logic and Applications, Vol. 2, Gordon & Breach Science Publishers, New York, 1989.

[122] V. G. Kac, Infinite root systems, representations of graphs and invariant theory, *Invent. Math.*, 56(1980), 57–92; II, *J. Algebra*, 78(1982), 141–162.

[123] S. Kawata, On Auslander-Reiten components and simple modules for finite group algebras, *Osaka J. Math.*, 34(1997), 681–688.

[124] S. Kawata, G. O. Michler and K. Uno, On simple modules in the Auslander-Reiten components of finite groups, *Math. Z.*, 234(2000), 375–398.

[125] B. Keller, Tilting theory and differential graded algebras, In: *Finite Dimensional Algebras and Related Topics*, NATO ASI Series, Series C: Mathematical and Physical Sciences, Kluwer Acad. Publ., Dordrecht, Vol. 424, 1994, pp. 183–190.

[126] B. Keller, Derived categories and their uses, *in Handbook of Algebra*, (Ed.: M. Hazewinkel), Vol. 1, North-Holland Elsevier, Amsterdam, 1996, pp. 671–701.

[127] O. Kerner, Tilting wild algebras, *J. London Math. Soc.*, 39(1989), 29–47.

[128] O. Kerner, Elementary stones, *Comm. Algebra*, 22(1994), 1797–1806.

[129] O. Kerner and F. Lukas, Elementary modules, *Math. Z.*, 223(1996) 421–434.

[130] O. Kerner and A. Skowroński, On module categories with nilpotent infinite radical, *Compositio Math.*, 77(1991), 313–333.

[131] O. Kerner, A. Skowroński, K. Yamagata and D. Zacharia, Finiteness of the strong global dimension of radical square zero algebras, *Centr. Eur. J. Math.*, 2(2004), 103–111 (electronic).

[132] S. M. Khorochkin, Indecomposable representations of Lorentz groups, *Funkc. Anal. i Priložen.*, 15(1981), 50–60 (in Russian).

[133] A. D. King, Moduli of representations of finite-dimensional algebras, *Quart. J. Math. Oxford*, 45(1994), 515–530.

[134] H. Kraft and C. Riedtmann, Geometry of representation of quivers, In: *Representations of Algebras*, London Math. Soc. Lecture Notes Series, 116(1986), pp. 109–145.

[135] H. Krause, Stable equivalence preserves representation type, *Comment. Math. Helvetici*, 72(1997), 266–284.

[136] H. Krause, The spectrum of a module category, *Memoirs Amer. Math. Soc.*, Vol. 707, 2001.

[137] H. Krause and G. Zwara, Stable equivalence and generic modules, *Bull. London Math. Soc.*, 32(2000), 615–618.

[138] H. Lenzing, A *K*-theoretic study of canonical algebras, In: *Representation Theory of Algebras*, Canad. Math. Soc. Conf. Proc., AMS, Vol. 18, 1996, pp. 433–454.

[139] H. Lenzing, Generic modules over tubular algebras, In: *Advances in Algebra and Model Theory*, (Eds. M. Droste and R. Göbel), Algebra, Logic and Applications Series, Vol. 9, Gordon & Breach Science Publishers, Australia, 1997, pp. 375–385.

[140] H. Lenzing, Hereditary noetherian categories with a tilting object, *Proc. Amer. Math. Soc.*, 125(1997), 1893–1901.

[141] H. Lenzing, Representations of finite dimensional algebras and singularity theory, In: *Trends in Ring Theory*, Canad. Math. Soc. Conf. Proc., AMS, Vol. 22, 1998, pp. 71–97.

[142] H. Lenzing and J. A. de la Peña, Concealed-canonical algebras and separating tubular families, *Proc. London Math. Soc.*, 78(1999), 513–540.

[143] H. Lenzing and J. A. de la Peña, Supercanonical algebras, *J. Algebra*, 282(2004), 298–348.

[144] H. Lenzing and A. Skowroński, Quasi-tilted algebras of canonical type, *Colloq. Math.*, 71(1996), 161–181.

[145] H. Lenzing and A. Skowroński, On selfinjective algebras of Euclidean type, *Colloq. Math.*, 79(1999), 71–76.

[146] H. Lenzing and A. Skowroński, Roots of Nakayama and Auslander-Reiten translations, *Colloq. Math.*, 86(2000), 209–230.

[147] Z. Leszczyński, On the representation type of tensor product algebras, *Fund. Math.*, 144(1994), 143–151.

[148] Z. Leszczyński, Representation-tame locally hereditary algebras, *Colloq. Math.*, 99(2004), 175–187.

[149] Z. Leszczyński and A. Skowroński, Auslander algebras of tame representation type, In: *Representation Theory of Algebras*, Canad. Math. Soc. Conf. Proc., AMS, Vol. 18, 1996, pp. 475–486.

[150] Z. Leszczyński and A. Skowroński, Tame tensor products of algebras, *Colloq. Math.*, 98(2003), 125–145.

[151] Z. Leszczyński and A. Skowroński, Tame generalized canonical algebras, *J. Algebra*, 273(2004), 412–433.

[152] S. Liu, The degrees of irreducible maps and the shapes of the Auslander–Reiten quivers, *J. London Math. Soc.*, 45(1992), 32–54.

[153] S. Liu, Semi-stable components of an Auslander–Reiten quiver, *J. London Math. Soc.*, 47(1993), 405–416.

[154] S. Liu, Tilted algebras and generalized standard Auslander–Reiten components, *Archiv Math. (Basel)*, 61(1993), 12–19.

[155] S. Liu, The connected components of the Auslander–Reiten quiver of a tilted algebra, *J. Algebra*, 161(1993), 505–523.

[156] S. Liu, Almost split sequences for non-regular modules, *Fund. Math.*, 143(1993), 183–190.

[157] S. Liu, Infinite radical in standard Auslander–Reiten components, *J. Algebra*, 166(1994), 245–254.

[158] S. Liu, On short cycles in a module category, *J. London Math. Soc.*, 51(1995), 62–74.

[159] S. Liu and R. Schultz, The existence of bounded infinite DTr-orbits, *Proc. Amer. Math. Soc.*, 122(1994), 1003–1005.

[160] Y. Liu and C. C. Xi, Constructions of stable equivalences of Morita type for finite dimensional algebras, I, *Trans. Amer. Math Soc.*, 358(2006), 2537–2560 (electronic).

[161] Y. Liu and C. C. Xi, Constructions of stable equivalences of Morita type for finite dimensional algebras, II, *Math. Z.*, 251(2005), 21–39.

[162] P. Malicki, On the composition factors of indecomposable modules in almost cyclic coherent Auslander-Reiten components, *J. Pure Appl. Algebra*, 207(2006), 469–490.

[163] P. Malicki and A. Skowroński, Almost cyclic coherent components of an Auslander-Reiten quiver, *J. Algebra*, 229(2000), 695–749.

[164] E. Marmolejo and C. M. Ringel, Modules of bounded length in Auslander–Reiten components, *Archiv Math. (Basel)*, 50(1988), 128–133.

[165] H. Meltzer, Auslander–Reiten components for concealed canonical algebras, *Colloq. Math.*, 71(1996), 183–202.

[166] H. Meltzer and A. Skowroński, Group algebras of finite representation type, *Math. Z.*, 182(1983), 129–148; Correction 187(1984), 563–569.

[167] K. Morita, Duality for modules and its applications to the theory of rings with minimum conditions, *Sci. Rep. Tokyo Kyoiku Daigaku*, A6(1958), 83–142.

[168] H. Nakajima, Varieties associated with quivers, In: *Representation Theory of Algebras and Related Topics*, Canad. Math. Soc. Conf. Proc., AMS, Vol. 19, 1996, pp. 139–157.

[169] L. A. Nazarova, Representations of quadruples, *Izv. Akad. Nauk SSSR*, 31(1967), 1361–1378 (in Russian).

[170] L. A. Nazarova, Representations of quivers of infinite type, *Izv. Akad. Nauk SSSR*, 37(1973), 752–791 (in Russian).

[171] L. A. Nazarova and A. V. Roiter, On a problem of I. M. Gelfand, *Funk. Anal. i Priložen.*, 7(1973), 54–69 (in Russian).

[172] L. A. Nazarova and A. V. Roiter, Kategorielle Matrizen-Probleme und die Brauer-Thrall-Vermutung, *Mitt. Math. Sem. Giessen*, 115(1975), 1–153.

[173] R. Nörenberg and A. Skowroński, Tame minimal non-polynomial growth strongly simply connected algebras, In: *Representation Theory of Algebras*, Canad. Math. Soc. Conf. Proc., AMS, Vol. 18, 1996, pp. 519–538.

[174] R. Nörenberg and A. Skowroński, Tame minimal non-polynomial growth simply connected algebras, *Colloq. Math.*, 73(1997), 301–330.

[175] S. A. Ovsienko, Integral weakly positive forms, In: *Schur Matrix Problems and Quadratic Forms*, Inst. Mat. Akad. Nauk Ukr. S.S.R., Preprint 78.25, 1978, pp. 3–17 (in Russian).

[176] S. A. Ovsienko, A bound of roots of weakly positive forms, In: *Representations and Quadratic Forms*, Akad. Nauk Ukr. S.S.R., Inst. Mat., Kiev, 1979, pp. 106–123 (in Russian).

[177] J. A. de la Peña, Algebras with hypercritical Tits form, In: *Topics in Algebra, Part I: Rings and Representations of Algebras*, (Eds: S. Balcerzyk, T. Józefiak, J. Krempa, D. Simson, W. Vogel), Banach Center Publications, Vol. 26, PWN Warszawa, 1990, pp. 353–369.

[178] J. A. de la Peña, Tame algebras with sincere directing modules, *J. Algebra,* 161(1993), 171–185.

[179] J. A. de la Peña, Coxeter transformations and the representation theory of algebras, In: *Finite Dimensional Algebras and Related Topics*, NATO ASI Series, Series C: Mathematical and Physical Sciences, Vol. 424, Kluwer Academic Publishers, Dordrecht, 1994, pp. 223–253.

[180] J. A. de la Peña, On the corank of the Tits form of a tame algebra, *J. Pure Appl. Algebra,* 107(1996), 89–105.

[181] J. A. de la Peña, The Tits form of a tame algebra, In: *Representation Theory of Algebras and Related Topics*, Canad. Math. Soc. Conf. Proc., AMS, Vol. 19, 1996, pp. 159–183.

[182] J. A. de la Peña and A. Skowroński, Geometric and homological characterizations of polynomial growth strongly simply connected algebras, *Invent. math.*, 126(1996), 287–296.

[183] J. A. de la Peña and A. Skowroński, The Tits and Euler forms of a tame algebra, *Math. Ann.,* 315(1999), 37–59.

[184] J. A. de la Peña and M. Takane, Spectral properties of Coxeter transformations and applications, *Archiv Math. (Basel)*, 55(1990), 120–134.

[185] L. Peng and J. Xiao, On the number of DTr-orbits containing directing modules, *Proc. Amer. Math. Soc.*, 118(1993), 753–756.

[186] R. S. Pierce, *Associative Algebras*, Springer-Verlag, New York, Heidelberg, Berlin, 1982.

[187] Z. Pogorzały, On the Auslander-Reiten periodicity of self-injective algebras, *Bull. London Math. Soc.*, 36(2004), 156–168.

[188] Z. Pogorzały and A. Skowroński, On algebras whose indecomposable modules are multiplicity-free, *Proc. London. Math. Soc.*, 47(1983), 463–479.

[189] Z. Pogorzały and A. Skowroński, Selfinjective biserial standard algebras, *J. Algebra,* 138(1991), 491–504.

[190] M. Reineke, Quivers, desingularizations and canonical bases, In: *Studies in memory of Issai Schur*, Progress Math., 210, Birkhäuser, Boston, 2003, pp. 325–344.

[191] I. Reiten and C. Riedtmann, Skew group algebras in the representation theory of artin algebras, *J. Algebra,* 92(1985), 224–282.

[192] I. Reiten and C. M. Ringel, Infinite dimensional representations of canonical algebras, *Canad. J. Math.*, 58(2006), 180–224.

[193] I. Reiten and A. Skowroński, Sincere stable tubes, *J. Algebra,* 232(2000), 64–75.

[194] I. Reiten and A. Skowroński, Characterizations of algebras with small homological dimensions, *Adv. Math.*, 179(2003), 122–154.

[195] I. Reiten and A. Skowroński, Generalized double tilted algebras, *J. Math. Soc. Japan,* 56(2004), 269–288.

[196] I. Reiten, A. Skowroński and S. Smalø, Short chains and short cycles of modules, *Proc. Amer. Math. Soc.*, 117(1993), 343–354.

[197] J. Rickard, Some recent advances in modular representation theory, In: *Algebras and Modules I*, Canad. Math. Soc. Conf. Proc., AMS, Vol. 23, 1998, pp. 157–178.

[198] C. Riedtmann, Algebren, Darstellungsköcher, Überlagerungen und zurück, *Comment. Math. Helvetici,* 55(1980), 199–224.

[199] C. Riedtmann, Representation-finite selfinjective algebras of class $\mathbb{A}_n$, In: *Representation Theory* II, Lecture Notes in Math., No. 832, Springer-Verlag, Berlin, Heidelberg, New York, 1980, pp. 449–520.

[200] C. Riedtmann, Representation-finite selfinjective algebras of class $\mathbb{D}_n$, *Compositio Math.*, 49(1983), 231–282.

[201] C. Riedtmann, Degenerations for quivers with relations, *Ann. Sci. École Norm. Sup.*, 4(1986), 275–301.

[202] C. Riedtmann and A. Schofield, On a simplicial complex associated with tilting modules, *Comment. Math. Helvetici*, 66(1991), 70–78.

[203] C. Riedtmann and G. Zwara, On the zero set of semi-invariants for quivers, *Ann. Sci. École Norm. Sup.*, 36(2003), 969–976.

[204] C. M. Ringel, The indecomposable representations of the dihedral 2-groups, *Math. Ann.*, 214(1975), 19–34.

[205] C. M. Ringel, Representations of K-species and bimodules, *J. Algebra,* 41(1976), 269–302.

[206] C. M. Ringel, Report on the Brauer-Thrall conjectures: Rojter's theorem and the theorem of Nazarova and Rojter, In: *Representation Theory I*, Lecture Notes in Math., No. 831, Springer-Verlag, Berlin, Heidelberg, New York, 1980, pp. 104–136.

[207] C. M. Ringel, Report on the Brauer-Thrall conjectures: Tame algebras, In: *Representation Theory I*, Lecture Notes in Math. No. 831, Springer-Verlag, Berlin, Heidelberg, New York, 1980, pp. 137–287.

[208] C. M. Ringel, The rational invariants of the tame quivers, *Invent. Math.*, 58(1980), 217–239.

[209] C. M. Ringel, Reflection functors for hereditary algebras, *J. London Math. Soc.*, 21(1980), 465–479.

[210] C. M. Ringel, Kawada's theorem, In: *Abelian Group Theory,* Lecture Notes in Math., No. 874, Springer-Verlag, Berlin, Heidelberg, New York, 1981, pp. 431–447.

[211] C. M. Ringel, Indecomposable representations of finite-dimensional algebras, In: *Proceedings Intern. Congress of Math.*, PWN - Polish Scientific Publishers, Warszawa, 1983, pp. 425–436.

[212] C. M. Ringel, Bricks in hereditary length categories, *Result. Math.,* 6(1983), 64–70.

[213] C. M. Ringel, Separating tubular series, In: *Séminaire d'Algèbre Paul Dubreil et Marie-Paul Malliavin,* Lecture Notes in Math., No. 1029, Springer-Verlag, Berlin, Heidelberg, New York, 1983, pp. 134–158.

[214] C. M. Ringel, Unzerlegbare Darstellungen endlich-dimensionaler Algebren, *Jber. Deutche Math. Verein*, 85(1983), 86–105.

[215] C. M. Ringel, *Tame Algebras and Integral Quadratic Forms*, Lecture Notes in Math., No. 1099, Springer-Verlag, Berlin, Heidelberg, New York, 1984, pp. 1–371.

[216] C. M. Ringel, The regular components of Auslander–Reiten quiver of a tilted algebra, *Chinese Ann. Math.*, 9B(1988), 1–18.

[217] C. M. Ringel, The canonical algebras, with an appendix by W. Crawley-Boevey, In: *Topics in Algebra, Part 1: Rings and Representations of Algebras*, (Eds: S. Balcerzyk, T. Józefiak, J. Krempa,

D. Simson, W. Vogel), Banach Center Publications, Vol. 26, PWN - Polish Scientific Publishers, Warszawa 1990, pp. 407–439.

[218] C. M. Ringel, Recent advances in the representation theory of finite dimensional algebras, In: *Representation Theory of Finite Groups and Finite-Dimensional Algebras*, Progress in Mathematics 95, Birkhäuser-Verlag, Basel, 1991, pp. 137–178.

[219] C. M. Ringel, The spectral radius of the Coxeter transformations for a generalized Cartan matrix, *Math. Ann.*, 300(1994), 331–339.

[220] C. M. Ringel, Cones, In: *Representation Theory of Algebras*, Canad. Math. Soc. Conf. Proc., AMS, Vol. 18, 1996, pp. 583–586.

[221] C. M. Ringel, The Liu-Schultz example, In: *Representation Theory of Algebras*, Canad. Math. Soc. Conf. Proc., AMS, Vol. 18, 1996, pp. 587–600.

[222] C. M. Ringel, Infinite length modules. Some examples as introduction, In: *Infinite Length Modules* (Eds: H. Krause and C.M. Ringel), Trends in Mathematics, Birkhäuser Verlag, Basel-Boston-Berlin, 2000, pp. 1–73.

[223] C. M. Ringel, Combinatorial representation theory. History and future, In: *Representations of Algebras*, Vol I, Proc. Conf. ICRA IX, Beijing 2000, (Eds: D. Happel and Y.B. Zhang), Beijing Normal University Press, 2002, pp. 122–144.

[224] C. M. Ringel and K. W. Roggenkamp, Diagrammatic methods in representation theory of orders, *J. Algebra*, 60(1979), 11–42.

[225] C. M. Ringel and H. Tachikawa, QF-3 rings, *J. reine angew. Math.*, 272(1975), 49–72.

[226] C. M. Ringel and D. Vossieck, Hammocks, *Proc. London Math. Soc.*, 54(1987), 216–246.

[227] A. V. Roiter, Unboundedness of the dimension of the indecomposable representations of an algebra which has infinitely many indecomposable representations, *Izv. Akad. Nauk. SSSR. Ser. Mat.*, 32(1968), 1275–1282.

[228] A. Schofield, Semi-invariants of quivers, *J. London Math. Soc.*, 43(1991), 385–395.

[229] A. Schofield, General representations of quivers, *Proc. London Math. Soc.*, 65(1992), 46–64.

[230] A. Schofield and M. Van den Bergh, Semi-invariants of quivers for arbitrary dimension vectors, *Indag. Math.*, (N.S.) 12(2001), 125–138.

[231] J. Schröer, On the infinite radical of a module category, *Proc. London Math. Soc.*, 81(2000), 651–674.

[232] V. V. Sergeichuk, Canonical matrices for linear matrix problems, *Linear Algebra Appl.*, 317(2000), 53–102.

[233] W. Sierpiński, *Elementary Theory of Numbers*, Warszawa, 1964.

[234] D. Simson, Partial Coxeter functors and right pure semisimple hereditary rings, *J. Algebra* 71(1981), 195-218.

[235] D. Simson, *Linear Representations of Partially Ordered Sets and Vector Space Categories*, Algebra, Logic and Applications, Vol. 4, Gordon & Breach Science Publishers, 1992.

[236] D. Simson, An endomorphism algebra realisation problem and Kronecker embeddings for algebras of infinite representation type, *J. Pure Appl. Algebra*, 172(2002), 293–303.

[237] D. Simson and A. Skowroński, The Jacobson radical power series of module categories and the representation type, *Bol. Soc. Mat. Mexicana*, 5(1999), 223–236.

[238] D. Simson and A. Skowroński, Hereditary stable tubes in module categories, Preprint, Toruń, 2007.

[239] A. Skowroński, The representation type of group algebras, CISM Courses and Lectures No. 287, pp. 517–531, Springer-Verlag, Wien - New York, 1984.

[240] A. Skowroński, Group algebras of polynomial growth, *Manuscr. Math.*, 59 (1987), 499–516.

[241] A. Skowroński, Selfinjective algebras of polynomial growth, *Math. Ann.*, 285 (1989), 177–199.

[242] A. Skowroński, Algebras of polynomial growth, In: *Topics in Algebra, Part 1: Rings and Representations of Algebras*, (Eds: S. Balcerzyk, T. Józefiak, J. Krempa, D. Simson, W. Vogel), Banach Center Publications, Vol. 26, PWN - Polish Scientific Publishers, Warszawa 1990, pp. 535–568.

[243] A. Skowroński, Generalized standard Auslander–Reiten components without oriented cycles, *Osaka J. Math.*, 30(1993), 515–527.

[244] A. Skowroński, Regular Auslander–Reiten components containing directing modules, *Proc. Amer. Math. Soc.*, 120(1994), 19–26.

[245] A. Skowroński, Cycles in module categories, In: *Finite Dimensional Algebras and Related Topics*, NATO ASI Series, Series C: Mathematical and Physical Sciences, Vol. 424, Kluwer Academic Publishers, Dordrecht, 1994, pp. 309–345.

[246] A. Skowroński, Generalized standard Auslander–Reiten components, *J. Math. Soc. Japan*, 46(1994), 517–543.

[247] A. Skowroński, On the composition factors of periodic modules, *J. London Math. Soc.*, 49(1994), 477–492.

[248] A. Skowroński, Minimal representation-infinite artin algebras, *Math. Proc. Cambridge Phil. Soc.*, 116(1994), 229-243.

[249] A. Skowroński, Module categories over tame algebras, In: *Representation Theory of Algebras and Related Topics*, Canad. Math. Soc. Conf. Proc., AMS, Vol. 19, 1996, pp. 281-313.

[250] A. Skowroński, On omnipresent tubular families of modules, In: *Proceedings of the Seventh International Conference on Representation Theory of Algebras*, Canad. Math. Soc. Conf. Proc., AMS, Vol. 18, 1996, pp. 641-657.

[251] A. Skowroński, Simply connected algebras of polynomial growth, *Compositio Math.*, 109(1997), 99-133.

[252] A. Skowroński, Tame module categories of finite dimensional algebras, In: *Trends in Ring Theory*, Canad. Math. Soc. Conf. Proc., AMS, Vol. 22, 1998, pp. 187-219.

[253] A. Skowroński, On the structure of periodic modules over tame algebras, *Proc. Amer. Math. Soc.*, 127(1999), 1941-1949.

[254] A. Skowroński, Generalized canonical algebras and standard stable tubes, *Colloq. Math.*, 90(2001), 77-93.

[255] A. Skowroński, Generically directed algebras, *Archiv Math. (Basel)*, 78(2002), 358-361.

[256] A. Skowroński, Selfinjective algebras: finite and tame type, In: *Trends in Representation Theory of Algebras and Related Topics*, Contemp. Math., 406(2006), 169-238.

[257] A. Skowroński and S. Smalø, Artin algebras with only preprojective or preinjective modules are of finite type, *Archiv Math. (Basel)*, 64(1995), 8-10.

[258] A. Skowroński, S. Smalø and D. Zacharia, On the finiteness of the global dimension of Artin rings, *J. Algebra*, 251(2002), 475-478.

[259] A. Skowroński and K. Yamagata, Socle deformations of self-injective algebras, *Proc. London Math. Soc.*, 72(1996), 545-566.

[260] A. Skowroński and K. Yamagata, Stable equivalence of selfinjective algebras of tilted type, *Archiv Math. (Basel)*, 70(1998), 341-350.

[261] A. Skowroński and K. Yamagata, On selfinjective artin algebras having nonperiodic generalized standard Auslander-Reiten components, *Colloq. Math.*, 96(2003), 235-244.

[262] A. Skowroński and K. Yamagata, Selfinjective artin algebras with all Auslander-Reiten components generalized standard, Preprint, Toruń, 2007.

[263] A. Skowroński and J. Waschbüsch, Representation-finite biserial algebras, *J. reine angew. Math.*, 345(1983), 172-181.

[264] A. Skowroński and M. Wenderlich, Artin algebras with directing indecomposable projective modules, *J. Algebra*, 165(1994), 507–530.

[265] A. Skowroński and J. Weyman, Semi-invariants of canonical algebras, *Manuscr. Math.*, 100(1999), 391–403.

[266] A. Skowroński and J. Weyman, The algebras of semi-invariants of quivers, *Transform. Groups*, 5(2000), 361–402.

[267] A. Skowroński and G. Zwara, On indecomposable modules without self extensions, *J. Algebra*, 195(1997), 151–169.

[268] S. Smalø, The inductive step of the second Brauer–Thrall conjecture, *Canad. J. Math.*, 2(1980), 342–349.

[269] D. Smith, On generalized standard Auslander-Reiten components having only finitely many non-directing modules, *J. Algebra*, 279(2004), 493–513.

[270] H. Strauss, On the perpendicular category of a partial tilting module, *J. Algebra*, 144(1991), 43–66.

[271] H. Tachikawa and T. Wakamatsu, Tilting functors and stable equivalence for self-injective algebras, *J. Algebra,* 109(1987), 138–165.

[272] L. Unger, The simplicial complex of tilting modules over quiver algebras, *Proc. London Math. Soc.*, 73(1996), 27–46.

[273] R. Vila-Freyer and W. W. Crawley-Boevey, The structure of biserial algebras, *J. London. Math. Soc.*, 57(1998), 41–54.

[274] B. Wald and J. Waschbüsch, Tame biserial algebras, *J. Algebra,* 95(1985), 480–500.

[275] J. Waschbüsch, Symmetrische Algebren vom endlichen Modultyp, *J. reine angew. Math.*, 321(1981), 78–98.

[276] P. Webb, The Auslander-Reiten quiver of a finite group, *Math. Z.*, 179(1982), pp. 79–121.

[277] C. C. Xi, On the finitistic dimension conjecture, I. Related to representation-finite algebras, *J. Pure Appl. Algebra,* 193(2004), 287–305; Erratum: 202(2005), 325–328.

[278] C. C. Xi, On the finitistic dimension conjecture, II. Related to finite global dimension, *Advances Math,* 201(2006), 116–142.

[279] K. Yamagata, Frobenius algebras, In: *Handbook of Algebra*, (Ed.: M. Hazewinkel), Vol. 1, North-Holland Elsevier, Amsterdam, 1996, pp. 841–887.

[280] T. Yoshi, On algebras of bounded representation type, *Osaka Math. J.,* 8(1956), 51–105.

[281] M. V. Zeldich, Sincere weakly positive unit quadratic forms, In: *Representations of Algebras*, Canad. Math. Soc. Conf. Proc., AMS, Vol. 14, 1993, pp. 453–461.

[282] Y. Zhang, The modules in any component of the AR-quiver of a wild hereditary algebra are uniquely determined by their composition factors, *Archiv Math. (Basel)*, 53(1989), 250–251.

[283] Y. Zhang, The structure of stable components, *Canad. J. Math.*, 43(1991), 652–672.

[284] B. Zimmermann-Huisgen, The finitistic dimension conjectures - a tale of 3.5 decades, In: *Abelian Groups and Modules*, Math. Appl., 343, Kluwer Acad. Publ., Dordrecht, 1995, pp. 501–517.

Index

acyclic quiver   2
annihilator of a component   35
annihilator of a module   35, 130
Auslander–Reiten translation   110

bound quiver   268
bound quiver algebra   268
brick   13
    orthogonal bricks   13

canonical algebras of Euclidean
    type   93
canonically oriented Euclidean
    quivers   145
canonical radical vector   102
canonical radical vector $\mathbf{h}_\Delta$ 150
category add $\mathcal{R}(A)$ of regular
    modules   52
concealed algebra of Euclidean
    type   69
concealed domain of $\mathcal{P}(H)$ 248
connected vector of $K_0(A)$ 133
coray   4
coray module   5
coray point of a component   5
coray point of a stable tube   5
Coxeter matrix of an algebra   54
Coxeter transformation of an
    algebra   54

defect of an algebra   58
defect number of a Euclidean
    algebra   58

defect number of a Euclidean
    quiver   58

$\mathcal{E}$-length of a module   16
enlarged Kronecker quiver   227
Euclidean graphs (diagrams)   52, 53
Euler characteristic of an algebra   69
Euler (non-symmetric) $\mathbb{Z}$-bilinear
    form of an algebra   54
Euler quadratic form of an
    algebra   53
exact subcategory   13
Ext-vanishing lemma   232

faithful module   35
four subspace quiver   197
four subspace problem   145, 197
frames   268

generalised standard component   32

hereditary family of modules   14
hereditary tube   26
homogeneous tube   3

idempotent embedding functors   107
incidence matrix of a finite
    quiver   88
infinite translation quiver   2
infinite radical of mod A 7
integral quadratic form   54, 228
    positive semidefinite   228
    weakly nonnegative   228
    weakly positive   228

Kronecker algebra 52, 77
Kronecker diagram 53
Kronecker graph 53
Kronecker problem 52
Kronecker quiver 52, 77

$\mathcal{E}$-length of a module 16

mesh category 21
mesh element 21
minimal representation-infinite
       algebras 231
mouth 4
module
    cogenerated by a module 130
    faithful 130
    generated by a module 130
mod A is controlled by q_A 133
mod A is link controlled by q_A 133
multiplicity-free postprojective
       tilting module 247

Nakayama functor 28

orbit space 2
orthogonal bricks 13
orthogonal components 63

path category 21
positive generator of the
       group rad q_B 75
positive semidefinite quadratic
       form 228
positive vector 228
projective line $\mathbb{P}_1(K)$ over K 80

radical of a module category 7
radical of a quadratic form 54, 73

ray 4, 5
ray module 5
ray point of a component 5
ray point of a stable tube 5
regular component 1
regular length of a regular
       module 63
regular module 1, 52
roots of an integral quadratic
       form 56

sectional path 4
self-hereditary component 22
self-hereditary family of
       modules 14
self-hereditary tube 26
serial category 63
separating family 125
    $\mathcal{C}$ separates $\mathcal{P}$ from $\mathcal{Q}$ 126
simple object 13
simple regular module 63
sincere vector 228
stable tube $\mathcal{T}$ of rank $r = r_{\mathcal{T}}$ 3
standard component 21
standard homomorphism 24

τ-cycle of a tube 3
τ_A-cycle of mouth A-modules 5, 11
τ-orbit 64
τ-periodic point of a stable
       tube 4
tubular type $(m_1, \ldots, m_s)$ 125
tubular type $\mathbf{m}_\Delta = (m_1, \ldots, m_s)$ 144

uniserial object 16

weakly positive quadratic form 228
weakly nonnegative quadratic
       form 228

List of symbols

$\mathcal{P}(A)$ 1, 60

$\mathcal{Q}(A)$ 1, 60

$\mathcal{R}(A)$ 1, 60

$Q = (Q_0, Q_1)$ 2

KQ 2

$\mathbb{A}_\infty$ 2

$\mathbb{Z}\mathbb{A}_\infty/(\tau^r)$ 2

τ-cycle $(x_1, \ldots, x_r)$ 3

$\mathrm{rad}_A, \mathrm{rad}_A^m, \mathrm{rad}_A^\infty$ 7

$\mathcal{E}_A = \mathcal{E}\mathcal{X}\mathcal{T}_A(E_1, \ldots, E_r)$ 13

$\ell_\mathcal{E}(U)$ 16

$K\mathcal{C}$ 21

$M_\mathcal{C}$ 21

m_x 21

$K(\mathcal{C}) = K\mathcal{C}/M_\mathcal{C}$ 21

$\mathrm{ind}\,\mathcal{C}$ 21

$\mathcal{T}_\mathcal{E}$ 22, 64

$\mathrm{Ann}_A M$ 35

$\mathrm{Ann}_A \mathcal{C}$ 35

$J_{m,0}$ 50

$\boldsymbol{\mathcal{T}}^B = \{\mathcal{T}_\lambda^B\}_{\lambda \in \Lambda}$ 51

$\widetilde{\mathbb{A}}_n$ 53

$\widetilde{\mathbb{D}}_n$ 53

$\widetilde{\mathbb{E}}_6$ 53

$\widetilde{\mathbb{E}}_7$ 53

$\widetilde{\mathbb{E}}_8$ 53

$\mathrm{add}\,\mathcal{R}(A)$ 53, 70, 71

$q_A : K_0(A) \longrightarrow \mathbb{Z}$ 53

$q_A(\mathbf{x}) = \mathbf{x}^t(\mathbf{C}_A^{-1})^t\mathbf{x}$ 54

$\mathbf{C}_A$ - Cartan matrix of A 54

$\mathrm{rad}\,q_A \subseteq K_0(A)$ 54

$\langle -, - \rangle_A : \mathbb{Z}^n \times \mathbb{Z}^n \longrightarrow \mathbb{Z}$ 54

$\boldsymbol{\Phi}_\mathbf{A}$ 54

$\mathrm{Gl}(n, \mathbb{Z})$ 54

$\overline{Q}$ 56

$\partial_A : K_0(A) \longrightarrow \mathbb{Z}$ 57, 150

$d_Q, d_A \in \mathbb{N}$ 58

$\mathrm{add}\,\mathcal{R}(A)$ 61

$r\ell(M)$ 63

$\boldsymbol{\mathcal{T}}^A = \{\mathcal{T}_\lambda^A\}_{\lambda \in \Lambda}$ 63

$\mathcal{O}(E)$ 64

$\chi_R : K_0(R) \times K_0(R) \longrightarrow \mathbb{Z}$ 69

$\mathrm{rad}\,q_B \subseteq K_0(B)$ 73

$\mathbf{h}_B \in K_0(B)$ 75

$\mathbb{P}_1(K) = K \cup \{\infty\}$ 80

$\mathcal{T}_\lambda^A$ 80

$C(p, q)$ 92

$\Delta(p, q)$ 92

$C(p, q, r)$ 92

$\Delta(p, q, r)$ 92

$I(p, q, r)$ 92

$\mathbf{h}_C \in K_0(C)$ 102

$\mathcal{P}(C), \mathcal{Q}(C), \mathcal{R}(C)$ 106

T_e, L_e, res_e 107

$\tau_C = D\mathrm{Tr}$ 110

$\boldsymbol{\mathcal{T}}^C, \mathcal{T}_\infty^C, \mathcal{T}_\lambda^C$ 111

$E_1^{(\infty)}, E_2^{(\infty)}$ 115

$E_j^{(0)}, E_k^{(1)}$ 115

R^{α_1} 116

R^{β_t} 116

R^{γ_m} 116

$E^{(\lambda)}$ 116

$\boldsymbol{\mathcal{T}}^C, \mathcal{T}_\infty^C, E^{(0)}, E^{(1)}, \mathcal{T}_\lambda^C$ 116

$\mathbf{m}_B, \mathbf{m}_Q$ 126

$\operatorname{Ann} N$ 130

$r\ell(X)$ 134

$\Delta(\widetilde{\mathbb{A}}_m)$, $\Delta(\widetilde{\mathbb{D}}_m)$ 145

$\Delta(\widetilde{\mathbb{E}}_6)$, $\Delta(\widetilde{\mathbb{E}}_7)$, $\Delta(\widetilde{\mathbb{E}}_8)$ 145

$d_\Delta = d_A \in \mathbb{N}$ 150

$\mathbf{h}_\Delta \in K_0(K\Delta)$ 150

$\boldsymbol{\mathcal{T}}^{\Delta(\widetilde{\mathbb{D}}_m)}$, $\mathcal{T}_\infty^{\Delta(\widetilde{\mathbb{D}}_m)}$, $\mathcal{T}_\lambda^{\Delta(\widetilde{\mathbb{D}}_m)}$ 159

$\boldsymbol{\mathcal{T}}^{\Delta(\widetilde{\mathbb{E}}_6)}$, $\mathcal{T}_\infty^{\Delta(\widetilde{\mathbb{E}}_6)}$, $\mathcal{T}_\lambda^{\Delta(\widetilde{\mathbb{E}}_6)}$ 168

$\boldsymbol{\mathcal{T}}^{\Delta(\widetilde{\mathbb{E}}_7)}$, $\mathcal{T}_\infty^{\Delta(\widetilde{\mathbb{E}}_7)}$, $\mathcal{T}_\lambda^{\Delta(\widetilde{\mathbb{E}}_7)}$ 177

$\boldsymbol{\mathcal{T}}^{\Delta(\widetilde{\mathbb{E}}_8)}$, $\mathcal{T}_\infty^{\Delta(\widetilde{\mathbb{E}}_8)}$, $\mathcal{T}_\lambda^{\Delta(\widetilde{\mathbb{E}}_8)}$ 187

$P(m,j)$ 202, 205

$I(m,j)$ 206

$\mathcal{T}_0^Q$, $\mathcal{T}_1^Q$, $\mathcal{T}_\infty^Q$ 221

$\mathcal{K}_m$ 227

$\mathcal{DP}(H)$ 248

Printed in the United States
by Baker & Taylor Publisher Services